Recombinant DNA and
Cell Proliferation

This is a volume in
CELL BIOLOGY
A series of monographs

Editors: D. E. Buetow, I. L. Cameron, G. M. Padilla, and A. M. Zimmerman

A complete list of the books in this series appears at the end of the volume.

Recombinant DNA and Cell Proliferation

Edited by

Gary S. Stein

Department of Biochemistry and Molecular Biology
University of Florida College of Medicine
Gainesville, Florida

Janet L. Stein

Department of Immunology and Medical Microbiology
University of Florida College of Medicine
Gainesville, Florida

1984

ACADEMIC PRESS, INC.

(Harcourt Brace Jovanovich, Publishers)

Orlando   San Diego   San Francisco   New York   London
Toronto   Montreal   Sydney   Tokyo   São Paulo

ACADEMIC PRESS, INC.
Orlando, Florida 32887

United Kingdom Edition published by
ACADEMIC PRESS, INC. (LONDON) LTD.
24/28 Oval Road, London NW1 7DX

Library of Congress Cataloging in Publication Data
Main entry under title:

Recombinant DNA and cell proliferation.

(Cell biology)
Includes index.
1. Recombinant DNA. 2. Cell proliferation. 3. Gene
expression. I. Stein, Gary S. II. Stein, Janet L.
III. Title: Recombinant DNA and cell proliferation.
IV. Series. [DNLM: 1. DNA, recombinant. 2. Gene
expression regulation. 3. Cell division. QH 605 R311]

QH442.R374 1984 547 574.87'62 83–21506
ISBN 0–12–665080–2 (alk. paper)

PRINTED IN THE UNITED STATES OF AMERICA

84 85 86 87   9 8 7 6 5 4 3 2 1

Contents

I Expression of Specific Genes during the Cell Cycle

1 Use of Cloned SV40 DNA Fragments to Study Signals for Cell Proliferation

Kenneth J. Soprano

v

2 Expression of Dihydrofolate Reductase and Thymidylate Synthase Genes in Mammalian Cells

Lee F. Johnson

3 Enzyme Expression during Growth and Cell Division in *Saccharomyces cerevisiae:* A Study of Galactose and Phosphorus Metabolism

H. O. Halvorson, K. A. Bostian, J. G. Yarger,
and J. E. Hopper

4 Gene Expression during the Cell Cycle of *Chlamydomonas reinhardtii*

David Herrin and Allan Michaels

II Expression of Specific Genes Associated with Proliferation and Differentiation

11 Regulation of Nonmuscle Actin Gene Expression during Early Development

Lewis J. Kleinsmith, N. Kent Peters, and Mary E. Zeigler

12 Functional Architecture at Telomeres of Linear DNA in Eukaryotes

Edward M. Johnson, Peter Bergold, and Gerald R. Campbell

III Overview

13 Recombinant DNA Approaches to Studying Control of Cell Proliferation: An Overview

Renato Baserga

Contributors

Numbers in parentheses indicate the pages on which the authors' contributions begin.

Renato Baserga (337), Department of Pathology and Fels Research Institute, Temple University School of Medicine, Philadelphia, Pennsylvania 19140

L. L. Baumbach (107), Department of Biochemistry and Molecular Biology, University of Florida College of Medicine, Gainesville, Florida 32610

Peter Bergold[1] (303), The Rockefeller University, New York, New York 10021

K. A. Bostian (49), Section of Biochemistry, Division of Biology and Medicine, Brown University, Providence, Rhode Island 02912

Gerald R. Campbell[2] (303), The Rockefeller University, New York, New York 10021

David Conkie (195), The Beatson Institute for Cancer Research, Garscube Estate, Glasgow G61 1BD, Scotland

Charles P. Emerson, Jr. (219), Department of Biology, University of Virginia, Charlottesville, Virginia 22901

Nelson Fausto (145), Department of Pathology, Division of Biology and Medicine, Brown University, Providence, Rhode Island 02912

Howard M. Fried (169), Department of Biochemistry and Nutrition, University of North Carolina School of Medicine, Chapel Hill, North Carolina 27514

H. O. Halvorson (49), Rosenstiel Basic Medical Sciences Research Center, and Department of Biology, Brandeis University, Waltham, Massachusetts 02254

Kenneth E. M. Hastings (219), Department of Biology, University of Virginia, Charlottesville, Virginia 22901

David Herrin (87), Department of Biology, University of South Florida, Tampa, Florida 33620

[1]Present address: Memorial Sloan-Kettering Cancer Center, 1275 York Avenue, New York, New York 10021.

[2]Present address: Department of Biochemistry, New York University Medical Center, New York, New York 10016.

J. E. Hopper (49), Department of Biological Chemistry, Hershey Medical Center, Hershey, Pennsylvania 17033

Edward M. Johnson (303), The Rockefeller University, New York, New York 10021

Lee F. Johnson (25), Department of Biochemistry, The Ohio State University, Columbus, Ohio 43210

Lewis J. Kleinsmith (273), Division of Biological Sciences, The University of Michigan, Ann Arbor, Michigan 48109

F. F. Marashi (107), Department of Biochemistry and Molecular Biology, University of Florida College of Medicine, Gainesville, Florida 32610

Oded Meyuhas (243), Developmental Biochemistry Research Unit, Institute of Biochemistry, Hebrew University–Hadassah Medical School, Jerusalem 91010, Israel

Allan Michaels (87), Department of Biology, University of South Florida, Tampa, Florida 33620

John Paul (195), The Beatson Institute for Cancer Research, Garscube Estate, Glasgow G61 1BD, Scotland

N. Kent Peters (273), Division of Biological Sciences, The University of Michigan, Ann Arbor, Michigan 48109

M. A. Plumb (107), Department of Biochemistry and Molecular Biology, University of Florida College of Medicine, Gainesville, Florida 32610

L. F. Sierra[3] (107), Department of Biochemistry and Molecular Biology, University of Florida College of Medicine, Gainesville, Florida 32610

Kenneth J. Soprano (3), Department of Microbiology and Immunology, Temple University School of Medicine, Philadelphia, Pennsylvania 19140

G. S. Stein (107), Department of Biochemistry and Molecular Biology, University of Florida College of Medicine, Gainesville, Florida 32610

J. L. Stein (107), Department of Immunology and Medical Microbiology, University of Florida College of Medicine, Gainesville, Florida 32610

Taiki Tamaoki (145), Department of Medical Biochemistry, University of Calgary, Calgary, Alberta T2N 4N1, Canada

Jonathan R. Warner (169), Departments of Cell Biology and Biochemistry, Albert Einstein College of Medicine, Bronx, New York 10461

J. G. Yarger[4] (49), Department of Biochemistry and Molecular Biology, Fairchild Biochemical Laboratory, Harvard University, Cambridge, Massachusetts 02139

Mary E. Zeigler[5] (273), Division of Biological Sciences, The University of Michigan, Ann Arbor, Michigan 48109

---

[3]Present address: Swiss Institute for Experimental Cancer Research, CH-1066 Epalinges, Switzerland.

[4]Present address: Department of Biosynthesis Research, Miles Laboratories, Inc., Elkhart, Indiana 46514.

[5]Present address: Department of Plant Pathology, Cornell University, Ithaca, New York 14853.

Preface

It is becoming increasingly apparent that in continuously dividing cells and following stimulation of nondividing cells to proliferate, DNA replication and mitosis are preceded and accompanied by a complex and interdependent series of biochemical events involving modifications in gene expression. Hence, understanding mechanisms operative in the control of the proliferative process necessitates knowledge of how gene expression is regulated. At the same time, the cell cycle offers an effective model system for studying the control of gene expression, in fact, one in which control of expression resides at several levels.

Recombinant DNA technology encompasses a series of powerful approaches for elucidating the structure, organization, and regulation of genetic sequences. Already recombinant DNA techniques have been effectively utilized to study the regulation of some specific genes during the cell cycle, the interrelationship among the various genetic sequences expressed during the cell cycle, and the functional implications of certain genetic sequences for the control of the proliferative process.

In preparing this volume we attempted to assemble a series of contributions dealing with applications of recombinant DNA technology to the study of the control of cell proliferation. They include gene transfer to assess the role of defined DNA sequences in triggering DNA replication, nucleic acid hybridization probes for examining the regulation of specific genes during the cell cycle, and cloned DNAs for analyzing genes expressed in conjunction with proliferation and differentiation. An overview which summarizes "where things are as well as where they are going" concludes the book.

Recombinant DNA technology is a timely topic, and to date no comparable work is available which is addressed specifically to the applications of recombinant DNA techniques to the investigation of cell proliferation. We therefore hope that this volume will be of value both to advanced students and to research scientists.

Gary S. Stein
Janet L. Stein

I

Expression of Specific Genes during the Cell Cycle

1

Use of Cloned SV40 DNA Fragments to Study Signals for Cell Proliferation

KENNETH J. SOPRANO

Department of Microbiology and Immunology
Temple University School of Medicine
Philadelphia, Pennsylvania

I. INTRODUCTION

One key to the understanding of the control of mammalian cell proliferation involves the identification of genes and gene products that regulate the orderly process of cell division. For cell division to occur, at least three fundamental processes must take place: (1) The cells must grow in mass, i.e., a cell must double all its cellular components; (2) the cell must replicate DNA; and (3) the

3

cell must undergo mitosis. It is usually taken for granted that these three processes are strictly interrelated to the point that cell DNA synthesis, for instance, is often equated with cell proliferation. However, it may not be necessarily so.

Through the use of recombinant DNA technology and gene transfer by microinjection, we have been able to show that increases in cell mass and cell DNA replication induced by DNA oncogenic viruses such as SV40 are under separate and distinct controls. In fact, the information for stimulation of these two fundamental processes by SV40 T antigen is encoded in separate and distinct gene sequences or domains of the *A* gene. By analysis of these viral sequences and comparison to gene sequences and proteins normally found in the mammalian cell, it is possible to determine if sequences or proteins of comparable structure and function are present and functioning in the normal, proliferating mammalian cell.

II. GENE TRANSFER BY MANUAL MICROINJECTION

During the last decade, the development of recombinant DNA cloning and sequencing technology has made it possible to investigate the structure of a variety of prokaryotic and eukaryotic genes (Proudfoot *et al.*, 1980; Tucker *et al.*, 1980). Likewise, development of methods to introduce recombinant DNA into a biologically functional host cell has permitted one to study the function of genes and gene fragments, including those that play a role in the process of cell proliferation.

The most commonly used methods of gene transfer are cell fusion, DNA transfection, liposomes, and manual microinjection (for review, see Baserga, 1980). Of all of these methods, we found the manual microinjection technique to be the most suitable for our studies on cell proliferation. This technique was developed by Diacumakos (1980; Anderson *et al.*, 1980) and Graessmann and co-workers (Graessmann and Graessmann, 1976; Graessmann *et al.*, 1977). The major advantage and most attractive feature of microinjection is that it permits one to deliver a known, purified molecule or macromolecule directly into the nucleus or the cytoplasm of a particular mammalian cell, preserving its integrity and viability. Thus, this technique not only permits the introduction of a single molecular species into the cell but also provides precise placement of these molecules, whether in an aqueous or a nonaqueous solvent, within different compartments of an individual cell (Stacey, 1980). Among the macromolecules that have been microinjected are DNA (Mueller *et al.*, 1978; Floros *et al.*, 1981; Soprano *et al.*, 1981; Galanti *et al.*, 1981), cRNA (Graessmann and Graessmann, 1976), mRNA (Richardson *et al.*, 1980; Bravo and Celis, 1980; Floros *et al.*, 1981), and proteins (Tjian *et al.*, 1978; Kreis *et al.*, 1979; Feramisco, 1979).

Once injected, expression of the introduced molecule can be followed at definite phenotypic (Mueller *et al.*, 1978; Tjian *et al.*, 1978; Soprano *et al.*, 1981) and genotypic (Graessmann *et al.*, 1978; Anderson *et al.*, 1980) end points. These include incorporation of [^{3}H]thymidine into cellular DNA (detectable by autoradiography), the appearance of a protein (detectable by either immunoprecipitation or direct/indirect immunofluorescence), and the synthesis of a specific RNA species (detected by gel electrophoresis and/or Northern blotting followed by hybridization). It is also possible to follow polypeptide synthesis in microinjected cells using two-dimensional gel electrophoresis. Finally, microinjection of specific antibodies can be used to inhibit some of these same processes. The major disadvantage of this technique is that it is a time-consuming, delicate technique in which only a limited number of cells can be microinjected.

Manual microinjection is performed under a phase-contrast microscope by guiding a glass micropipette (tip diameter is approximately 0.5 μm) into a cell with a micromanipulator. The cells are grown on cover slips and must be subconfluent. Approximately 10^{-11} ml of a very concentrated solution (for example, 1 mg/ml plasmid DNA) is delivered into each cell. Thus, with this technique only very minute amounts of material are required and, after delivery to a specific location within the cell, the microinjected cells can be analyzed either immediately or at any time after microinjection.

With respect to the fate of the microinjected DNA, results from a typical experiment are presented in Table I. In this experiment a recombinant plasmid, pSV2G (Soprano *et al.*, 1981; Galanti *et al.*, 1981), containing the entire SV40 *A*

TABLE I

Expression of a Cloned Fragment of SV40 after Microinjection[a]

Hours after microinjection	% Microinjected cells T antigen (+)
4	0
8	40
12	70
24	75
48	400
72	450
96	392
120	0
144	0

[a] Plasmid pSV2G was manually microinjected at time 0 into the nucleus of 100 *ts*13 cells. At various times after microinjection, cover slips were harvested, fixed in absolute methanol (−20°C), and stained by indirect immunofluorescence for detection of large T antigen. Presence of more than 100% T antigen (+) cells at 48, 72, and 96 hr is likely to reflect cell division.

gene inserted into pBR322, was microinjected into the nucleus of *ts*13 cells, a Syrian hamster cell line. For each time point, 100 cells on one cover slip were microinjected. At 4, 8, 12, 24, 48, 72, 96, and 120 hr after microinjection, one cover slip was harvested, fixed, and stained by indirect immunofluorescence to test for the expression of large T antigen, the major product of the SV40 *A* gene (see Section III). Expression of the microinjected DNA is first detectable at 8 hr and progressively increases with time. In fact the number of cells that express the injected DNA doubles for the first 3 days. This suggests that enough copies of the gene are introduced to permit expression even after several cell divisions. However, by day 4 or 5 the number of cells expressing the gene declines such that by day 6 or 7, only one or two cells express the viral gene. If the cultures are kept for longer periods the number again begins to increase, indicating that integration of the injected DNA into the host chromosome has occurred, leading to a stable genotypic and phenotypic change in the cell.

Similar studies have been performed with a variety of cloned genes and a number of different cell lines (Shen *et al.*, 1982). In all cases, expression occurs very rapidly after microinjection of the DNA into the nucleus. However, it should be noted that poor expression results when the DNA is injected into the cytoplasm (Capecchi, 1980; Galanti *et al.*, 1981). Also, expression is virtually eliminated if the injected DNA is linear. This suggests that the DNA is degraded by nucleases in the cytoplasm prior to entering the nucleus. Despite this limitation we still concluded that this technique was the easiest and most efficient procedure to use in our studies of the identification of viral genes and gene fragments involved in the induction of cell proliferation in resting cells.

III. SV40 AND CELL PROLIFERATION

It has been known for some time that certain DNA oncogenic viruses such as SV40, polyoma, and adenovirus can induce cell proliferation in mammalian cells. The use of such viruses to study genes involved in cell proliferation offers several advantages. For example, SV40 is a small DNA virus with a circular genome of only 5,243 base pairs (Tooze, 1980). It has been fully sequenced by Fiers *et al.* (1978) and Weissman and his associates (Reddy *et al.*, 1978; for a review, see Lebowitz and Weissman, 1979). Obviously, a genome of this size is much more amenable to analysis than the approximately 3×10^9 base pairs of the mammalian genome (Maniatis *et al.*, 1978). Also, SV40 is a mitogenic virus (Weil, 1978), i.e., it not only induces cell DNA replication (Henry *et al.*, 1966; Tjian *et al.*, 1978) but it also produces a coordinate increase in ribosomal RNA, messenger RNA, and proteins (May *et al.*, 1976; Khandjian *et al.*, 1980), as well as mitosis (Weil *et al.*, 1977). SV40 infection, therefore, causes both an increase in cell mass and cell DNA replication.

The SV40 genome is divided into two genes. One extends from the origin of replication [nucleotide 5235, 0.66 map units (mu)] counterclockwise to nucleotide 2693 (0.17 mu). This is the "early" or A gene. The other region or "late" gene is found between the origin of replication, clockwise to nucleotide 2591 (0.16 mu). When SV40 infects a cell, expression of its genome occurs in two phases (Tooze, 1980). Immediately after infection, two mRNAs are synthesized through different splicing of a common 16 S precursor RNA transcribed from the early gene. One of these mRNAs codes for a protein of approximately 100,000 daltons, called large T antigen. The other mRNA codes for a protein of 17,000 daltons, small t antigen. The late gene, which codes for the structural proteins of the virus, is not expressed until 10–12 hr after infection (Tooze, 1980). Genetic and biochemical evidence has indicated that it is the early region of the SV40 genome that is responsible for many of the biological activities exhibited by SV40, including stimulation of cell DNA and RNA synthesis.

The biological activity of the two products of the A gene, large T antigen and small t antigen, has been studied not only by infection but also by manual microinjection (Graessmann and Graessmann, 1976). Graessmann et al. (1977) showed that microinjection of the SV40 A gene was sufficient to induce cell DNA replication in quiescent cells in culture. Tjian et al. (1978) induced cell DNA synthesis in mouse kidney cells microinjected with a partially purified large T protein. Thus, formal demonstration was given that the SV40 large T antigen contains all the necessary information to induce cell DNA replication in resting cells. Therefore, we hypothesized that a study of the means by which this protein acts on the cell could provide important information about the control processes or signals of cell proliferation which act within the cell itself. Such a hypothesis is based on a model of control of cell proliferation which requires that a cell (1) grows in mass, i.e., doubles its cellular components; and (2) replicates DNA. It also requires that SV40, and the large T antigen in particular, can induce each of these processes. Let us consider each of these points.

IV. A GROWING CELL RECEIVES SIGNALS FOR GROWTH IN SIZE

When a cell goes from one mitosis to the next one, it must double in size. If it did not, each subsequently produced daughter cell would become progressively smaller and would eventually vanish. This requirement has been shown experimentally (for a review, see Hartwell, 1978) by assaying a variety of parameters in a number of systems. For example, the protein content (Skog et al., 1979; Baxter and Stanners, 1978), total RNA (Ashihara et al., 1978a; Darzynkiewicz et al., 1979), and even nucleus size (Yen and Pardee, 1979) have been shown to double as the cell progresses from early G_1 through mitosis.

This increase in cell size is regularly accompanied by an increase in the amount of rRNA. This is not surprising, since nucleic acids and proteins constitute about 50% of the total dry weight of the cell, and rRNA constitutes 70% of the total nucleic acids of the cell (Galanti *et al.*, 1981). Furthermore, ribosomes are the structures on which proteins are made. Indeed, an increase in the synthesis and/or accumulation of rRNA is a well-established characteristic of a variety of cell and cell systems stimulated to proliferate by a variety of means [e.g., partial hepatectomy (Organtini *et al.*, 1975; Tsukada and Liebermann, 1964; Schmid and Sekaris, 1975); serum stimulation of WI38 (Zardi and Baserga, 1974), 3T6 cells (Mauck and Green, 1973), and CV-1 monkey cells (Rovera *et al.*, 1975); and SV40 and polyoma virus infection of mouse kidney cells (Benjamin, 1966; Weil *et al.*, 1975; May *et al.*, 1976; Whelly *et al.*, 1978].

Thus, mammalian cells that are actively proliferating do increase their various components and increase in size as a consequence. It also seems reasonable to use increases in rRNA as an assay of this aspect of cell growth.

V. A GROWING CELL RECEIVES SIGNALS TO REPLICATE DNA

In most cases, there is a temporal gap between completion of mitosis and onset of DNA synthesis, a gap which is called G_1 (Howard and Pelc, 1951) and is markedly lengthened in quiescent cells stimulated to proliferate. This is not to deny or ignore the reports that certain cells can do without a G_1 phase (Baserga, 1963; Lala and Patt, 1966; Robbins and Scharff, 1967; Liskay and Prescott, 1978; Liskay, 1978). But most cells, after mitosis, go through a phase during which no DNA synthesis occurs and during which something occurs that is a prerequisite for entry into S. Formal evidence for such prerequisites derived from the existence of *ts* mutants of the cell cycle which specifically arrest in G_1 at the nonpermissive temperature (Basilico, 1977; Siminovitch and Thompson, 1978). Such G_1 *ts* mutants, when shifted to the nonpermissive temperature, go through late G_1, S, G_2, and M regularly, but they arrest in mid-G_1 when shifted up after mitosis (Ashihara *et al.*, 1978b; Smith and Wigglesworth, 1974) or if made quiescent, after serum stimulation (Ashihara *et al.*, 1978b). Therefore, it is clear that some functions must be completed before cells can enter into S, and that in most cells, these functions are carried out in the interval between M and S, i.e., G_1.

Supporting evidence for the existence of these G_1 functions comes from the identification of a number of biochemical events that occur in G_1. These include synthesis of actin (Riddle *et al.*, 1979), phosphorylation of ribosomal protein S6 (Haselbacher *et al.*, 1979), synthesis of non-histone chromosomal proteins

(Tsuboi and Baserga, 1972), and a requirement for RNA polymerase II (Rossini et al., 1979).

It has been known for a number of years that infection with certain DNA oncogenic viruses (including SV40 and polyoma) induces cellular DNA synthesis in the host cell (as described previously in this chapter) in a manner similar to growth factor or serum stimulation of quiescent cultures, i.e., after a lag period and after the appearance of large T antigen.

VI. EVIDENCE THAT SIGNALS FOR GROWTH IN SIZE CAN BE SEPARATED FROM SIGNALS FOR CELL DNA REPLICATION

As mentioned previously, there is evidence that a proliferating cell does receive signals for growth in cell mass and replication of its DNA and that DNA oncogenic viruses such as SV40 and polyoma are capable of providing these signals. The question that arises is whether these signals are independent, i.e., can we find certain stimuli that provide only the signal for DNA replication and not the signal for increase in cell size?

The literature contains a number of reports that show that these signals can indeed be separable.

1. It is clear from the studies of Ross and collaborators (1978) and of Pledger et al. (1977) that there are at least two signals for the entry of cells into S. The first signal comes from the PDGF that primes the cell, stimulates RNA synthesis, but does not by itself cause cells to enter into S. A second signal comes from a factor in platelet-poor plasma that makes cells, made competent by PDGF, enter into S. The simplest explanation is that a plasma factor interacts with a PDGF-induced cellular product to initiate DNA synthesis. It appears that PDGF gives the signal for doubling the size of the cells, whereas the signal for DNA replication comes from the plasma factor plus a PDGF-induced cellular product (Scher et al., 1979).

2. Originally isolated from BHK cells by Meiss and Basilico (1972), tsAF8 cells are bona fide G_1 mutants. They have been identified as a mutant of RNA polymerase II by biochemical and genetic evidence (Rossini and Baserga, 1978; Shales et al., 1980). Although tsAF8 cells, at the nonpermissive temperature, are incapable of entering the S phase of the cell cycle, RNA polymerase I activity is not affected. Studies by flow cytophotometry have demonstrated that in tsAF8 cells, RNA accumulates just as effectively at the nonpermissive temperature as at the permissive (Ashihara et al., 1978b). These findings suggest that the requirements for cell DNA replication may be different from those for growth in size.

However, it should be made clear that cell size in itself is not sufficient for entry into S. Indeed, in human aged diploid fibroblasts (in culture), an inverse relationship appears to exist between growth rate and cell volume (Mitsui and Schneider, 1976).

3. 422E cells, a mutant of BHK that fails to accumulate ribosomes at the nonpermissive temperature (Toniolo *et al.*, 1973), will enter DNA synthesis even under nonpermissive conditions (Grummt *et al.*, 1979) although they fail to divide (Mora *et al.*, 1980).

4. Adenovirus 2 stimulates DNA synthesis in infected cells (Laughlin and Strohl, 1976) but does not cause an increase in rRNA synthesis (Eliceiri, 1973). Infection by adenovirus also fails to increase the accumulation of cellular RNA (Pochron *et al.*, 1980). This is at variance with SV40, which stimulates rRNA synthesis (Soprano *et al.*, 1979) and accumulation of cellular RNA (Weil *et al.*, 1977; Petralia *et al.*, 1980).

VII. THE INFORMATION FOR GROWTH IN SIZE AND FOR CELL DNA REPLICATION IS ENCODED IN DIFFERENT DNA SEQUENCES OF THE SV40 *A* GENE

Since adenovirus 2 contains information for the stimulation of cell DNA synthesis in quiescent mammalian cells but not the information for growth in size (Laughlin and Strohl, 1976; Rossini *et al.*, 1979; Eliceiri, 1973; Soprano *et al.*, 1980; Pochron *et al.*, 1980), it is conceivable that the information for cell DNA replication may be encoded in DNA sequences that do not contain the information for growth in size. This hypothesis has been supported by a series of experiments which we have performed with SV40.

As mentioned previously, SV40 infection can stimulate cell DNA replication and mitosis in quiescent mammalian cells (Weil, 1978). We have investigated whether the capacity to stimulate growth in size can be dissociated from the capacity to stimulate cell DNA replication by using adeno–SV40 hybrid viruses and mutants and fragments of SV40 that have been cloned in plasmids and that can be microinjected into the nuclei of quiescent cells.

A. Ability of Portions of the SV40 *A* Gene to Reactivate rRNA Genes

To determine SV40 stimulation of growth in cell size, we investigated the ability of this virus to act on rRNA genes on the reasonable assumption that rRNA genes constitute a target of any signal for growth in size of the cell. We felt that this was not an unreasonable assumption because of the aforementioned role of rRNA synthesis and accumulation in cell proliferation.

To determine the effect of SV40 on rRNA genes, we used the system of reactivation of silent rRNA genes (Soprano et al., 1979). This method is based on the principle that hybrids between human and rodent cells express only the rRNA of the dominant species (the dominant species is the one whose chromosomes are not segregated). The rRNA genes of the recessive species, though often present, are not expressed (Croce, 1976; Croce et al., 1977). The two species of rRNA can be distinguished because human 28 S rRNA migrates in gels at a slightly slower speed than 28 S rRNA of rodents (Eliceiri and Green, 1969). For our studies we selected 55-54 cells, a hybrid cell line between human fibrosarcoma HT1080 cells and BALB/c macrophages, in which the human species is dominant. These hybrid cells have retained all human chromosomes and 18 mouse chromosomes, including chromosomes 12, 15, and 18, where mouse genes for rRNA are located (Croce, 1976). These hybrid cells express human 28 S rRNA but not mouse 28 S rRNA, although mouse rRNA genes are present (Croce et al., 1977). In addition, restriction endonuclease analysis of the gene products (Perry et al., 1979) and silver staining of the nucleolus organizers (Miller et al., 1976) have shown that it is the synthesis and not the processing of the rRNA of the recessive species that is inhibited. Thus, one can say that in 55-54 cells, even when growing exponentially in 10% serum, the expression of mouse rRNA genes is repressed.

We previously determined that treatment of such hybrid cells with tetradecanoyl phorbol-13 acetate (Soprano and Baserga, 1980) or SV40 (Soprano et al., 1979, 1980, 1981) reactivates the silent rRNA genes, but infection with adenovirus 2 does not. This supported our idea that the SV40 genome contains information for growth in size as well as for cell DNA replication, whereas the genome of adenovirus 2 contains information only for the latter. These findings allowed us to determine the sequences in the SV40 genome that are necessary and sufficient for the reactivation of silent rRNA genes by the use of a number of adeno–SV40 hybrid viruses. These are adenoviruses containing hybrid DNA which consists of all or part of the SV40 genome and all or part of the adenovirus 2 genome (Lewis, 1977). The size and location of the SV40 fragment inserted within these hybrids has been well characterized (see Table II) along with the specific SV40 RNA and antigens they produce (Morrow et al., 1973; Kelly and Lewis, 1973; Lebowitz et al., 1974; Walter and Martin, 1975). In general, these viruses contain a single SV40 substitution inserted in place of a single adenovirus 2 deletion. They produce hybrid mRNAs which code for hybrid proteins. These proteins, however, are functional despite the fact that they may be coded partially by adenovirus and partially by SV40 (Khoury et al., 1973; Levine et al., 1973; Martin et al., 1974; Westphal et al., 1979). For example, the 23K hybrid protein synthesized by Ad2+ND1, which contains only about 175 SV40 coded amino acids, is capable of enhancing adenovirus growth in monkey kidney cells (adenovirus helper function of larger T antigen) (Fey et al., 1979). Since ade-

TABLE II

Sequences of SV40 Genome Required for the Reactivation of Ribosomal RNA Genes[a]

Virus or plasmid	Linear map of SV40 sequences (—— A Gene ——)	SV40 map units	rRNA reactivation
SV40	1.0 ———— 0.67 ———— 0.17 —— 0	1.00/0	+
pSV3B	1.0 ———————————————— 0	1.00/0	+
pSV2G	1.0 ———————————— 0.144	1.00/0.144	+
pSV-Pst	0.73 ———— 0.27	0.73/0.27	+
pSV1058	1.0 ———— 0.288 0.26 ———— 0	Δ[b] 0.288–0.260*[c]	+
pSV1061	1.0 ———— 0.231 0.215	Δ 0.231–0.215*	+
pSV-PvuII	0.70 ———— 0.32	0.70/0.32	+
pSV-HpaI	0.73 ———— 0.37	0.73/0.37	–
pSV1139	1.0 ———— 0.339 0.332 ———— 0	Δ 0.339–0.332*	–

pSV1151	1.0 —— 0.385	0.322 —— 0		Δ 0.385–0.322		–
pSV1055	1.0 —— 0.351	0.340 —— 0		Δ 0.351–0.340*		–
pVR200	0.72 —— 0.42			0.72/0.42		–
pSV1135	1.0 —— 0.639	0.636 —— 0		Δ 0.639–0.636		+
dl 2005	1.0 —— 0.59	0.54 —— 0		0.59/0.54		+
Ad2+ND2	0.44 —— 0.11			0.44/0.11		+
Ad2+ND5	0.39 —— 0.11			0.39/0.11		+
Ad2+ND1	0.28 —— 0.11			0.28/0.11		–
Adenovirus 2	None			—		–
pBR322	None			—		–

[a] The plasmids were manually microinjected into the nucleus of approximately 300 55-54 cells. The effect of the viruses was determined by infection of 55-54 cells using a multiplicity of infection (MOI) of at least 10. The SV40 sequences are given counterclockwise. Experimental details are given in the papers by Soprano et al. (1979, 1980, 1981). Some of the clones were constructed by Dr. J. Pipas in the laboratory of Dr. D. Nathans.

[b] Δ, Deletion of the sequences between these map units.

[c] *, Predicted T antigen polypeptide will terminate at the deletion.

13

novirus 2 by itself does not stimulate cellular RNA synthesis (Eliceiri, 1973; Soprano et al., 1980), it is possible to correlate this function of SV40 with the segment of SV40 DNA contained in those hybrids capable of rRNA gene reactivation.

One of these hybrid viruses contains a small deletion of the adenovirus genome that has been replaced by a segment of the SV40 genome that extends from nucleotide residue 232 to nucleotide residue 2385 of the conventional SV40 map. In this chapter we follow the numbering system of nucleotide residues given by Tooze (1980, p. 799). The numbers are always given in counterclockwise order. This fragment of the SV40 genome contains the whole early region of SV40 (nucleotides 5230–2586), and indeed cells infected with this hybrid virus, called Ad2-D1 (Hassell et al., 1978), become positive by indirect immunofluorescence for the SV40 T antigen. When 55-54 cells were infected with this hybrid virus they became positive by indirect immunofluorescence for SV40 T antigen, and the silent mouse rRNA genes were reactivated (Soprano et al., 1980). These experiments formally demonstrated that the information in the SV40 genome that causes the reactivation of silent mouse rRNA genes is encoded in the A gene of the virus, specifically, that portion that codes for the T antigens. By extrapolation, we can therefore say that SV40 T antigen induces growth in size in cells, a statement confirmed by the reports that infection with SV40 causes a coordinate increase of cellular RNA (Khandjian et al., 1980; Petralia et al., 1980).

By a combination of infection with other adeno–SV40 hybrid viruses and microinjection of a number of SV40 DNA fragments or deletion mutants of SV40, we have been able to localize the sequences in the SV40 genome that are critical for reactivation of rRNA synthesis. The results of these experiments are summarized in Table II. It should be remembered that the SV40 A gene, which codes both the large T and small t antigens, maps from 0.67 to 0.17 mu, counterclockwise. From Table II it is apparent that the large T antigen is sufficient for the stimulation of rRNA genes, i.e., the small t is not required (dl 2005 makes a large T but no small t). It is also apparent that the sequences from 0.67 to 0.39 of the SV40 A gene (the sequences deleted in Ad2+ND5, Ad2+ND4, and pSV1135) are not necessary for the reactivation of rRNA genes, and neither are the sequences from 0.33 to 0.17 (the sequences absent from pSV-Pst, pSV-Pvu, pSV1058, and pSV1061). However, deletion of sequences between 0.39 and 0.33 (e.g., pSV-HpaI, pSV1139, pSV1055, and pSV1151) results in loss of this activity. Therefore, the conclusion is that in the SV40 A gene, the sequences critical for reactivation of rRNA genes are those mapping between 0.39 and 0.33 mu. These results should be interpreted with caution. They do not mean that the 100 or so amino acids, coded by those 300 nucleotides between 0.39 and 0.33 mu, are sufficient for the reactivation of ribosomal RNA genes. Obviously, tertiary structure of proteins is important for protein function; but what these data convey is that the critical sequences for reactivation of ribosomal

RNA genes are indeed specified by this nucleotide sequence. An analogy could be made with the active site of an enzyme or with the antigen-binding site of an antibody.

B. Ability of SV40 Fragments and Deletion Mutants to Induce Cell DNA Replication

To study the induction of cell DNA synthesis by the various cloned fragments and deletion mutants of SV40, quiescent NIH 3T3, and ts13 cells were microinjected with intact, supercoil, plasmid DNA. Control cells were microinjected with either pBR322, pSV3B (a clone containing the entire SV40 genome), or pSV2G (a clone containing the entire A gene of SV40). 3T3 cells were incubated at 37°C, and ts13 cells at 34°C in conditioned Dulbecco's medium, supplemented with 1% serum, in the presence of [^{3}H]thymidine. Twenty-four hours after microinjection the cells were fixed, stained for T antigen, and processed for autoradiography.

The results are summarized in Table III, which gives the map coordinates of the SV40 clones microinjected into cells, as well as their ability to stimulate cell DNA replication. From Table III it is apparent that deletion of any sequences distal (or 3') to nucleotide 4001 (0.42 mu) (pVR200, pSV1055, 1151, 1139, 1061, 1058, pSV-Pst, -Hpa, and -PvuII) does not affect the ability of the microinjected DNA to stimulate cell DNA synthesis in either ts13 or NIH 3T3 cells. Deletions that lead to presumptive termination at a site proximal (or 5') to nucleotide 4147 (0.45 mu) (pSV1138, 1137, and 1046) are unable to induce cell DNA synthesis. We conclude, therefore, that the sequences between 0.45 and 0.42 mu are essenial for stimulation of cell DNA synthesis by T antigen. Of particular interest is the observation that clones pVR200, pSV1055, 1151, and 1139 have all the necessary capacity for stimulating cell DNA replication, although these mutants are incapable of reactivating rRNA genes. Therefore, the information for the stimulation of cell DNA replication is clearly separable from the information for the reactivation of rRNA genes.

VIII. DISCUSSION

The ability to reduce functional, critical portions of a gene to a minimum number of sequences opens new possibilities. In fact, one can hypothesize that similar gene sequences are present in noninfected, nontransformed cells and that a fundamental difference between transformed and nontransformed cells is that these genes are under tighter control in normal cells than in transformed cells. In other words, it may be possible that similar gene sequences are expressed in nontransformed cells only under optimal conditions of nutrition and growth

TABLE III

Stimulation of Cell DNA Replication by Recombinant SV40 Plasmids[a]

Plasmid	Linear map of SV40 sequences (A Gene)	SV40 map units	Stimulation of cell DNA synthesis
SV40	1.0 — 0.67 — 0.17 — 0	1.00/0	+
pSV3B	1.0 — 0	1.00/0	+
pSV2G	1.0 — 0.144	1.00/0.144	+
pSV-Pst	0.73 — 0.27	0.73/0.27	+
pSV1058	1.0 — 0.288 0.26 — 0	Δ[b] 0.288–0.26*[c]	+
pSV1061	1.0 — 0.231 0.21 — 0	Δ 0.231–0.215*	+
pSV-PvuII	1.0 — 0.70 — 0.32	0.70/0.32	+

Plasmid					+ / −
pSV-HpaI		0.73 ——— 0.37		0.73/0.37	+
pSV1139	1.0	0.339 0.332	0	Δ 0.339–0.332*	+
pSV1151	1.0	0.385 0.322	0	Δ 0.385–0.322	+
pSV1055	1.0	0.351 0.34	0	Δ 0.351–0.34*	+
pVR200		0.72 ——— 0.42		0.72/0.42	+
pSV1137	1.0	0.51 0.504	0	Δ 0.51–0.504*	−
pSV1138	1.0	0.49 0.482	0	Δ 0.49–0.482*	−
pSV1046	1.0	0.46 0.42	0	Δ 0.46–0.42*	−
pBR322		None		—	−

[a] The plasmids were manually microinjected into the nucleus of mouse 3T3 cells and hamster ts13 cells. Cell DNA synthesis was determined by labeling with [3H]thymidine for 24 hr after microinjection followed by autoradiography. Background levels of labeling were always below 7%. SV40 map units are given counterclockwise. Some of the deletion plasmids were constructed by Dr. J. Pipas in the laboratory of Dr. D. Nathans.

[b] Δ, Deletion of the sequences between these map units.

[c] *, Predicted T antigen polypeptide will terminate at the deletion.

Baserga, R. (1963). Mitotic cycle of ascites tumor cells. *Arch. Pathol.* **75,** 156–161.

Baserga, R. (1980). Introduction of RNA and proteins into viable cells. *In* "Introduction of Macromolecules into Viable Mammalian Cells" (R. Baserga, C. Croce, and G. Rovera, eds.), pp. 79–85. Alan R. Liss, Inc., New York.

Basilico, C. (1977). Temperature-sensitive mutations in animal cells. *Adv. Cancer Res.* **24,** 223–266.

Baxter, G. C., and Stanners, C. P. (1978). The effect of protein degradation on cellular growth characteristics. *J. Cell. Physiol.* **96,** 139–146.

Benjamin, T. L. (1966). Virus specific RNA in cells productively infected or transformed by polyoma virus. *J. Mol. Biol.* **16,** 359–373.

Bravo, R., and Celis, J. E. (1980). Direct microinjection of rabbit globin mRNA into mouse 3T3 cells. *Exp. Cell Res.* **126,** 481–485.

Capecchi, M. R. (1980). High efficiency transformation by direct microinjection of DNA into cultured mammalian cells. *Cell* **22,** 479–488.

Conrad, S., and Botchan, M. R. (1982). Isolation and characterization of human DNA fragments with nucleotide sequences homologies with the simian virus 40 regulatory region. *Mol. Cell. Biol.* **2,** 949–965.

Crawford, L., Leppard, K., Lane, D., and Harlon, E. (1982). Cellular proteins reactive with monoclonal antibodies directed against simian virus 40 T. Antigen. *J. Virol.* **42,** 612–620.

Croce, C. M. (1976). Loss of mouse chromosomes in somatic cell hybrids between HT-1080 human fibrosarcoma cells and mouse peritoneal macrophages. *Proc. Natl. Acad. Sci. U.S.A.* **73,** 3248–3252.

Croce, C. M., Talavera, A., Basilico, C., and Miller, O. J. (1977). Suppression of production of mouse 28S ribosomal RNA in mouse-human hybrids segregating mouse chromosomes. *Proc. Natl. Acad. Sci. U.S.A.* **74,** 694–697.

Darzynkiewicz, Z., Evanson, D. P., Staiano-Coico, L., Sharpless, T. K., and Melamed, M. L. (1979). Correlation between cell cycle duration and RNA content. *J. Cell. Physiol.* **100,** 425–438.

Deppert, W., Gurney, E., and Harrison, R. (1981). Monoclonal antibodies against simian virus 40 tumor antigens: Analysis of antigenic binding sites, using adenovirus type 2-simian virus 40 hybrid viruses. *J. Virol.* **37,** 478–482.

Diacumakos, E. (1980). Introduction of macromolecules into viable mammalian cells by precise physical microinjection. *In* "Introduction of Macromolecules into Viable Mammalian Cells" (R. Baserga, C. Croce, and G. Rovera, eds.), pp. 85–99. Alan R. Liss, Inc., New York.

Eliceiri, G. L. (1973). Ribosomal RNA synthesis after infection with adenovirus type 2. *Virology* **56,** 604–607.

Eliceiri, G. L., and Green, H. (1969). Ribosomal RNA synthesis in human-mouse hybrid cells. *J. Mol. Biol.* **41,** 253–260.

Feramisco, J. R. (1979). Microinjection of fluorescently labeled α-actin into living fibroblasts. *Proc. Natl. Acad. Sci. U.S.A.* **76,** 3967–3971.

Fey, G., Lewis, J. B. Grodzicker, T., and Bothwell, A. (1979). Characterization of a fused protein specified by the adenovirus type 2-simian virus 40 hybrid Ad2$^+$ND1 dp2. *J. Virol.* **30,** 201–217.

Fiers, W., Contreras, R., Haegeman, G., Rogiers, R., Van de Veorde, A., Van Heuversuyn, H., Van Herreweghe, G., Volckaert, G., and Ysebaert, M. (1978). Complete nucleotide sequence of SV40 DNA. *Nature (London)* **273,** 113–120.

Floros, J., Jonak, G., Galanti, N., and Baserga, R. (1981). Induction of cell DNA replication of G_1 specific ts mutants by microinjection of SV40 DNA. *Exp. Cell Res.* **132,** 215–223.

Galanti, N., Jonak, G. J., Soprano, K. J., Floros, J., Kaczmarek, L., Weissman, S., Reddy, V. B.,

Tilghman, S. M., and Baserga, R. (1981). Characterization and biological activity of cloned simian virus 40 DNA fragments. *J. Biol. Chem.* **256**, 6469–6474.

Graessmann, A., and Graessmann, M. (1976). "Early" simian virus-40-specific RNA contains information for tumor antigen formation and chromatin replication. *Proc. Natl. Acad. Sci. U.S.A.* **73**, 366–370.

Graessmann, A., Graessmann, M., and Mueller, C. (1977). Regulatory function of simian virus 40 DNA replication for late viral gene expression. *Proc. Natl. Acad. Sci. U.S.A.* **74**, 4831.

Graessmann, A., Graessmann, M., Guhl, E., and Mueller, C. (1978). Quantitative correlation between simian virus 40 T-antigen synthesis and late viral gene expression in permissive and non-permissive cells. *J. Cell Biol.* **77**, R1.

Grummt, F., Grummt, I., and Mayer, E. (1979). Ribosome biosynthesis is not necessary for initiation of DNA replication. *Eur. J. Biochem.* **97**, 37–42.

Hartwell, L. (1978). Cell division from a genetic perspective. *J. Cell Biol.* **77**, 627–637.

Haselbacher, G. K., Humble, R. E., and Thomas, G. (1979). Insulin like growth factor: Insulin or serum increase phosphorylation of ribosomal protein S6 during transition of stationary chick embryo fibroblasts into early G_1 phase of the cell cycle. *FEBS Lett.* **100**, 185–190.

Hassell, J. A., Lukanindin, E., Fey, G., and Sambrook, J. (1978). The structure and expression of two defective adenovirus 2 simian virus 40 hybrids. *J. Mol. Biol.* **120**, 209–247.

Henry, P., Black, P. H., Oxman, M. N., and Weissman, S. M. (1966). Stimulation of DNA synthesis in mouse cell line 3T3 by simian virus 40. *Proc. Natl. Acad. Sci. U.S.A.* **56**, 1170–1176.

Howard, A., and Pelc, S. R. (1951). A nuclear incorporation of P^{32} as demonstrated by autoradiographs. *Exp. Cell Res.* **2**, 178–187.

Kelly, T. J., Jr., and Lewis, A. M., Jr. (1973). Use of nondefective adenovirussimian virus 40 hybrids for mapping the simian virus 40 genome. *J. Virol.* **12**, 643–652.

Khandjian, E. W., Matter, J., Leonard, N., and Weil, R. (1980). Simian virus 40 and polyoma virus stimulate overall cellular RNA and protein synthesis. *Proc. Natl. Acad. Sci. U.S.A.* **77**, 1476–1480.

Khoury, G., Lewis, A. M., Jr., Oxman, M. N., and Levine, A. S. (1973). Strand orientation of SV40 transcription of cells infected by non-defective adenovirus 2-SV40 hybrid viruses. *Nature (London) New Biol.* **246**, 202–205.

Kreis, T. E., Winterhalter, K. H., and Birchmeier, W. (1979). *In vivo* distribution and turnover of fluorescently labeled actin microinjected into human fibroblasts. *Proc. Natl. Acad. Sci.* **76**, 3814–3818.

Lala, P. K., and Patt, H. M. (1966). Cytokinetic analysis of tumor growth. *Proc. Natl. Acad. Sci. U.S.A.* **56**, 1735–1742.

Lane, D., and Hoeffler, W. K. (1980). SV40 large T shares an antigenic determinant with a cellular protein of molecular weight 68,000. *Nature (London)* **288**, 167–170.

Laughlin, C., and Strohl, W. A. (1976). Factors regulating cellular DNA synthesis induced by adenovirus infection. 2. The effects of actinomycin D on productive virus cell systems. *Virology* **74**, 44–56.

Lebowitz, P., and Weissman, S. (1979). Organization and transcription of the simian virus 40 genome. *Curr. Top. Microbiol. Immunol.* **87**, 44–172.

Lebowitz, P., Kelly, T. J., Jr., Nathans, D., Lee, T. N. H., and Lewis, A. M., Jr. (1974). A colinear map relating the simian virus 40 (SV40) DNA segments of six adenovirus-SV40 hybrids to the DNA fragments produced by restriction endonuclease cleavage of SV40 DNA. *Proc. Natl. Acad. Sci. U.S.A.* **71**, 441–445.

Levine, A. S., Levin, M. J., Oxman, M. N., and Lewis, A. M., Jr. (1973). Studies of nondefective adenovirus 2-simian virus-40 hybrid viruses. VII. Characterization of the simian virus 40 RNA species induced by five nondefective hybrid viruses. *J. Virol.* **11**, 672–681.

Lewis, A. M. (1977). Defective and nondefective Ad2-SV40 hybrids. *Prog. Med. Virol.* **23,** 96–139.

Liskay, R. M. (1978). Genetic analysis of a Chinese hamster cell line lacking G_1 phase. *Exp. Cell Res.* **114,** 69–77.

Liskay, R. M., and Prescott, D. M. (1978). Genetic analysis of the G_1 period: Isolation of mutants (or variants) with the G_1 period from a Chinese hamster cell like lacking G_1. *Proc. Natl. Acad. Sci. U.S.A.* **75,** 2873–2877.

McCutchan, T., and Singer, M. (1981). DNA sequences similar to those around the simian virus 40 origin of replication are present in the monkey genome. *Proc. Natl. Acad. Sci. U.S.A.* **78,** 95–99.

Maniatis, T., Hardison, R., Lacy, E., Lauer, J. O'Connell, C., and Quon, D. (1978). The isolation of Structural genes from libraries of eucaryotic DNA. *Cell* **15,** 687–701.

Martin, R. G., Chou, J. Y., Avila, J., and Saral, R. (1974). The semi-autonomous replicon: A molecular model for the oncogenicity of SV40. *Cold Spring Harbor Symp. Quant. Biol.* **39,** 17–24.

Mauck, J. C., and Green H. (1973). Regulation of RNA synthesis in fibroblasts during transition from resting to growing state. *Proc. Natl. Acad. Sci. U.S.A.* **70,** 2819–2822.

May, P., May, E., and Borde, J. (1976). Stimulation of cellular RNA synthesis in mouse kidney cell cultures infected with SV40 virus. *Exp. Cell Res.* **100,** 433–436.

Meiss, H. K., and Basilico, C. (1972). Temperature-sensitive mutants of BHK 21 cells. *Nature (London), New Biol.* **239,** 66–68.

Miller, O. J., Miller, D. A., Dev, V. G., Tantravahi, R., and Croce, C. M. (1976). Expression of human and suppression of mouse nucleolus organizer activity in mouse-human somatic cell hybrids. *Proc. Natl. Acad. Sci. U.S.A.* **73,** 4531–4535.

Mitsui, Y., and Schneider, E. L. (1976). Relationship between cell replication and volume in senescent human diploid fibroblasts. *Mech. Ageing Dev.* **5,** 45–56.

Mora, M., Darzynkiewicz, Z., and Baserga, R. (1980). DNA synthesis and cell division in a mammalian cell mutant temperature-sentitive for the processing of ribosomal RNA. *Exp. Cell Res.* **125,** 241–249.

Morrow, J. F., Berg, P., Kelly, T. J., Jr., and Lewis, A. M., Jr. (1973). Mapping of simian virus 40 early functions on the viral chromosome. *J. Virol.* **12,** 653–658.

Mueller, G., Graessmann, A., and Graessmann, M. (1978). Mapping of early SV40 specific functions by microinjection of different early viral DNA fragments. *Cell* **15,** 579–585.

Organtini, J. E., Joseph, C. R., and Farber, J. L. (1975). Increase in the activity of the solubilized rat liver nuclear polymerases following partial hepatectomy. *Arch. Biochem. Biophys.* **170,** 485–491.

Perry, R. P., Kelley, D. E., Schibler, U., Huebner, K., and Croce, C. (1979). Selective suppression of the transcription of ribosomal genes in mouse-human hybrid cells. *J. Cell. Physiol.* **98,** 553–560.

Petralia, S., Soprano, K. J., Pochron, S., Darzinkiewicz, Z., Traganos, F., and Baserga, R. (1980). Increases in cellular RNA in cells infected with DNA oncogenic viruses. *Cancer Res.* **40,** 3177–3180.

Pledger, W. J., Stiles, C. D., Antoniades, H. N., and Scher, C. D. (1977). Induction of DNA synthesis in Balb C 3T3 cells by serum components: Reevaluation of the commitment process. *Proc. Natl. Acad. Sci. U.S.A.* **74,** 4481–4485.

Pochron, S., Rossini, M., Darzynkiewicz, Z., Traganos, F., and Baserga, R. (1980). Failure of accumulation of cellular RNA in hamster cells stimulated to synthesize DNA by infection with adenovirus 2. *J. Biol. Chem.* **255,** 4411–4413.

Proudfoot, N. J., Shander, M., Mancey, J. L., Gefter, M. L., and Maniatis, T. (1980). Structure and in vitro transcription of human globin genes. *Science* **209,** 1129–1136.

Queen, C., Loro, S. T., McCutchan, T., and Singer, M. (1981). Three segments from the monkey genome that hybridize to simian virus 40 have common structural elements. *Mol. Cell. Biol.* **1,** 1061–1068.

Reddy, V. B., Thimmappaya, B., Dhar, R., Subramanian, N., Zain B. S., Pan, J., Ghosh, P. K., Celma, M. L., and Weissman, S. (1978). The genome of simian virus 40. *Science* **200,** 494–502.

Richardson, W. D., Carter, B. J., and Westphal, H. (1980). Vero cells injected with adenovirus type 2 mRNA produce authentic viral polypeptide patterns: Early mRNA promotes growth of adenovirus-associated virus. *Proc. Natl. Acad. Sci. U.S.A.* **77,** 931–935.

Riddle, V. G. H., Dubrow, R., and Pardee, A. B. (1979). Changes in the synthesis of actin and other cell proteins after stimulation of serum arrested cells. *Proc. Natl. Acad. Sci. U.S.A.* **76,** 1298–1302.

Robbins, E., and Scharff, M. D. (1967). The absence of a detectable G_1 phase in a cultured strain of Chinese hamster lung cells. *J. Cell Biol.* **34,** 684.

Ross, R., Nist, C., Kariya, B., Rivest, M. J., Raines, E., and Callis, J. (1978). Physiological quiescence in plasma derived serum. Influence of platelet derived growth factor on cell growth in culture. *J. Cell. Physiol.* **97,** 497–508.

Rossini, M., and Baserga, R. (1978). RNA synthesis in a cell cycle-specific temperature sensitive mutant from a hamster cell line. *Biochemistry* **17,** 858–863.

Rossini, M., Weinmann, R., and Baserga, R. (1979). DNA synthesis in temperature-sensitive mutants of the cell cycle infected by polyoma virus and adenovirus. *Proc. Natl. Acad. Sci. U.S.A.* **76,** 4441–4445.

Rovera, G., Mehta, S., and Maul, G. (1975). Ghost monolayers in the study of the modulation of transcription in cultures of CV1 Fibroblasts. *Exp. Cell Res.* **89,** 295–305.

Scher, C. D., Stone, M. E., and Stiles, C. D. (1979). Platelet derived growth factor prevents G_1 growth arrest. *Nature (London)* **281,** 390–392.

Schmid, W., and Sekaris, C. E. (1975). Nucleolar RNA synthesis in the liver of partially hepatectomized and Cortisol treated rats. *Biochim. Biophys. Acta* **402,** 244–252.

Shales, M., Bergsagel, J., and Ingles, C. J. (1980). Defective RNA polymerase II in the G_1 specific temperature sensitive hamster cell mutant tsAF8. *J. Cell. Physiol.* **105,** 527–532.

Shen, Y., Hirschhern, R., Mercer, W., Surmacz, E., Tsutsui, Y., Soprano, K. J., and Baserga, R. (1982). Gene transfer: DNA microinjection compared with DNA transfection with a very high efficiency. *Mol. Cell. Biol.* **2,** 1145–1154.

Siminovitch, L., and Thompson, L. H. (1978). The nature of conditionally lethal temperature-sensitive mutations in somatic cells. *J. Cell Physiol.* **95,** 361–366.

Skog, S., Eliasson, E., and Eliasson, E. (1979). Correlation between cell size and position within the division cycle in suspension cultures of Chang liver cells. *Cell Tissue Kinet.* **12,** 501–511.

Smith, B. J., and Wigglesworth, N. M. (1974). Studies on the Chinese hamster line that is temperature-sensitive for the commitment to DNA synthesis.

Soprano, K. J., and Baserga, R. (1980). Reactivation of ribosomal RNA gene in human-mouse hybrid cells by 12-0 tetradecanoyl phorbol-13 acetate (TPA) *Proc. Natl. Acad. Sci. U.S.A.* **77,** 1566–1569.

Soprano, K. J., Dev. V. G., Croce, C. M., and Baserga, R. (1979). Reactivation of silent rRNA genes by simian virus 40 in human-mouse hybrid cells. *Proc. Natl. Acad. Sci. U.S.A.* **76,** 3885–3889.

Soprano, K. J., Rossini, M., Croce, C., and Baserga, R. (1980). The role of large T antigen in simian virus 40-induced reactivation of silent rRNA genes in human-mouse hybrid cells. *Virology* **102,** 317–326.

Soprano, K. J., Jonak, G. J., Galanti, N., Floros, J., and Baserga, R. (1981). Identification of an

SV40 DNA sequence related to the reactivation of silent rRNA genes in human-mouse hybrid cells. *Virology* **109**, 127–136.

Stacey, D. (1980). Behavior of microinjected molecules and recipient cells. In "Introduction of Macromolecules into Viable Mammalian Cells" (R. Baserga, C. Croce, and G. Rovera, eds.), pp. 125–134. Alan R. Liss, Inc., New York.

Tjian, R., Fey, G., and Graessmann, A. (1978). Biological activity of purified simian virus 40 T antigen proteins. *Proc. Natl. Acad. Sci. U.S.A.* **75**, 1279–1283.

Toniolo, D., Meiss, H. K., and Basilico, C. (1973). A temperature-sensitive mutation affecting 28S ribosomal RNA production in mammalian cells. *Proc. Natl. Acad. Sci. U.S.A.* **70**, 1273–1277.

Tooze, J. (1980). "DNA Tumor Viruses." Cold Spring Harbor Lab., Cold Spring Harbor, New York.

Tsuboi, A., and Baserga, R. (1972). Synthesis of nuclear acidic proteins in density inhibited fibroblasts stimulated to proliferate. *J. Cell. Physiol.* **80**, 107–118.

Tsukada, K., and Lieberman, I. (1964). Synthesis of ribonucleic acid by liver nuclear and nucleolar preparations after partial hepatectomy. *J. Biol. Chem.* **239**, 2952–2956.

Tucker, P. W., Liu, C., Mushinski, J. F., and Blattner, F. R. (1980). Mouse immunoglobulin D: Messenger RNA and genomic DNA sequences. *Science* **209**, 1353–1360.

Walter, G., and Martin, H. (1975). Simian virus 40-specific proteins in HeLa cells infected with nondefective adenovirus 2-simian virus 40 hybrid viruses. *J. Virol.* **16**, 1236–1247.

Walter, G., Scheidtmann, K. H., Carbonne, A., Laudano, A. P., and Doolittle, R. F. (1980). Antibodies specific for the carboxy- and amino-terminal regions of simian virus 40 large tumor antigen. *Proc. Natl. Acad. Sci. U.S.A.* **77**, 5197–5200.

Weil, R. (1978). Viral 'tumor antigens' a novel type of mammalian regulatory protein. *Biochim. Biophys. Acta* **516**, 301–388.

Weil, R., Salomon, C., May, E., and May, P. (1975). A simplifying concept in tumor virology: Virus-specific pleiotropic affectors. *Cold Spring Harbor Symp. Quant. Biol.* **39**, 381–395.

Weil, R., Turler, H., Leonard, N., and Ahman-Zadeh, C. (1977). Dissociation of lytic and mitogenic action of SV40 in permissive monkey kidney cells. INSERM **69**, 263–280.

Westphal, H., Lai, S., Lawrence, C., Hunter, T., and Walter, G. (1979). Mosaic adenovirus-SV40 RNA specified by the non-defective hybrid virus Ad2+ND$_4$. *J. Mol. Biol.* **130**, 337–351.

Whelly, S., Ide, T., and Baserga, R. (1978). Stimulation of RNA synthesis in isolated nucleoli by preparations of Simian virus 40 T antigen. *Virology* **88**, 82–91.

Yen, A., and Pardee, A. B. (1979). Exponential 3T3 cells escape in mid-G$_1$ from their high serum requirement. *Exp. Cell Res.* **116**, 103–113.

Zardi, L., and Baserga, R. (1974), Ribosomal RNA synthesis in WI38 cells stimulated to proliferate. *Exp. Mol. Pathol.* **20**, 69–77.

2

Expression of Dihydrofolate Reductase and Thymidylate Synthase Genes in Mammalian Cells

LEE F. JOHNSON

Department of Biochemistry
The Ohio State University
Columbus, Ohio

25

RECOMBINANT DNA AND
CELL PROLIFERATION

I. OVERVIEW

Much of our knowledge of the mechanisms for controlling the expression of genes for proteins in higher eukaryotes is based on studies of genes for abundant proteins (e.g., globin and ovalbumin) in highly differentiated cells. The reason for this is quite simple: The high level of expression of these genes facilitates the purification of the proteins and their messages, the cloning of DNA sequences corresponding to the mRNAs and the genes for the proteins, and the analysis of mRNA metabolism and gene transcription.

In contrast, it has been far more difficult to conduct detailed studies of the mechanisms for regulating the expression of genes for enzymes and other "housekeeping" proteins in higher eukaryotes since these normally represent a very small fraction of total cellular protein. Although many of these genes are probably expressed in a constitutive manner, others are known to be controlled over a wide range in response to various stimuli.

It is likely that there are fundamental differences in the mechanisms for controlling the expression of the genes for housekeeping versus abundant proteins. In particular, it is probable that there are differences in the mechanisms for controlling the synthesis and/or processing of the hnRNA corresponding to these two classes of genes since the number of mRNA molecules produced per gene copy differs enormously (Perry *et al.*, 1979). Therefore, it is important to develop procedures that will permit direct and detailed analysis of the mechanisms for controlling the expression of genes for enzymes.

One class of enzymes that is regulated over a wide range under well-defined conditions is the set of "S-phase enzymes" that is necessary for DNA replication. It has been recognized for many years that the activities of these enzymes are much greater in cells that are synthesizing DNA than in those that are not (for review, see Mitchison, 1971; Prescott, 1976). The activities of these enzymes are regulated at several different levels. First, the amounts and rates of production of many of the enzymes appear to be carefully controlled. Second, the activities of some of the enzymes are regulated by end product inhibition (e.g., ribonucleotide reductase). Third, many of the enzymes participate in the formation of a nuclear multienzyme complex ("replitase") that is formed only during S phase and may be responsible for supplying DNA precursors near the site of DNA replication (Reddy and Pardee, 1980). An important problem in the analysis of the cell cycle is to determine how the cell coordinates the production of the individual enzymes and regulates their assembly into the replitase complex during S phase.

In this chapter, I describe recent progress toward elucidating the mechanisms by which mammalian cells control the rates of production of two different (but metabolically related) enzymes, dihydrofolate reductase (DHFR; tetrahydrofolate dehydrogenase) and thymidylate synthase (TS). Both of these enzymes are present at much lower levels in resting (G_0) or in G_1-phase cells than in S-phase cells. Detailed analysis of DHFR and TS gene expression has been facilitated by the isolation of drug-resistant cell lines that overproduce these enzymes. Since DHFR and TS gene expression appear to be regulated in the same manner in the overproducing cells as in the parental cells, they provide a convenient model system for studying the content and metabolism of these two enzymes and their messages.

II. REGULATION OF DHFR, TS, AND TK ENZYME LEVELS IN NORMAL CELLS

Thymidylic acid (TMP) is synthesized by two different biochemical pathways, as shown in Fig. 1. In the *de novo* synthesis pathway, deoxyuridylic acid (dUMP) is converted to TMP in a reductive methylation reaction catalyzed by thymidylate synthase (TS). The methyl donor is N^5,N^{10}-methylenetetrahydrofolate (me-THFA), which is oxidized to dihydrofolate (DHFA) during the course of the reaction. DHFA is converted back to tetrahydrofolate (THFA) by dihydrofolate reductase (DHFR). The salvage pathway involves the transfer of the γ-phosphate of ATP to the 5' position of thymidine and is catalyzed by thymidine kinase (TK).

A. Enzyme Levels in Growing and Nongrowing Cells

Since TMP is synthesized at high rates only in cells that are synthesizing DNA, it is not surprising that the levels of TS and TK are much higher in rapidly proliferating cells than in nongrowing cells (e.g., Maley and Maley, 1960; Conrad and Ruddle, 1972; Littlefield, 1965; Kit, 1970, 1976). Studies with synchronized cultures have shown that the activities of these enzymes are low

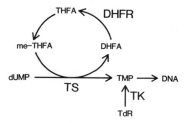

Fig. 1. Enzymes involved in the biosynthesis of thymidylic acid. Thymidylate synthase (TS) catalyzes the conversion of deoxyuridylic acid (dUMP) to thymidylic acid (TMP). Thymidine kinase (TK) catalyzes the phosphorylation of thymidine (TdR) to TMP. Dihydrofolate reductase (DHFR) catalyzes the reduction of dihydrofolic acid (DHFA) to tetrahydrofolic acid (THFA), which is then converted to N^5,N^{10}-methylene-THFA (me-THFA).

during G_1 phase but increase as the cells traverse S phase (e.g., Stubblefield and Murphree, 1967; Conrad, 1971; Kit and Jorgensen, 1976; Piper et al., 1980; Rode et al., 1980).

The situation for DHFR is not quite as obvious. THFA is required for a variety of reactions besides the production of TMP. However, THFA is oxidized to DHFA only during the de novo synthesis of TMP. Consequently, the requirement for DHFR activity is greatest when cells are actively synthesizing DNA. Therefore, it is also reasonable to expect that the DHFR level might be regulated, at least to some extent, during the cell cycle. This possibility is consistent with observations that the specific activity of DHFR is several times greater in growing than in stationary phase or resting cells (e.g., Hillcoat et al., 1967; Chello et al., 1977; Johnson et al., 1978a).

B. Studies with Serum-Stimulated 3T6 Cells

For the past few years, my laboratory has been studying the mechanisms for controlling DHFR, TS, and (to a lesser extent) TK gene expression. Our studies have been conducted with cultured mouse 3T6 fibroblasts (Todaro and Green, 1963). These cells are able to rest at confluent cell density for prolonged periods in the G_0 state of the cell cycle when maintained in medium containing 0.5% calf serum. The resting cells can be induced to reenter the cell cycle by increasing the serum concentration to 10% (Mauck and Green, 1973; Johnson et al., 1974). The rate of DNA synthesis is very low in resting cells, but increases sharply beginning about 10 hr following serum stimulation, and reaches a maximum at about 18 hr, at which time 60–80% of the cells are incorporating [^{3}H]thymidine into DNA. Cell division begins about 20 hr following stimulation, although the cell population does not double until about 30 hr after serum addition.

Serum-stimulated 3T6 cells are an excellent model system for studying the biochemical events leading to DNA replication. Since synchrony is achieved without exposing the cells to toxic drugs or starving them for essential metabolites, there is less need for concern about artifacts that may be introduced by the synchronization procedure. Furthermore, since the cells are maintained in the resting state for 1 week (with regular feeding) prior to stimulation, proteins and mRNAs that are synthesized at high rates in S-phase cells, but not in resting or G_1-phase cells, are present at very low, basal levels at the time of stimulation. This is especially important when studying the expression of genes coding for rather stable proteins or mRNAs.

To determine how closely the cell coordinates the expression of the genes for DHFR, TS, and TK, we have measured the levels of these three enzymes in 3T6 cells as a function of time following serum stimulation (Johnson et al., 1978b, 1982; Navalgund et al., 1980). Figure 2 shows that the level of these enzymes was low in resting cells and remained at the resting level until at least 10 hr following stimulation. Thereafter, the rate of accumulation of enzyme activity

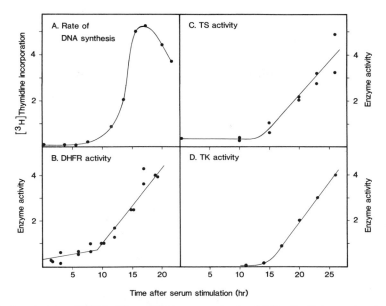

Fig. 2.   Activity of DHFR, TS, and TK in serum-stimulated 3T6 cells. Cultures of resting 3T6 cells were serum stimulated at time = 0. At the indicated times, cultures were assayed for: A, the rate of incorporation of [³H]thymidine into DNA; B, the amount of newly synthesized DHFR activity (resting cultures were incubated briefly prior to stimulation with unlabeled MTX to inactivate preexisting DHFR so that the time of increase in DHFR activity could be determined more accurately); C, TS activity; D, TK activity. Enzyme activities and rates of thymidine incorporation are all expressed in arbitrary units. The figures are adapted from data presented in Johnson *et al.* (1978b) (A and B), Navalgund *et al.* (1980) (C), and Johnson *et al.* (1982) (D).

increased sharply as the cells proceeded through S phase. The increase in specific activities of the enzymes in exponentially growing, as compared to resting, 3T6 cells ranged from about 5-fold for DHFR, to greater than 40-fold for TK.

The increase in the activity of each enzyme was blocked by the addition of inhibitors of protein synthesis, suggesting that the increase is due to *de novo* synthesis of the enzyme rather than activation of a previously existing enzyme. Furthermore, the increase of each enzyme was blocked by 5 μg/ml actinomycin D if the drug was added at about 8 hr following serum stimulation. This suggests that the increase in enzyme activity requires transcription, presumably (although not necessarily) of the gene for the enzyme. We also found that addition of actinomycin D at 16 hr following stimulation had little effect on the increase in enzyme level for at least 6 hr. This shows that the drug does not inhibit the translation of the mRNA for the enzyme. It also suggests that the majority of the mRNA is produced prior to 16 hr following stimulation, and that the mRNA is not extremely labile, at least in the presence of the drug.

The increase in activity of the three enzymes occurs at about the same time that the cells enter S phase. However, the increase in enzyme activities is not tightly linked with DNA replication since the activities increase in essentially the same manner in cells stimulated in the presence of DNA synthesis inhibitors as in control stimulated cells. Thus, the control of expression of the genes for these enzymes differs in this respect from the control of histone gene expression. Histones are also synthesized primarily during S phase. However, histone synthesis is blocked very rapidly following inhibition of DNA synthesis (e.g., Borun *et al.,* 1967). It is important to note that even though the increase in the expression of the genes for the three enzymes is not linked directly with DNA replication, the two events may still be coordinated in some manner by a common regulatory signal.

The only significant difference in the mechanisms for regulating the levels of these enzymes was in the stability of the enzymes. The activity of TK decreased fairly rapidly (half-life = 4 hr) in cells treated with cycloheximide, whereas the activities of the other two enzymes decreased slowly (half-life greater than 24 hr) under these conditions.

III. ISOLATION AND CHARACTERIZATION OF DHFR-OVERPRODUCING CELL LINES

To understand thoroughly the mechanisms for controlling enzyme gene expression, one must be able to quantitate the content and metabolism of both the protein and its mRNA. Due to the small amount of most enzymes in normal mammalian cells, even the isolation of an enzyme in pure form is an involved process. Therefore, the isolation and quantitation of the mRNA for an enzyme, to say nothing of the mRNA precursor or its gene, would be exceedingly difficult. Fortunately, a remarkable series of observations in several different laboratories has greatly reduced this problem for at least certain enzymes.

A. Selection of MTX-Resistant Cells

DHFR is the target enzyme for the cancer chemotherapeutic agent methotrexate (MTX). This drug is a substrate analog that binds extremely tightly at the active site of the enzyme, thereby inactivating it (Blakley, 1969). Since DHFR activity is essential during DNA replication, exposure of proliferating (but not resting) cells to concentrations of MTX as low as 10^{-8} M leads to cell death (Hryniuk *et al.,* 1969; Bertino, 1971, 1979; Johnson *et al.,* 1978a).

A number of laboratories have isolated cell lines that have developed resistance to high concentrations of MTX. These are usually selected by culturing cells for prolonged periods in the presence of gradually increasing concentrations of the drug. Resistant lines have been isolated that are able to proliferate at normal rates in concentrations of MTX as high as $10^{-4}-10^{-3}\,M$ (e.g., Hakala *et al.*, 1961; Littlefield, 1969; Milbrandt *et al.*, 1981).

B. Mechanism of MTX Resistance

The mechanisms by which cells develop resistance to MTX have been studied in detail in the past few years. Although decreased transport of MTX and reduced affinity of DHFR for the drug may be responsible for resistance to low concentrations of MTX, cells usually develop resistance to high concentrations of the drug by overproducing DHFR several hundredfold (for review, see Schimke *et al.*, 1978). The level of the enzyme is so high that even though most of the enzyme is inhibited by MTX, enough active enzyme remains to allow cell proliferation. DHFR represents several percent of total cell protein in these drug-resistant cells (Hakala *et al.*, 1961; Alt *et al.*, 1976). In some cases, MTX resistance has been achieved by more than one mechanism, such as overproduction of DHFR with reduced affinity for MTX (Flintoff *et al.*, 1976; Goldie *et al.*, 1980; Haber *et al.*, 1981).

Overproduction of DHFR is the result of a corresponding increase in the rate of synthesis of the enzyme (Alt *et al.*, 1976; Hanggi and Littlefield, 1976), which is due to a parallel increase in the amount of DHFR mRNA (Kellems *et al.*, 1976; Chang and Littlefield, 1976). Remarkably, the overproduction of DHFR mRNA is due to amplification of the DHFR gene (Alt *et al.*, 1978). Several other examples of amplification of genes for proteins in cells resistant to toxic agents have been observed (e.g., Wahl *et al.*, 1979; Beach and Palmiter, 1981).

The mechanism responsible for gene amplification remains to be determined. It has been suggested that the initial even might be one or more unscheduled duplications of the replicon containing the DHFR gene. Since this gene is known to replicate early in S phase (Kellems *et al.*, 1982; Milbrandt *et al.*, 1981) it may be more likely than other genes to experience disproportionate replication. This process may be stimulated by agents that damage DNA, that interrupt normal DNA replication, or by known tumor promoters (Schimke *et al.*, 1978; Schimke 1982; Varshavsky, 1981; Brown *et al.*, 1983). Disproportionate replication of the DHFR gene (and perhaps many other genes as well) is believed to occur spontaneously. However, the duplicated DHFR genes are rapidly lost unless they confer a selective advantage to the cells, such as the ability to grow in the presence of MTX. Cells containing still higher levels of DHFR (and a higher number of copies of the DHFR gene) are selected by increasing the concentration of MTX in the culture medium.

The amplified genes are localized either to a homogeneously staining chromosomal region (Biedler and Spengler, 1976; Nunberg *et al.*, 1978; Dolnick *et al.*, 1979) or to double minute chromosomes (Kaufman *et al.*, 1979; Haber and

Schimke, 1981; Brown *et al.*, 1981), which lack centromeres and apparently distribute randomly during cell division. The overproduction trait is stable in the absence of selective pressure in the former, but unstable in the latter.

The structure of the DHFR gene in normal and MTX-resistant cells has been studied in some detail. The DHFR gene is about 32 kilobases (kb) in length, which is quite large for a protein with a molecular weight of 21,000. The coding sequences are interrupted by five intervening sequences, up to about 16 kb in length (Nunberg *et al.*, 1980). The length of the total amplification unit is believed to be many times greater than the length of the gene (Nunberg *et al.*, 1978; Milbrandt *et al.*, 1981).

C. Isolation and Characterization of MTX-Resistant 3T6 Cells

We reasoned that if DHFR gene expression were regulated in DHFR-over-producing cells in the same manner as in normal cells, the overproducing cells would be an ideal model system for studying DHFR gene expression. In such cells it would be possible to measure directly the content, rate of synthesis, and stability of DHFR mRNA and hnRNA, as well as the transcription of the DHFR gene. Therefore, we set out to isolate a MTX-resistant 3T6 cell line that would overproduce DHFR and its mRNA several hundredfold, retain the ability to rest in the G_0 state when maintained in medium containing 0.5% serum, and regulate DHFR gene expression in the same manner as normal 3T6 cells.

After exposing 3T6 cells for many months to gradually increasing concentrations of MTX, we isolated a clone (designated M50L3) that was resistant to 50 μM MTX. DHFR was overproduced about 300-fold in these cells and represented about 4% of total cytoplasmic protein (Wiedemann and Johnson, 1979). A similar cell line was isolated independently by Kellems *et al.* (1979). The M50L3 cells were able to rest reasonably well (although not quite as well as normal 3T6 cells) when kept in medium containing low concentrations of serum. The overproduction trait was gradually lost when M50L3 cells were grown in the absence of selective pressure, suggesting that the amplified DHFR genes were localized on double minute chromosomes. The level of DHFR appeared to be regulated in exactly the same manner in the M50L3 cells as in normal 3T6 cells. We repeated all of the experiments performed originally with 3T6 cells on M50L3 cells and found that the results were virally identical (Wiedemann and Johnson, 1979). Therefore, the M50L3 cells appear to be an excellent model system for studying the biochemical mechanisms for controlling DHFR gene expression.

It is interesting that the ability of the cell to control the expression of the multiple gene copies was not lost as a result of the amplification process. This suggests that if specific regulatory molecules are responsible for controlling DHFR gene expression, they must either be amplified in parallel with the structural gene or be present in great excess in normal cells.

IV. DHFR GENE EXPRESSION IN OVERPRODUCING CELLS

Detailed studies of the content and metabolism of DHFR and its mRNA and the transcription of the DHFR gene required specific probes (i.e., antibodies and cloned cDNAs) for these molecules. Antibodies to the enzyme were obtained by injecting rabbits with DHFR purified from overproducing cells by affinity chromatography (Alt *et al.*, 1976; Wu *et al.*, 1982). The nucleic acid probe was prepared by cloning cDNA corresponding to DHFR mRNA into the *Pst*I site of pBR322 by the poly(dG)–poly(dC) tailing procedure (Chang *et al.*, 1978). The probe used in our studies (pDHFR-21) contained DNA corresponding to the entire length of the 1.6-kb DHFR mRNA molecule, except for about 100 nucleotides at the 5′ end, and was kindly provided by Dr. Robert Schimke.

A. Regulation of Enzyme Level

Previous studies showed that DHFR was a particularly stable enzyme and that the cellular level of DHFR was determined by the rate of synthesis of the enzyme (Alt *et al.*, 1976; Kellems *et al.*, 1979). To confirm that the increase in DHFR enzyme level in serum-stimulated M50L3 cells was due to an increase in the rate of synthesis of the enzyme, cultures of cells were pulse-labeled with [^{3}H]leucine at various times following serum stimulation. The rate of synthesis of DHFR relative to total cell protein was determined by immunoprecipitation. Figure 3A shows that the rate of synthesis of DHFR relative to the rate of synthesis of total proteins increased about fourfold by 20 hr following stimulation (Wiedemann and Johnson, 1979; Wu *et al.*, 1982). This is very close to the increase in the specific activity of the enzyme in growing, compared to resting, cells.

B. Regulation of DHFR mRNA Content and Metabolism

1. Properties of DHFR mRNA

The properties of DHFR mRNA have been analyzed in normal and overproducing cells, and several unusual features have been observed. Although most poly(A)$^{+}$ DHFR mRNA is about 1600 nucleotides in length, several other size classes, ranging from about 750 to 5600 nucleotides in length, have also been observed. All of these messages appear to be functional (Setzer *et al.*, 1980; Lewis *et al.*, 1981; Dolnick and Bertino, 1981). The different sizes appear to result from polyadenylation at different sites within the 3′ untranslated region of the mRNA (Setzer *et al.*, 1982). The physiological significance of the different size classes is not known.

The percentage of the DHFR hnRNA molecule that corresponds to intron sequences is unusually large. Assuming that the initial transcription product is the same size as the gene (32 kb), greater than 95% of the DHFR hnRNA

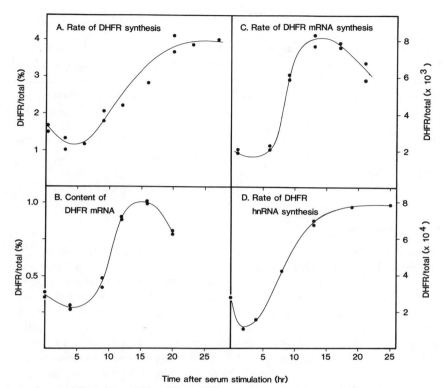

Fig. 3. DHFR mRNA production in M50L3 cells. Cultures of resting M50L3 cells were serum stimulated at time = 0. At the indicated times, cultures were assayed for: A, incorporation of [3H]leucine into DHFR relative to total protein during a 1-hr labeling period as determined by immunoprecipitation; B, the content of DHFR poly(A)+ mRNA relative to total poly(A)+ mRNA as determined by filter hybridization of RNA labeled to equilibrium with $^{32}PO_4$; C, incorporation of [3H]uridine into DHFR poly(A)+ mRNA relative to total poly(A)+ mRNA during a 2-hr labeling period as determined by filter hybridization; D, incorporation of [3H] uridine into DHFR hnRNA relative to total hnRNA during a 20-min labeling period as determined by filter hybridization. The figures are adapted from data presented in Wu *et al.* (1982) (A), Hendrickson *et al.* (1980) (B and C), and Wu and Johnson (1982) (D).

molecule is removed during the splicing reactions leading to mature-sized DHFR mRNA.

2. DHFR mRNA Content in Serum-Stimulated Cells

The increase in rate of DHFR synthesis following serum stimulation could be due to an increase in the content of DHFR mRNA or to an increase in the efficiency of translation of the message. Direct quantitation of DHFR mRNA (by *in vitro* translation or by RNA-excess hybridization) in other cell systems and under other physiological conditions showed that there was always a close correlation between the content of the message and the amount and rate of synthesis of the enzyme (Chang and Littlefield, 1976; Kellems *et al.*, 1976, 1979). We wanted to determine if this was true in serum-stimulated M50L3 cells as well.

We decided to use the technique of DNA-excess filter hybridization in our analyses. This would allow us to measure not only the content, but also the rates, of synthesis and turnover of both nuclear and cytoplasmic DHFR RNA sequences. The availability of recombinant plasmids containing DHFR cDNA sequences permitted the isolation of the large quantities of DNA complementary to DHFR mRNA that are required for this procedure.

In the filter hybridization procedure excess DNA, corresponding to DHFR mRNA, is bound to nitrocellulose filters which are incubated with solutions containing labeled DHFR RNA until all hybridizable sequences are bound to the immobilized DNA. The filters are then washed extensively with buffers and incubated with ribonuclease to remove unhybridized RNA sequences. Using this procedure, an RNA species can be quantitated that represents as little as 0.003% (30 ppm) of total labeled RNA incubated with the filter.

We used this procedure to determine the content of DHFR mRNA in M50L3 cells at various times following serum stimulation. Resting cells were labeled to equilibrium with $^{32}PO_4$ at low specific activity. Cultures were harvested at various times following stimulation, and the amount of labeled poly(A)$^+$ DHFR mRNA relative to total poly(A)$^+$ mRNA was determined. Figure 3B shows that there was good agreement between the increase in the relative amount of DHFR mRNA and the increase in the rate of synthesis of the enzyme (Hendrickson *et al.*, 1980). We also found that about 85% of the poly(A)$^+$ DHFR mRNA was associated with polysomes under all conditions examined (Wu *et al.*, 1982). These results are consistent with the idea that the rate of synthesis of DHFR in serum-stimulated cells is determined primarily by the content of DHFR mRNA and that the translation of the mRNA is not being controlled to any significant extent under these conditions.

3. DHFR mRNA Metabolism in Serum-Stimulated Cells

To determine if the increase in DHFR mRNA content was the result of an increase in the rate of production of the message, we measured the rate of labeling of cytoplasmic DHFR mRNA at various times following stimulation (Hendrickson *et al.*, 1980). As shown in Fig. 3C, the relative rate of production of DHFR mRNA also increased about fourfold following serum stimulation. The increase in rate of production preceeded the increase in content, as would be expected.

We also determined the stability of DHFR mRNA in a pulse-chase experiment. We found that the half-life of the message was the same (about 7.5 hr) in both resting and exponentially growing M50L3 cells (Hendrickson *et al.*, 1980). Therefore, it appears that the increase in DHFR mRNA content is due primarily to an increase in the rate of production of the message.

C. Regulation of Transcription and Processing

The rate of production of cytoplasmic DHFR mRNA could be controlled by regulating the rate of DHFR gene transcription and/or by regulating the processing of DHFR hnRNA and the export of the mature mRNA to the cytoplasm. Our initial studies with normal 3T6 cells suggested that the control was at the transcriptional level (Johnson *et al.*, 1978b). This could be studied directly with the overproducing cells which synthesize enough DHFR hnRNA to allow detection by filter hybridization.

The transcription of the DHFR gene was studied in serum-stimulated M50L3 cells by pulse-labeling the cells with [^{3}H]uridine for very brief (5–20 min) periods, and determining the rate of labeling of DHFR hnRNA relative to total hnRNA by filter hybridization. Figure 3D shows that the relative rate of synthesis of DHFR hnRNA began increasing at the same time (6 hr), and increased to approximately the same extent, as the relative rate of production of DHFR mRNA (Wu and Johnson, 1982).

More recently, we have studied the relative rate of transcription of the DHFR gene *in vitro* using nuclei isolated at various times following serum stimulation. The results were very similar to the *in vivo* labeling results (Santiago *et al.*, 1983).

It should be noted that the increase in DHFR hnRNA synthesis and mRNA production occurs about 6 hr prior to entry into S phase, which is certainly consistent with the observation that there is no direct coupling between the increase in DHFR gene expression and DNA replication. It also shows that DHFR gene duplication is not necessary for the increase in the level of transcription of this gene.

To study the efficiency of processing of DHFR hnRNA into DHFR mRNA, we determined the kinetics of labeling of the two RNA species in resting and exponentially growing cells. Figure 4 shows that the rate of labeling of DHFR mRNA relative to the rate of labeling of DHFR hnRNA is nearly identical in resting and exponentially growing cells. This is probably best illustrated by the fact that the nuclear and cytoplasmic labeling curves intersect at 60 min in both resting and growing cells. Although we cannot determine from these data alone if all potential DHFR mRNA sequences that are synthesized are subsequently exported to the cytoplasm, the data are consistent with the idea that the efficiency as well as the kinetics of processing are the same in cells resting in 0.5% serum and in exponentially growing cells (Collins *et al.*, 1983).

D. DHFR Gene Expression in Cells Synchronized by Mitotic Selection or Amino Acid Starvation

Changes in DHFR gene expression occur not only in cells undergoing a serum-induced transition from the G_0 to the growing state, but also in cells synchronized by other mechanisms. Mariani *et al.* (1981) found that the level and

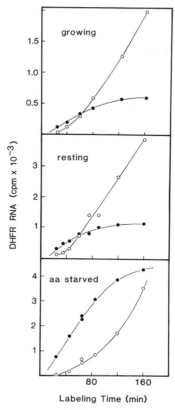

Fig. 4. Comparison of DHFR hnRNA and mRNA labeling kinetics. Cultures of exponentially growing, resting (in 0.5% serum), and amino acid (aa) starved (maintained in medium lacking isoleucine and glutamine for 36 hr) M50L3 cells were labeled with [³H]uridine at time = 0. At the indicated times, cultures were harvested and separated into nuclear and cytoplasmic compartments. Total RNA was purified from each, and the amount of radioactivity in nuclear (●) and cytoplasmic (○) DHFR RNA was determined by filter hybridization. Note that the sequences detected by the hybridization probe correspond only to DHFR mRNA sequences; the filters were treated with ribonuclease during the washing procedure, which would remove the intervening sequences in DHFR hnRNA. Therefore, the rates of labeling of DHFR mRNA sequences in the nuclear and cytoplasmic compartments may be directly compared. However, since different numbers of cells and specific activities of [³H]uridine were used in the various experiments, it is not possible to compare the rates of incorporation into nuclear (or cytoplasmic) DHFR RNA from one experiment to another.

rate of synthesis of DHFR varied over a twofold range during the cell cycle of MTX-resistant Chinese hamster cells that had been synchronized by mitotic selection. The increase also occurred primarily during the S phase, as in serum-stimulated cells. Since DHFR mRNA is rather stable, it is possible that the rate of production of the mRNA might vary by much more than a factor of 2 during the cell cycle in exponentially growing cells.

We examined the metabolism of DHFR mRNA in M50L3 cells synchronized in G_1 following a 36-hr starvation for isoleucine and glutamine (Collins *et al.*, 1983). When these cells are fed with complete medium, they began DNA replication about 8 hr later. Greater than 60% of the cells were in S phase by 16 hr after feeding. The rate of labeling of DHFR mRNA relative to total poly(A)$^+$ mRNA increased about fourfold by 6 hr following refeeding. However, the rate of synthesis of DHFR hnRNA changed very little during this period. Furthermore, the content of DHFR mRNA was the same in amino acid-starved and exponentially growing cells. This suggested that there was a fundamental difference in the processing of DHFR hnRNA in amino acid-starved cells as compared to growing cells.

To study this further, we compared the kinetics of labeling of DHFR hnRNA to that of mRNA in exponentially growing cells and in cells deprived of isoleucine and glutamine. Figure 4 shows that the rate of labeling of DHFR mRNA relative to DHFR hnRNA was about three- to fourfold lower in the amino acid-starved cells than in exponentially growing cells or in cells resting due to serum limitation. Further experiments showed that this was the result of a much lower rate of processing of DHFR RNA sequences rather than turnover of DHFR hnRNA sequences within the nuclei of the amino acid-starved cells (Collins *et al.*, 1983).

Leys and Kellems (1981) found evidence for posttranscriptional control of DHFR mRNA production in MTX-resistant sarcoma 180 cells. They found that when stationary phase cells were replated in fresh medium at a lower cell density, the rate of synthesis of DHFR hnRNA changed very little, even though the content and rate of production of DHFR mRNA were about three- to fourfold greater by 25 hr following replating. Although the stationary phase sarcoma cells are probably starving for a variety of nutrients, the physiological state of the cells may be similar to that of the M50L3 cells starving for isoleucine and glutamine.

E. Effects of cAMP and Polyoma Virus on DHFR Gene Expression

The biochemical signals responsible for turning on the expression of the DHFR gene, whether at the level of gene transcription or processing and export of the message, are not known. However, Kellems and co-workers (1979; Gudewicz *et al.*, 1981) observed that high intracellular concentrations of cAMP prevented the increase in the synthesis of DHFR and its mRNA as well as entry into S phase in growth-stimulated 3T6 cells. We made similar observations in serum-stimulated M50L3 cells (Wu *et al.,* 1982). The effect of cAMP on DHFR gene expression may either be direct (e.g., cAMP may stimulate the phosphorylation of, or directly interact with, proteins involved in the synthesis or processing of DHFR hnRNA) or be indirect (e.g., cAMP may block progression through the cell cycle, thereby preventing the formation of the biochemical signal for stimulating DHFR gene expression). We found that addition of dibutyryl-cAMP during S phase resulted in an immediate decrease in the rate of synthesis of DHFR hnRNA relative to total hnRNA, even though progression through the cell cycle was not blocked until the cells had completed S phase (Wu and Johnson, 1982). This suggests that the effect may be rather direct.

Kellems *et al.* (1979) found that infection of quiescent cells with polyoma virus also stimulated DHFR mRNA production. However, this increase was not blocked by high concentrations of cAMP. This interesting observation suggests that the stimulation of DHFR gene expression by serum factors or by polyoma virus infection may occur by two different pathways.

F. Future Directions

Although much progress has been made, it is clear that more work is required to understand fully the mechanisms for controlling DHFR gene expression. It will be very useful to clone the entire DHFR gene, especially the 5' end of the gene, and to determine the sequence of the DNA upstream from the transcriptional initiation site. The promoter and other control sites are likely to be located in this region, and the sequences of these regulatory sites may be significantly different from those of the genes for abundant proteins that have been studied in detail previously. The cloned 5' end can also be used in *in vitro* transcription experiments to study proteins and DNA sequences that are necessary for DHFR gene transcription. If specific proteins regulate DHFR gene transcription by binding near the promoter sites, it may eventually be possible to purify them (e.g., by chromatography of nuclear proteins on affinity columns containing the cloned 5' end of the DHFR gene). The purified proteins and cloned gene fragments might then be used to reconstruct transcriptional regulation of the DHFR gene *in vitro* using defined components.

It will also be important to determine how the processing of DHFR hnRNA is controlled in amino acid starved cells. For example, is the splicing or polyadenylation of the precursor delayed? Is nuclear-to-cytoplasmic transport rate-limiting? Are potential DHFR mRNA sequences destroyed in the nucleus? Finally, is DHFR mRNA production controlled at the transcriptional or posttranscriptional level during the cell cycle of exponentially growing cells?

V. ISOLATION AND CHARACTERIZATION OF TS-OVERPRODUCING CELLS

A. Isolation of FdUrd-Resistant Cells

The above studies show that an overproducing cell line can be used as a convenient model system to study the control of gene expression in mammalian cells. We also wanted to isolate a cell line that would overproduce TS, which might provide a model system for detailed studies of TS gene expression.

It has been known for many years that TS is specifically and irreversibly inactivated by the substrate analog 5-fluorodeoxyuridylic acid (FdUMP) (Heidelberger, 1965; Santi *et al.*, 1974; Walsh, 1979). The analog inhibits the enzyme by binding covalently at the active site, apparently as a result of an abortive attempt to methylate the 5 position of the pyrimidine ring. The enzyme can be inactivated in living cells by exposing them to fluorodeoxyuridine (FdUrd), which is transported across the cell membrane more readily than the nucleotide form of the drug. FdUrd is converted to FdUMP by TK. Inhibition of TS activity in rapidly proliferating cells leads to starvation for TMP, inhibition of DNA synthesis, and eventually, cell death. Because of their ability to selectively kill rapidly proliferating cells, FdUrd and similar derivatives have been widely used as antineoplastic drugs (reviewed in Blakley, 1969; Danenberg, 1977).

We reasoned that rapidly growing cells might be able to develop resistance to FdUrd by overproducing TS in much the same manner that cells develop resistance to MTX by overproducing DHFR. Earlier studies demonstrated a low level of overproduction (less than tenfold) in some cell lines selected for resistance to FdUrd (Baskin *et al.*, 1975; Baskin and Rosenberg, 1975; Priest *et al.*, 1980; Wilkinson *et al.*, 1977). However, we were aware that cells might also develop resistance to FdUrd by becoming TK⁻ (Baskin *et al.*, 1977), by decreasing the rate of transport of the inhibitor into the cell, by increasing the rate of breakdown of the drug, or by altering the gene for TS so that the enzyme would have a decreased affinity for the drug. Therefore, we set out to isolate a TK⁺ 3T6 cell line that was able to grow in high concentrations of FdUrd and that had a high level of overproduction of TS enzyme activity. To be useful as a model system in our studies, the cell line would also have to retain the ability to rest in medium containing low concentrations of serum and regulate TS gene expression in the same manner as normal 3T6 cells.

Cultures of 3T6 cells were selected for the ability to grow in medium containing gradually increasing concentrations of FdUrd. The medium was also supplemented with 1 mM uridine and 1 mM cytidine to prevent toxicity from possible metabolic products of FdUrd. TK⁻ cells were eliminated from the population by periodic selection in HAT medium (Littlefield, 1964). Unfortunately, 3T6 cells were never able to grow in FdUrd concentrations greater than 0.3 μM. Nevertheless, some clones isolated from this population were found to overproduce TS up to 20-fold (Rossana *et al.*, 1982). We also subjected the M50L3 cells to the FdUrd selection procedure and found that they were able to adapt much more rapidly, and to much higher drug concentrations, than 3T6 cells (Rossana *et al.*, 1982). The reason the M50L3 cells adapt more readily is not known at present.

Several clones of cells were selected that were able to grow at normal rates in medium containing 3 μM FdUrd, and these were characterized further (Rossana *et al.*, 1982). The maximum TS specific activity was found in a clone designated LU3-7 and was about 50 times greater than that found in normal 3T6 cells. The molecular weight of the overproduced enzyme was the same as that in normal 3T6 cells (38,000). Furthermore, it was inactivated by FdUMP at the same concentration as the normal enzyme. Therefore, it appears that the cells are overproducing an enzyme that is the same as, or very similar to, the normal enzyme. We estimate, based on analysis of proteins by SDS gel electrophoresis, that TS represents between 0.1 and 1% of total cytoplasmic protein in the LU3-7 cells. TS from the overproducing cell line was purified by affinity chromatography on MTX–polyacrylamide and injected into rabbits to stimulate the production of TS antibodies. This reagent will be useful for studies of the synthesis of TS *in vivo* or *in vitro*.

To determine if the LU3-7 clone retains the ability to rest, we plated the cells in medium containing 0.5% serum and maintained them for 7 days with alternate-day feedings. Very few mitotic cells were detected, and the cells had the flat morphology of normal resting 3T6 cells. The rate of incorporation of [^{3}H]thymidine into DNA was determined in resting cells and at various times following stimulation. The results were very similar to those observed with serum-stimulated 3T6 or M50L3 cells (Rao *et al.*, 1984).

B. Regulation of Enzyme Level

To determine if the LU3-7 cells regulate TS gene expression in the same manner as 3T6 cells, we repeated all of our previous experiments on the overproducing cell line. The results were quite similar to those obtained with 3T6. TS activity began increasing at about the same time as the cells began DNA replication. The increase in enzyme activity was blocked by the addition of cycloheximide or actinomycin D if the drugs were added prior to entry of the cells into S phase. We did note that the fold increase in TS activity during S phase was somewhat less than that observed in stimulated 3T6 cells. This might be due to the fact that the LU3-7 cells do not rest quite as well as the 3T6 cells. To determine if the increase in TS activity was due to an increase in the rate of synthesis of the enzyme, cultures of cells were pulse-labeled with [^{35}S]methionine at various times following serum stimulation. Analysis of the labeled cell extracts by SDS–polyacrylamide gel electrophoresis showed that the rate of synthesis of TS relative to total protein increased by at least a factor of 7 by 20 hr following stimulation (Rao *et al.*, 1984).

C. TS cDNA Cloning

Analysis of the products of *in vitro* translation of poly(A)$^+$ mRNA isolated from the LU3-7 cell line revealed that TS mRNA was also overproduced in the fluorodeoxyuridine-resistant cell line. Based on this observation, we developed a strategy for cloning TS cDNA. Our approach was similar to one used for the cloning of cDNA corresponding to argininosuccinate synthase mRNA, which represents roughly 0.1% of total mRNA in canavanine-resistant human cells (Su *et al.*, 1981). DNA complementary to total cytoplasmic poly(A)$^+$ mRNA from the overproducing cell line was synthesized using reverse transcriptase primed with oligo(dT). The second strand was also synthesized using the same enzyme. The DNA was then treated with S1 nuclease to yield a population of blunt-ended, double-stranded DNA molecules representing the mRNA of the overproducing cell line. The DNA was fractionated on a sucrose gradient, and DNA greater than 500 nucleotides in length was isolated. This was tailed with poly(dC) using terminal transferase, annealed with poly(dG)-tailed pBR322, and transformed into *Escherichia coli*. Approximately 30,000 tetracycline-resistant colonies were isolated.

Replica plates of the colonies were screened with [^{32}P]cDNA synthesized from mRNA isolated from either the overproducing cell line or the parental 3T6 cells. Colonies that contained TS cDNA sequences would be expected to give a strong hybridization signal with the former probe but a weak signal with the latter probe. Plasmids were isolated from such colonies and immoblized on nitrocellulose filters. RNA that hybridized to the DNA was eluted from the filter and translated *in vitro* and immunoprecipitated with TS antiserum to confirm the identity of the cloned TS cDNA sequences. The largest TS cDNA insert identified so far is about 1200 nucleotides in length, which is about 200–300 nucleotides shorter than the size of the mature TS mRNA (Geyer and Johnson, 1984).

D. Future Directions

The TS cDNA clone will be used to study the production and metabolism of TS mRNA during the cell cycle. These studies will parallel our studies on DHFR mRNA metabolism to determine if the two messages are controlled by similar mechanisms. Since TS enzyme level is regulated over a much wider range than DHFR enzyme level, we expect that the changes in TS mRNA content and metabolism will be much more pronounced than those for DHFR mRNA. The cDNA clone will also be used as a hybridization probe to determine if the overproduction of TS is due to TS gene amplification or to some other mechanism (e.g., a "promoter up" mutation). It will also be used to study the structure of the TS gene and to facilitate the isolation of genomic clones containing portions of the TS gene. If both TS and DHFR gene expression are controlled, at least under some conditions, at the level of transcription, there may be similar (or even identical) sequences at the 5' ends of the two genes. Furthermore, the same chromosomal proteins may be responsible for regulating and coordinating the transcription of the two genes.

ACKNOWLEDGMENTS

I thank M. Collins, P. Geyer, C. Rossana, A. Delisle, and C. Santiago for comments on tbe manuscript. Our studies have been supported by a Basil O'Connor research grant from the March of Dimes Birth Defects Foundation, by grants from the National Cancer Institute (CA 26470) and the National Institute of General Medical Sciences (GM 29356), and by Faculty Research Award (FRA 210) from the American Cancer Society.

REFERENCES

Alt, F. W., Kellems, R. E., and Schimke, R. T. (1976). Synthesis and degradation of folate reductase in sensitive and methotrexate-resistant lines of S-180 cells. *J. Biol. Chem.* **251**, 3063–3074.

Alt, F. W., Kellems, R. E., Bertino, J. R., and Schimke, R. T. (1978). Selective multiplication of dihydrofolate reductase genes in methotrexate-resistant variants of cultured murine cells *J. Biol. Chem.* **253**, 1357–1370.

Baskin, F., and Rosenberg, R. N. (1975). A comparison of thymidylate synthetase activities from 5-fluorodeoxyuridine sensitive and resistant variants of mouse neuroblastoma. *J. Neurochem.* **25**, 233–238.

Baskin, F., Carlin, S. C., Kraus, P., Friedkin, M., and Rosenberg, R. N. (1975). Experimental chemotherapy of neuroblastoma. II. Increased thymidylate synthetase activity in a 5-fluorodeoxyuridine-resistant variant of mouse neuroblastoma. *Mol. Pharmacol.* **11**, 105–117.

Baskin, F., Davis, R., and Rosenberg, R. N. (1977). Altered thymidine kinase or thymidylate synthetase activities in 5-fluorodeoxyuridine resistant variants of mouse neuroblastoma. *J. Neurochem.* **29**, 1031–1037.

Beach, L. R., and Palmiter, R. D. (1981). Amplification of the metallothionein-1 gene in cadmium-resistant mouse cells. *Proc. Natl. Acad. Sci. U.S.A.* **78**, 2110–2114.

Bertino, J. R. (1971). "Folate Antagonists as Chemotherapeutic Agents." N.Y. Acad. Sci., New York.

Bertino, J. R. (1979). Toward improved selectivity in cancer chemotherapy. *Cancer Res.* **39**, 293–304.

Biedler, J. L., and Spengler, B. A. (1976). Metaphase chromosome anomaly: Association with drug resistance and cell-specific products. *Science* **191**, 185–187.

Blakley, R. L. (1969). "The Biochemistry of Folic Acid and Related Pteridines." North-Holland Publ., Amsterdam.

Borun, T. W., Scharff, M. D., and Robbins, E. (1967). Rapidly labeled polyribosome-associated RNA having the properties of histone messenger. *Proc. Natl. Acad. Sci. U.S.A.* **58**, 1977–1982.

Brown, P. C., Beverly, S. M., and Schimke, R. T. (1981). Relationship of amplified dihydrofolate reductase genes to double minute chromosomes in unstably resistant mouse fibroblast cell lines. *Mol. Cell. Biol.* **1**, 1077–1083.

Brown, P. C., Tlsty, T. D., and Schimke, R. T. (1983). Enhancement of methotrexate resistance and dihydrofolate reductase gene amplification by treatment of mouse 3T6 cells with hydroxyurea. *Mol. Cell. Biol.* **3**, 1097–1107.

Chang, A. C. Y., Nunberg, J. H., Kaufman, R. J., Erlick, H. A., Schimke, R. T., and Cohen, S. N. (1978). Phenotypic expression in E. coli of a DNA sequence coding for mouse dihydrofolate reductase. *Nature (London)* **275**, 617–624.

Chang, S. E., and Littlefield, J. W. (1976). Elevated dihydrofolate reductase messenger RNA levels in methotrexate-resistant BHK cells. *Cell* **7**, 391–396.

Chello, P. L., McQueen, C. A., DeAngelis, L. M., and Bertino, J. R. (1977). Comparative effects of folate antagonists versus enzymatic folate depletion on folate and thymidine enzymes in cultured mammalian cells. *Cancer Treat. Rep.* **61**, 539–548.

Collins, M. L., Wu, J.-S. R., Santiago, C. L., Hendrickson, S. L., and Johnson, L. F. (1983). Delayed processing of dihydrofolate reductase hnRNA in amino acid-starved mouse fibroblasts. *Mol. Cell. Biol.* **3**, 1792–1802.

Conrad, A. H. (1971). Thymidylate synthetase activity in cultured mammalian cells. *J. Biol. Chem.* **246**, 1318–1323.

Conrad, A. H., and Ruddle, F. H. (1972). Regulation of thymidylate synthetase activity in cultured mammalian cells. *J. Cell Sci.* **10**, 471–486.

Danenberg, P. V. (1977). Thymidylate synthetase—a target enzyme in cancer chemotherapy. *Biochim. Biophys. Acta* **473**, 73–92.

Dolnick, B. J., and Bertino, J. R. (1981). Multiple messenger RNAs for dihydrofolate reductase. *Arch. Biochem. Biophys.* **210**, 691–697.

Dolnick, B. J., Berenson, R. J., Bertino, J. R., Kaufman, R. J., Nunberg, J. H., and Schimke, R. T. (1979). Correlation of dihydrofolate reductase elevation with gene amplification in a homogeneously staining chromosomal region in L5178Y cells. *J. Cell Biol.* **83**, 394–402.

Flintoff, W. E., Davidson, S. V., and Siminovitch, L. (1976). Isolation and partial characterization of three methotrexate-resistant phenotypes from Chinese hamster ovary cells. *Somatic Cell Genet.* **2**, 245–261.

Geyer, P. K., and Johnson, L. F. (1984). In preparation.

Goldie, J. H., Krystal, G., Hartley, D., Gudauskas, G., and Dedhar, S. (1980). A methotrexate insensitive variant of folate reductase present in two lines of methotrexate-resistant L5178Y cells. *Eur. J. Cancer* **16**, 1539–1546.

Gudewicz, T. M., Morhenn, V. B., and Kellems, R. E. (1981). The effect of polyoma virus, serum factors, and dibutyryl cyclic AMP on dihydrofolate reductase synthesis and the entry of quiescent cells into S-phase. *J. Cell. Physiol.* **108**, 1–8.

Haber, D. A., and Schimke, R. T. (1981). Unstable amplification of an altered dihydrofolate reductase gene associated with double-minute chromosomes. *Cell* **26**, 355–362.

Haber, D. A., Beverly, S. M., Kiely, M. L., and Schimke, R. T. (1981). Properties of an altered dihydrofolate reductase encoded by amplified genes in cultured mouse fibroblasts. *J. Biol. Chem.* **256**, 9501–9510.

Hakala, M. T., Zakrzewski, S. F., and Nichol, C. A. (1961). Relation of folic acid reductase to amethopterin resistance in cultured mammalian cells. *J. Biol. Chem.* **236**, 952–958.

Hanggi, U. J., and Littlefield, J. W. (1976). Altered regulation of the rate of synthesis of dihydrofolate reductase in methotrexate-resistant hamster cells. *J. Biol. Chem.* **251**, 3075–3080.

Heidelberger, C. (1965). Fluorinated pyrimidines. *Prog. Nucleic Acid Res. Mol. Biol.* **4**, 1–50.

Hendrickson, S. L., Wu, J.-S. R., and Johnson, L. R. (1980). Cell cycle regulation of dihydrofolate reductase mRNA metabolism in mouse fibroblasts. *Proc. Natl. Acad. Sci. U.S.A.* **77**, 5140–5144.

Hillcoat, B. L., Swett, V., and Bertino, J. R. (1967). Increase of dihydrofolate reductase activity in cultured mammalian cells after exposure to methotrexate. *Proc. Natl. Acad. Sci. U.S.A.* **58**, 1632–1637.

Hryniuk, W. M., Fischer, G. A., and Bertino, J. R. (1969). S-phase cells of rapidly growing and resting populations. Differences in response to methotrexate. *Mol. Pharmacol.* **5**, 557–564.

Johnson, L. F., Abelson, H. T., Green, H., and Penman, S. (1974). Changes in RNA in relation to growth of the fibroblast. I. Amounts of mRNA, rRNA and tRNA in resting and growing cells. *Cell* **1**, 95–100.

Johnson, L. F., Fuhrman, C. L., and Abelson, H. T. (1978a). Resistance of resting 3T6 mouse fibroblasts to methotrexate cytotoxicity. *Cancer Res.* **38**, 2408–2412.

Johnson, L. F., Fuhrman, C. L., and Wiedmann, L. M. (1978b). Regulation of dihydrofolate reductase gene expression in mouse fibroblasts during the transition from the resting to growing state. *J. Cell. Physiol.* **97**, 397–406.

Johnson, L. F., Rao, L. G., and Muench, A. J. (1982). Regulation of thymidine kinase enzyme level in serum-stimulated mouse 3T6 fibroblasts. *Exp. Cell Res.* **138**, 79–85.

Kaufman, R. J., Brown, P. C., and Schimke, R. T. (1979). Amplified dihydrofolate reductase genes in unstably methotrexate-resistant cells are associated with double minute chromosomes. *Proc. Natl. Acad. Sci. U.S.A.* **76**, 5669–5673.

Kellems, R. E., Alt, F. W., and Schimke, R. T. (1976). Regulation of folate reductase synthesis in sensitive and methotrexate resistant sarcoma 180 cells. *J. Biol. Chem.* **251**, 6987–6993.

Kellems, R. E., Morhenn, V. B., Pfendt, E. A., Alt, F. W., and Schimke, R. T. (1979). Polyoma virus and cyclic AMP-mediated control of dihydrofolate reductase mRNA abundance in methotrexate-resistant mouse fibroblasts. *J. Biol. Chem.* **255**, 309–318.

Kellems, R. E., Harper, M. E., and Smith, L. M. (1982). Amplified dihydrofolate reductase genes are located in chromosome regions containing DNA that replicates during the first half of S-phase. *J. Cell Biol.* **92**, 531–539.

Kit, S. (1970). Nucleotides and nucleic acids. *In* "Metabolic Pathways" (D. M. Greenberg, ed.), 3rd ed., Vol. 4, pp. 69–257. Academic Press, New York.

Kit, S. (1976). Thymidine kinase, DNA synthesis and cancer. *Mol. Cell. Biochem.* **11**, 161–182.

Kit, S., and Jorgensen G. N. (1976). Formation of thymidine kinase and deoxycytidylate deaminase in synchronozed cultures of Chinese hamster cells temperature-sensitive for DNA synthesis. *J. Cell. Physiol.* **88**, 57–64.

Lewis, J. A., Kurtz, D. T., and Melera, P. W. (1981). Molecular cloning of Chinese hamster dihydrofolate reductase-specific cDNA and the identification of multplie dihydrofolate reductase mRNAs in antifolate-resistant Chinese hamster lung fibroblasts. *Nucleic Acids Res.* **9**, 1311–1322.

Leys, E. J., and Kellems, R. E. (1981). Control of dihydrofolate reductase messenger ribonucleic acid production. *Mol. Cell. Biol.* **1**, 961–971.

Littlefield, J. W. (1964). Selection of hybrids from matings of fibroblasts *in vitro* and their presumed recombinants. *Science* **145**, 709–710.

Littlefield, J. W. (1965). Studies on thymidine kinase in cultured mouse fibroblasts. *Biochim. Biophys. Acta* **95**, 14–22.

Littlefield, J. W. (1969). Hybridization of hamster cells with high and low folate reductase activity. *Proc. Natl. Acad. Sci. U.S.A.* **62**, 88–95.

Maley, F., and Maley, G. F. (1960). Nucleotide interconversions. II. Elevation of deoxycytidylate deaminase and thymidylate synthetase in regenerating rat liver. *J. Biol. Chem.* **235,** 2968–2970.

Mariani, B. D., Slate, D. L., and Schimke, R. T. (1981). S phase-specific synthesis of dihydrofolate reductase in Chinese hamster ovary cells. *Proc. Natl. Acad. Sci. U.S.A.* **78,** 4985–4989.

Mauck, J. C., and Green, H. (1973). Regulation of RNA synthesis in fibroblasts during the transition from resting to growing state. *Proc. Natl. Acad. Sci. U.S.A.* **70,** 2819–2822.

Milbrandt, J. D., Heintz, N. H., White, W. C., Rothman, S. M., and Hamlin, J. L. (1981). Methotrexate-resistant Chinese hamster ovary cells have amplified a 135-kilobase-pair region that includes the dihydrofolate reductase gene. *Proc. Natl. Acad. Sci. U.S.A.* **78,** 6043–6047.

Mitchison, J. M. (1971). "The Biology of the Cell Cycle." Cambridge Univ. Press, London and New York.

Navalgund, L. G., Rossana, C., Muench, A. J., and Johnson, L. F. (1980). Cell cycle regulation of thymidylate synthetase gene expression in cultured mouse fibroblasts. *J. Biol. Chem.* **255,** 7386–7390.

Nunberg, J. H., Kaufman, R.J., Schimke, R. T., Urlaub, G., and Chasin, L. A. (1978). Amplified dihydrofolate reductase genes are localized to a homogeneously staining region of a single chromosome in a methotrexate-resistant Chinese hamster ovary cell line. *Proc. Natl. Acad. Sci. U.S.A.* **75,** 5553–5556.

Nunberg, J. H., Kaufman, R. J., Chang, A. C. Y., Cohen, S. N., and Schimke, R. T. (1980). Structure and genomic organization of the mouse dihydrofolate reductase gene. *Cell* **19,** 355–364.

Perry, R. P., Schibler, U., and Meyuhas, O. (1979). The processing of messenger RNA and the determination of its relative abundance. *In* "From Gene to Protein: Information Transfer in Normal and Abnormal Cells" (T. R. Russell, K. Brew, H. Faber, and J. Schultz, eds.), pp. 187–206. Academic Press, New York.

Piper, A. A., Tattersall, M. H. N., and Fox, R. M. (1980). The activities of thymidine metabolising enzymes during the cell cycle of a human lymphocyte cell line LAZ-007 synchronized by centrifugal elutriation. *Biochim. Biophys. Acta* **633,** 400–409.

Prescott, D. M. (1976). "Reproduction of Eukaryotic Cells." Academic Press, New York.

Priest, D. G., Ledford, B. E., and Doig, M. T. (1980). Increased thymidylate synthetase in 5-fluorodeoxyuridine resistant cultured hepatoma cells. *Biochem. Pharmacol.* **29,** 1549–1553.

Rao, L. G., Jenh, C.-H., and Johnson, L. F. (1984). In preparation.

Reddy, G. P. V., and Pardee, A. B. (1980). Multienzyme complex for metabolic channeling in mammalian DNA replication. *Proc. Natl. Acad. Sci. U.S.A.* **77,** 3312–3316.

Rode, W., Scanlon, K. J.,Moroson, B. A., and Bertino, J. R. (1980). Regulation of thymidylate synthetase in mouse leukemia cells (L1210). *J. Biol. Chem.* **255,** 1305–1311.

Rossana, C., Rao, L. G., and Johnson, L. F. (1982). Thymidylate synthetase overproduction in 5-fluorodeoxyuridine-resistant mouse fibroblasts. *Mol. Cell. Biol.* **2,** 1118–1125.

Santi, D. V., McHenry, C. S., and Sommer, H. (1974). Mechanism of interaction of thymidylate synthetase with 5-fluorodeoxyuridylate. *Biochemistry* **13,** 471–481.

Santiago, C., Collins, M., and Johnson, L. F. (1983). In vitro and in vivo analysis of the control of dihydrofolate reducatse gene transcription in serum stimulated mouse fibroblasts. *J. Cell. Physiol.* (in press).

Schimke, R. T. (1982). "Gene Amplification." Cold Spring Harbor Lab., Cold Spring Harbor, New York.

Schimke, R. T., Kaufman, R. J., Alt, F. W., and Kellems, R. E. (1978). Gene amplification and drug resistance in cultured murine cells. *Science* **202,** 1051–1055.

Setzer, D. R., McGrogan, M., Nunberg, J. H., and Schimke, R. T. (1980). Size heterogeneity in the 3′ end of dihydrofolate reductase messenger RNAs in mouse cells. *Cell* **22,** 361–370.

Setzer, D. R., McGrogan, M., and Schimke, R. T. (1982). Nucleotide sequence surrounding multiple polyadenylation sites in the mouse dihydrofolate reductase gene. *J. Biol. Chem.* **257,** 5143–5147.

Stubblefield, E., and Murphree, S. (1967). Synchronized mammalian cell cultures. II. Thymidine kinase activity in colcemid synchronized fibroblasts. *Exp. Cell Res.* **48,** 652–656.

Su, T.-S., Bock, H.-G. O., O'Brien, W. E., and Beaudet, A. L. (1981). Cloning of cDNA for argininosuccinate synthetase mRNA and study of enzyme overproduction in a human cell line. *J. Biol. Chem.* **256,** 11826–11831.

Todaro, G. J., and Green H. (1963). Quantitative studies of the growth of mouse embryo cells in culture and their development into established lines. *J. Cell Biol.* **17,** 299–313.

Varshavsky, A. (1981). Phorbol ester dramatically increases incidence of methotrexate-resistant mouse cells: Possible mechanisms and relevance to tumor promotion. *Cell* **25,** 561–572.

Wahl, G. M., Padgett, R. A., and Stark, G. R. (1979). Gene amplification causes overproduction of the first three enzymes of UMP synthesis in N-(phosphonacetyl)-L-aspartate-resistant hamster cells. *J. Biol. Chem.* **254,** 8679–8689.

Walsh, C. (1979). "Enzymatic Reaction Mechanisms." Freeman, San Francisco, California.

Wiedemann, L. M., and Johnson, L. F. (1979). Regulation of dihydrofolate reductase synthesis in an overproducing 3T6 cell line during transition from resting to growing state. *Proc. Natl. Acad. Sci. U.S.A.* **76,** 2818–2822.

Wilkinson, D. S., Solomonson, L. P., and Cory, J. G. (1977). Increased thymidylate synthetase activity in 5-fluorodeoxyuridine-resistant Novikoff hepatoma cells. *Proc. Soc. Exp. Biol. Med.* **154,** 368–371.

Wu, J.-S. R., and Johnson, L. F. (1982). Regulation of dihydrofolate reductase gene transcription in methotrexate-resistant mouse fibroblasts. *J. Cell. Physiol.* **110,** 183–189.

Wu, J.-S. R., Wiedemann, L. M., and Johnson, L. F. (1982). Inhibition of dihydrofolate reductase gene expression following serum withdrawal or db-cAMP addition in methotrexate-resistant mouse fibroblasts. *Exp. Cell Res.* **141,** 159–169.

3

Enzyme Expression during Growth and Cell Division in *Saccharomyces cerevisiae:* A Study of Galactose and Phosphorus Metabolism

H. O. HALVORSON,[1] K. A. BOSTIAN,[2] J. G. YARGER,[3] AND J. E. HOPPER[4]

[1]Rosenstiel Basic Medical Sciences Research Center, and Department of Biology, Brandeis University, Waltham, Massachusetts.

[2]Section of Biochemistry, Division of Biology and Medicine, Brown University, Providence, Rhode Island.

[3]Department of Biochemistry and Molecular Biology, Fairchild Biochemical Laboratory, Harvard University, Cambridge, Massachusetts. Present address: Department of Biosynthesis Research, Miles Laboratories, Inc., Elkhart, Indiana.

[4]Department of Biological Chemistry, Hershey Medical Center, Hershey, Pennsylvania.

49

RECOMBINANT DNA AND
CELL PROLIFERATION

I. INTRODUCTION

Numerous cellular events have been identified which occur at specific periods within the cell cycle in eukaryotes (Mitchison, 1971; Lloyd *et al.*, 1982). These morphologically or biochemically distinct processes are related temporally and functionally, and define landmark events during cell growth and division. The regulation and coordination of these events and control of cell proliferation are central questions of cell biology. The studies reported here have employed the simple eukaryotic microorganism *Saccharomyces cerevisiae* to examine the relationship between cell growth and mechanisms regulating macromolecular synthesis within the cell division cycle. The temporal sequence of landmark events during the cell cycle of this budding yeast is shown in Fig. 1. Like the cell cycle of higher eukaryotes, replication of chromosomal DNA occurs during the S period, preceded and followed by periods of growth, which are G_1 and G_2, respectively. The G_2 period is followed by mitosis, nuclear division, cytokinesis, and cell separation. The asymmetric mode of cell growth through the formation of a bud precedes nuclear division. The mitotic spindle is formed during the earliest phase of bud emergence. However, major differences with higher eukaryotes occur during mitosis. For example, as in other fungi, the nuclear envelope does not break down during nuclear division, nor does the chromatin condense.

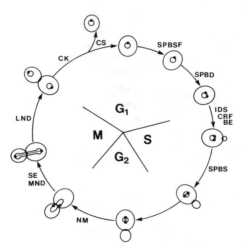

Fig. 1. Temporal sequence of the cell cycle in *Saccharomyces cerevisiae*. Abbreviations are as follows: spindle pole body satellite formation (SPBSF), spindle pole body duplication (SPBD), chitin ring formation (CRF), initiation of DNA synthesis (IDS), bud emergence (BE), spindle pole body separation (SPBS), nuclear migration (NM), medical nuclear division (MND), spindle elongation (SE), late nuclear division (LND), cytokinesis (CK), and cell separation (CS). (Adapted from Pringle and Hartwell, 1982.)

Much of our knowledge of the stage-specific functioning of the yeast cell cycle, in particular with DNA replication and mitosis, has come from genetic and biochemical analyses of conditional, lethal, cell division cycle mutants (Newlon *et al.*, 1974). Analysis of these mutants has shown that bud emergence and development can occur in the absence of nuclear replication and that these processes are largely independent of one another (Slater, 1973). Both, however, are dependent upon a common control point in G_1, termed "start." Control at "start" is responsive to nutrient limitation and to conjugation by mating pheromones. Many studies in which the rate of progression through the cell cycle has been varied suggest that traversal of "start" requires the attainment of a critical size in a semideterministic fashion upon which cells progress to division in a constant time, relatively independent of continued growth and biomass (Pringle and Hartwell, 1982). The degree to which the constituent processes of the growth cycle and the DNA division cycle are coordinated appears to result from a rate-limiting increase in cell mass, regulated by controls exerted in the variable G_1 period of the cell cycle. Whereas very little is known about the nature of these controls over cell growth and division, numerous models have been proposed (Halvorson *et al.*, 1971), and considerable data have been accumulated on the cell cycle timing of macromolecular synthesis.

Clarification of macromolecular synthesis during the cell cycle requires as a first step the generation of cell populations representing different stages of the cell cycle. Three experimental approaches have been employed for such analysis: Synchronous cultures have been generated by induction or size selection; asynchronous cultures have been age fractionated based on size; and single cells have been assayed during growth in asynchronous cultures. Inherent problems exist with all three analytical methods.

Synchronous division can be induced by release from reversible, stage-specific cell cycle poisons such as α-factor, periodic heat shocks, use of temperature-sensitive *cdc* mutants, chemical inhibition of DNA synthesis, and G_1 arrest due to nutrient limitation followed by addition of fresh medium. There is concern that each of these inducing agents may affect cell growth conditions after metabolic release, resulting in abnormal cell cycles (Elliott and McLaughlin, 1983). For example, synchronous cultures induced by use of a *cdc* mutant to block DNA synthesis show both a shortened cell life cycle and an increase in cell size at division (Slater, 1974; Fantes, 1976). Moreover, often the block in DNA synthesis is not accompanied by a block in cell growth. Cells may increase in size by as much as fourfold before arresting (Pringle and Hartwell, 1982).

The use of centrifugal elutriators, in which asynchronously growing cells are size fractionated to obtain populations of cells undergoing synchronous cellular growth, is less stressful than the use of metabolic poisons but may also introduce perturbations. One potential cause for perturbations in this instance arises from the analysis of only a unique subset of the total asynchronous cell population.

Small, unbudded cells are used in these experiments, which may not be entirely representative of the remainder of the asynchronous population.

The use of zonal rotors to fractionate asynchronous cultures by size presumably avoids the introduction of artifacts observed in synchronous culture experiments since the entire population of cells is analyzed after certrifugation, and cells are not required to resume normal, balanced growth. However, difficulties arise in setting cycle limits at the tails of the size distribution and also with dealing with the complexities of unequal cell division and size variations in mother–daughter cells (Hartwell and Unger, 1977).

The method that introduces the least number of potential perturbations involves direct enzyme measurements on single cells. The obvious limitation in such experiments is the inability to perform molecular analysis of transcriptional and translational events on single cells of yeast.

From the research generated using the various techniques, general disagreement remains over both the patterns and underlying controls of individual macromolecular synthesis. There is agreement that total RNA, tRNA, and total cellular protein increase continuously throughout the cell cycle (Wain and Staatz, 1973; Fraser and Moreno, 1976; Williamson and Scopes, 1960; Halvorson *et al.,* 1964; Elliot and McLaughlin, 1978, 1979), whereas DNA replication is confined to a brief S period. There is disagreement, however, over the pattern of synthesis of individual macromolecules. In our own laboratory all three of these analytical approaches have been employed to examine enzyme accumulation during the cell cycle. In contrast to the continuous accumulation of total proteins, it has been generally observed that specific enzymes are accumulated in a periodic fashion during the cell cycle. In total, over 30 enzymes examined displayed periodic, step increases in activities (Gorman *et al.,* 1964; Saunders *et al.,* 1975), not confined to a single period of the cell cycle (Saunders *et al.,* 1975; Matur and Berry, 1978).

These data, however, have become equivocal for several reasons. First, similar observations of step enzymes in the fission yeast *Schizosaccharomyces pombe,* using induced synchrony techniques, have been reinterpreted to be the result of perturbations induced by pregrowth conditions that appeared in both synchronous and asynchronous cultures (Mitchison and Carter, 1975). More recently, Mitchison (1977) has analyzed enzyme levels in synchronously growing cells derived from asynchronously growing cell cultures by size selection with an elutriator. With the exception of TMP kinase (EC 2.7.4.9), 18 out of 19 enzymes reexamined did not show periodic fluctuations in activity. The question has been raised whether the step increases observed in budding yeasts are the result of methods imposed to grow, fractionate, or analyze cells, or whether they reflected intrinsic differences between the two microorganisms. Recent experiments have implied the former may be true. Mitchison and colleagues (J. M. Mitchison, personal communication) have studied activity changes in several enzymes throughout the cell cycle. Whereas α-glucosidase in *Saccharomyces cerevisiae* (Gorman *et al.*, 1964; Halvorson *et al.*, 1964; Tauro and Halvorson, 1966; Tauro *et al.*, 1968; Sebastian *et al.*, 1971) and β-galactosidase in *Kluyveromyces lactis* (Halvorson *et al.*, 1964; Sebastian *et al.*, 1971; Yashpe and Halvorson, 1976) have previously been shown to increase periodically during the cell cycle, neither enzyme showed periodic changes in activity during the cell cycle in synchronous cultures generated by selection from an elutriator rotor (J. M. Mitchison, personal communication). Ludwig and colleagues (1982) showed similar results with α-glucosidase enzyme activity using cells that were size fractionated in an elutriatior.

Additional data suggesting a lack of periodic enzyme synthesis have been reported by Elliott and McLaughlin (1978, 1979). Cells of *S. cerevisiae* were pulse labeled during exponential growth and then size fractionated by centrifugal elutriation. Rates of radioisotope incorporation into individual proteins were examined by two-dimensional gel electrophoresis. Over 200 different proteins were examined, including 31 ribosomal proteins, 8 basic proteins, and over 150 other cellular proteins. All showed an exponential synthesis during the cell cycle with the exception of histone proteins, which were synthesized periodically, with the peak at the beginning of S phase (Elliott and McLaughlin, 1978, 1979). Lorincz *et al.* (1982) used a similar two-dimensional electrophoretic system to analyze protein synthesis in both synchronously growing cells and in size-fractionated cells. A computer-coupled autoradiogram scanning method was used to quantitate autoradiograms. They observed periodicity for only 17 nonabundant proteins (in addition to the histones H2A, H2B, and H4) out of 900 polypeptides analyzed. At least 8 of these proteins were regulated at the level of *de novo* synthesis and 6 were unstable. There is a general consensus that a class of proteins varies in concentration through the cell cycle by periodic degradation or modification. As reviewed by Elliott and McLaughlin (1983), a labile protein is required for cell cycle initiation (Shilo *et al.*, 1979), protease activation of a zymogen for chitin synthesis in bud morphogenesis (Cabib, 1975), proteins synthesized prior to DNA synthesis, and a thermolabile protein involved in mitosis (Polanshek, 1977; Bullock and Coakley, 1976).

There are a number of possible explanations for the paradox between the protein and enzyme data. First, the pattern of enzyme activity or of polypeptide synthesis is misleading, or the patterns of enzyme and/or polypeptide accumulation are easily altered depending upon growth conditions. Alternatively, step enzymes are not among the major protein spots that have been analyzed on two-dimensional gels. Finally, step increases in enzyme activities may be due to posttranslational controls. One approach to answering this dilemma is to reexamine the pattern of protein synthesis in *S. cerevisiae* with an emphasis towards determining what specific controls regulate the observed step enzyme systems. We have therefore undertaken an analysis of two abundant enzymes in *S. cere-*

visiae that showed periodic increases in enzyme activity [galactokinase and acid phosphatase (APase)], and posed the following question. Under the experimental conditions used in our laboratory for cell growth and zonal rotor cell fractionation, resulting in sharp steps in enzyme activity, what mechanism(s) operate to regulate the periodic increases in enzyme activities, and to what extent do our experimental techniques of cell cycle analysis affect the observed patterns?

We conclude in this chapter that periodic increases in APase enzyme activity observed by zonal rotor analysis may be due initially to a periodic transcription of APase mRNA as well as some posttranscriptional controls. In contrast, periodic increases in galactokinase activity measured by this technique are not due to periodic transcription of the galactokinase gene, but rather to a posttranscriptional mechanism. In addition, we suggest that growth of some strains of *S. cerevisiae* in the types of nutrient-poor media used during *in vivo* labeling experiments prior to two-dimensional gel analysis, for example, may alter the length of the G_1 and S periods of the cell cycle to the point that periodic increases in *GAL1* enzyme activity could be interpreted as reflecting a linear increase in enzyme activity. Thus, the length of the G_1 period within the cell cycle and the patterns of macromolecular synthesis during the cell cycle can vary dramatically, depending upon the particular set of experimental conditions used to grow and analyze cell populations.

II. REGULATION IN THE GALACTOSE METABOLIC PATHWAY

A. The Galactose System

The galactose utilization pathway in *S. cerevisiae* consists of one constitutive structural gene (*GAL5*) and four structural genes that are inducible by galactose (*GAL2, GAL1, GAL7,* and *GAL10*). Regulation of these inducible structural genes is under the control of three regulatory genes (*GAL4, GAL80,* and *GAL3*). Transcription of the inducible structural genes is also subject to catabolite repression (Adams, 1972). The pathway of galactose utilization in *S. cerevisiae* is shown in Fig. 2.

Active transport of galactose into the cell is facilitated by an inducible galactose permease (*GAL2*). Internal galactose is converted to glucose 1-phosphate, a substrate for the glycolytic pathway, through the action of the "Leloir" enzymes (Leloir, 1951; Kosterlitz, 1943; Kalckar *et al.,* 1953; Douglas and Hawthorne, 1964). *GAL2* maps to chromosome XII (Mortimer and Hawthorne, 1966). All three of the Leloir enzymes are unlinked to *GAL2* and map as a tightly linked trio near the centromere on chromosome II. The first enzyme of the pathway, galactokinase (*GAL1* or "kinase," EC 2.7.1.6), catalyzes the phosphorylation of

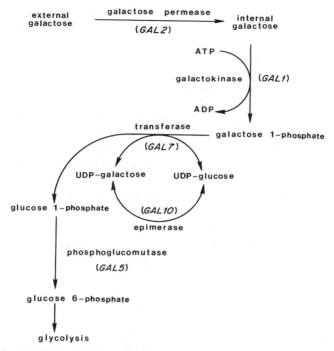

Fig. 2. The galactose metabolic pathway in *S. cerevisiae*.

galactose to galactose 1-phosphate. Galactose 1-phosphate is toxic to cells. In wild-type cells it is quickly converted to glucose 1-phosphate by the next inducible enzyme, galactose 1-phosphate uridylyltransferase (*GAL7* or "transferase," EC 2.7.7.10). The third inducible enzyme in the pathway, uridine diphospho-galactose-4-epimerase (*GAL10* or "epimerase," EC 5.1.3.2, UDPglucose-4-epimerase) recycles UDP-galactose, a by-product in the formation of glucose 1-phosphate, to UDP-glucose. The end product produced by these Leloir enzymes (glucose 1-phosphate) is converted to the substrate for the glycolytic pathway, glucose 6-phosphate, by the action of the constitutive enzyme phosphoglucomutase (*GAL5* or "mutase," EC 2.7.5.1) (Douglas and Hawthorne, 1964; Bevan and Douglas, 1969; Douglas, 1961). All three of the structural genes for the Leloir enzymes have been cloned and characterized by St. John and Davis (1981). The genes are clustered on a 6.5 kilobase (kb) region of DNA in which *GAL1* and *GAL10* are on separate strands and transcribed divergently. *GAL7* is downstream from the GAL10 gene on the same DNA strand. Analysis of *in vitro* synthesized deletions of the cloned DNA region has shown that all three genes have their own promoters (St. John and Davis, 1981).

The regulation of these structural genes has been shown by Hopper *et al.* (1978) and by St. John and Davis (1979) to occur at the transcriptional level. Transcription is dependent upon the presence of a wild-type *GAL4* gene (a positive effector of transcription). Another regulatory gene, *GAL80*, has been identified by mutations leading to either a constitutive or uninducible phenotype. A GAL80[s] uninducible phenotype is dominant to a wild-type *GAL80* allele and to the recessive *gal80* allele of the gene.

The *GAL4* protein is constitutively synthesized (Perlman and Hopper, 1979; Matsumoto *et al.*, 1978), and its activity regulated, through the action of *GAL80* protein (Matsumoto *et al.*, 1980). In a modification of the Douglas and Hawthorne model for regulation of the *GAL7*, *GAL10*, and *GAL1* gene cluster, Perlman and Hopper (1979) and Matsumoto *et al.* (1980) have proposed the existence of a *GAL4*, *GAL80* protein complex to explain the effect of the GAL80 repressor on the activity of the *GAL4* product. In the absence of inducer the complex is normally inactive, but in the presence of galactose (the presumed inducer) the *GAL4* protein becomes disassociated, allowing it to effect structural gene transcription. Alternatively, galactose could activate the entire complex which then promotes structural gene transcription. An alternative model which has not been excluded proposes that *GAL80* protein binds directly to structural gene promoters, thus inhibiting the binding of *GAL4* protein to DNA by exclusion.

A third regulatory gene, *GAL3*, has been defined by mutations characterized by a delay of 24–36 hr in the onset of normal galactose fermentation, in contrast to the 8–10 min required in wild-type cells. It has been postulated by Tsuyuma and Adams (1973) that the *GAL3* gene might specify a function required for the establishment but not the maintenance of the induced state.

B. Zonal Rotor Analysis of Galactokinase Cell Cycle Expression

For the experiments reported here, we have used the zonal rotor to fractionate yeast cells according to size by centrifugation in sucrose gradients at 4°C. The medium used for cell growth was the nutrient-rich medium yeast extract/peptone–galactose (YEP–galactose) unless noted otherwise. To minimize differences in cell size due to age, cell inocula of Y185 were made daily from fresh single colonies growing on YEP–dextrose agar plates.

The fractions from each zonal rotor centrifugation were analyzed by a number of standard cell cycle control assays to assess the accuracy of each cell cycle fractionation. Cell concentrations per fraction were determined by diluting cells in saline and then counting the samples in a Coulter electronic particle counter (Coulter Electronics, Inc.). Typical cell distributions are shown in Fig. 3. The size distribution of cells in each fraction (mean cell volume) was examined using

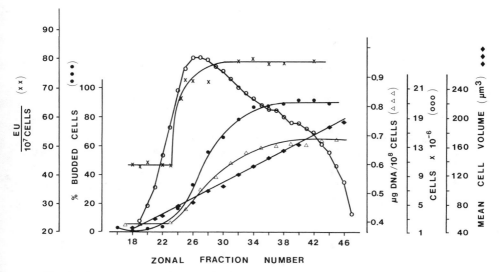

Fig. 3. Zonal rotor centrifugation of cells grown in YEP–galactose medium. The cell cycle control assays [% budded cells (●), µg DNA/cell (△), cell number (○), and mean cell volume (♦)] and the galactokinase enzyme assay, expressed in enzyme units (EU), (✕) are as described elsewhere (Sebastian *et al.*, 1971). Polystyrene beads having mean cell volumes of 72 and 64 µm³ were used as size standards for the Coulter counter analysis (Coulter Electronics, Inc.).

a particule size distribution plotter attached to a Coulter electronic channelyzer. As shown in Fig. 3, the mean cell volume increased linearly during the cell cycle. The mean cell volume increase is typically 3- to 4.5-fold. The explanation for this large divergence from a 2-fold increase expected for cells in balanced growth will be discussed in the following paragraphs.

In Fig. 3 we show that the percentage of budded cells per fraction began at 0% early in the early fractions, which would correspond with the G_1 period of the cell cycle, and increased to 90% bud emergence in the late fractions, which would correspond to the late S period and early G_2 period. Finally, DNA content per cell was examined for each zonal rotor fraction. The data in Fig. 3 confirmed that DNA synthesis is periodic, with synthesis being initiated just after bud emergence. The increase in DNA content per cell was nearly 2-fold (1.9✕) during the cell cycle.

The increase in *GAL1* enzyme activity was examined during the cell cycle. As shown in Fig. 3 a sharp step in *GAL1* enzyme activity began prior to step increase in percentage of budded cells and was completed by the time 50% of the cells were budded.

In Fig. 3 we show that the increase in mean cell volume across the rotor was routinely greater than the anticipated 2-fold increase. This phenomenon is explained by the model in Fig. 4. Hartwell and Unger (1977) have shown that yeast

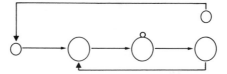

Fig. 4. Unequal cell size at division in *S. cerevisiae*. (Redrawn from Hartwell and Unger, 1977.)

cells undergo unequal cell division. Daughter cells are generally smaller than the mother cell. After cell division, the mother cell is capable of immediately re-initiating the cell cycle. In contrast, the daughter cell must first undergo a period of cell growth to obtain a minimal cell size prior to initiating the cell cycle (Pringle and Hartwell, 1982). Therefore, an increase in mean cell volume of 3- to 4.5-fold in zonal rotor fractionated cells (see Fig. 3) is expected from a culture of cells in which daughter cells are approximately 50% of the minimal size required to start the cell cycle (see Fig. 4) and in which mean cell volumes double during the cell cycle.

We examined the correlation between increase in mean cell volume and the increase in total cellular RNA content or total cellular protein content per cell by

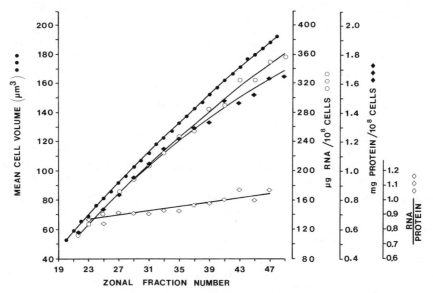

Fig. 5. Correlation between mean cell volume, RNA content per cell, and protein content per cell in cells size fractionated by zonal rotor centrifugation: mean cell volume (●), RNA (○), protein (◆), and the ratio of RNA/protein (◇). RNA was determined by the standard orcinol method and protein by the method of Lowry.

a similar zonal rotor analysis. The mean cell volume of the cells increased 3.15-fold during the cell cycle (Fig. 5). Also, the RNA content increased 3.1-fold during the cell cycle. Protein content per cell increased 3-fold (Fig. 5). Thus, the increase in RNA content and protein content per cell was proportional to the increase in mean cell volume.

There are two potential concerns when cells are fractionated according to size by gradient centrifugation. It is important to determine both whether cells at the top as well as the bottom of the gradient represent viable cells. To determine whether cells at the top of the gradient represented small, nonviable cells, we remove cells from the fractions shown as numbers 21 and 22 in Fig. 3, washed the cells free from sucrose and cycloheximide, and then resuspended the cells in fresh YEP–galactose growth medium. The culture was grown at 30°C in a rotatory water bath while aliquots of the culture were periodically monitored for mean cell volume and cell number. Figure 6 shows that the cell concentration remained constant for 2.5 hr prior to exhibiting a 2-fold, synchronous increase in cell number. Figure 6 also shows that the mean cell volume increased 2.6-fold prior to the synchronous round of cell division.

A different problem arises in the analyis of cells at the bottom of the gradient. Some of the large, double cells at the bottom may dissociate during subsequent handling and analysis. Therefore, cell concentrations at the bottom of the gradient, as determined by Coulter counter analysis, may be nonrepresentative of

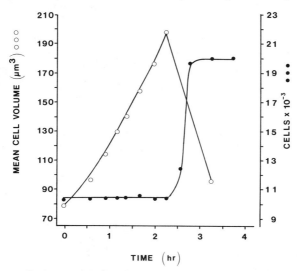

Fig. 6. Viability of cells in early fractions from the zonal rotor. Yeast cells were grown and fractionated as in Figs. 3 and 5. Cells from the earliest fractions of the rotor were used to start cultures grown at 30°C in YEP–galactose medium. The change in both mean cell volume (○) and cell concentration (●) with time are presented.

the original concentration of double cells. In our experiments, this problem occurred in only the last few fractions and represented a minority of the population. Any dissociation of these large, double cells was represented by an unexpected decrease in the mean cell volume.

C. Galactokinase mRNA Accumulation during the Cell Cycle

The simplest explanation for periodicity of *GAL1* enzyme synthesis is control at the level of transcription. To examine whether the periodic increase in *GAL1* enzyme activity was due to a periodic transcription of *GAL1* RNA, total yeast RNA was extracted from cells in the zonal rotor fractions shown in Fig. 7A. RNA from each cell fraction was electrophoresed on formaldehyde–agarose gels and transferred to nitrocellulose paper prior to hybridization analysis as described elsewhere (Yarger and Halvorson, 1983). An internal control was required in these hybridizations against which to standardize the amount of RNA added to each lane, the efficiency of RNA transfer to the nitrocellulose, etc. For this purpose we obtained from Lynna Hereford a plasmid DNA, prot1, which encodes for an abundant, unidentified RNA. The relative specific activity of prot1 RNA has been previously shown to remain essentially constant throughout the cell cycle (Hereford *et al.*, 1981). The data in Fig. 7B confirmed that the concentration of prot1 RNA, as determined by hybridization with ^{32}P-labeled prot1 DNA and densitometry of the autoradiogram, remained essentially constant throughout the rotor fractions. To determine the accuracy of quantitating RNA concentrations by standardizing to prot1 RNA concentrations on the RNA–nitrocellulose paper, we hybridized ^{32}P-labeled histone H2A plasmid DNA (a gift from Lynna Hereford) to the same set of filters. Histone H2A RNA, when standardized to prot1 RNA concentrations in this experiment, first appeared during middle G_1 period of the cell cycle with the maximum H2A RNA concentrations present at the onset of DNA synthesis in S period (Fig. 7B). The concentration of histone H2A RNA then rapidly declined as DNA synthesis continued (Fig. 7B). The histone H2A RNA data were virtually identical to the data previously published by Hereford and colleagues (1981). Therefore, standardizing RNA concentrations to prot1 RNA concentrations on the RNA–nitrocellulose paper represented an accurate quantitation of RNA concentrations in these cell fractions.

We examined the concentrations of *GAL1* RNA and *GAL10* RNA during the cell cycle by separately hybridizing ^{32}P-labeled plasmid pSC4817 DNA and ^{32}P-labeled plasmid pSC4911 DNA, respectively (kindly provided by Tom St. John), to the same nitrocellulose filters used for the prot1 and histone H2A hybridization analysis. The autoradiographic bands were again quantitated by densitometry and standardized to prot1 RNA. The data in Fig. 7B show that *GAL1* RNA did not increase in a step fashion prior to the increase in *GAL1* enzyme activity.

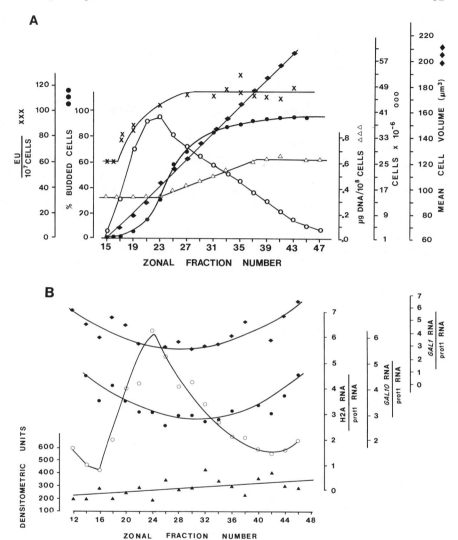

Fig. 7. Concentrations of specific RNAs in total yeast RNA extracted from cells fractionated by zonal centrifugation. (A) The parameters of the zonal centrifugation are as in Fig. 3. (B) The data are as follows: prot1 RNA (▲), H2A RNA/prot1 RNA (○), *GAL10* RNA/prot1 RNA (●), and *GA1* RNA/prot1 RNA (♦).

Neither did *GAL1* RNA increase linearly during the cell cycle. Instead, *GAL1* RNA smoothly oscillated in concentration over a twofold range. Variations in *GAL10* RNA concentration were similar to *GAL1* RNA (Fig. 7B). Because *GAL1* RNA at steady state represents 0.5–1.0% of the poly(A) RNA (St. John and Davies, 1981), at the point of least amount of *GAL1* RNA (mid-S period), *GAL1* RNA still represented approximately 0.25–0.50% of the cellular poly(A)-containing RNA. These data are not representative of a transcriptional control mechanism regulating periodic *GAL1* enzyme accumulation.

We examined the possibility that the translatability of *GAL1* mRNA varied over the cell cycle to produce the periodic increases in *GAL1* enzyme activity. Under the assumption that the translatability of total yeast RNA *in vitro* reflects the translational capacity *in vivo*, total RNA from the zonal rotor fractions shown in Fig. 7A was translated in a wheat germ *in vitro* translation system as previously described (Yarger and Halvorson, 1983). *In vitro* synthesized enolase polypeptide was immunoprecipitated from each *in vitro* translation reaction and electrophoresed on a 10% polyacrylamide gel as an internal control. The intensity of the enolase bands was then quantitated after autoradiography by densitometry. In Fig. 8 we show that the specific activity of enolase polypeptide, as analyzed by linear regression, remained essentially constant with only a marginal increase throughout the zonal rotor fractions.

Since glyceraldehyde-3-phosphate dehydrogenase (GPDH) synthesis was

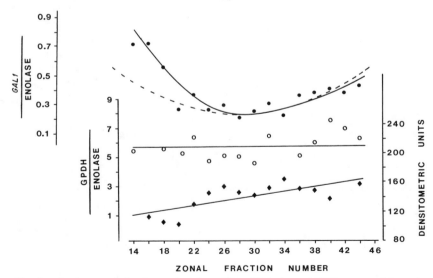

Fig. 8. *In vitro* translational capacity of specific mRNAs in total RNAs prepared from size-fractionated yeast cells: *GAL1* RNA (●), GPDH RNA (○), and enolase RNA (◆). The dashed line is the physical concentration of *GAL1* RNA. For methods, see Hopper *et al.* (1978).

known to be constant during the cell cycle (K. A. Bostian, L. M. Lemire and H. O. Halvorson, unpublished results), GPDH polypeptide was similarly immunoprecipitated and electrophoresed on a 10% polyacrylamide gel as an additional control. Autoradiographs were quantitated by densitometry, and the ratio of GPDH to enolase polypetide determined (Fig. 8). By linear regression analysis, the data yielded a line of slope 0. The *in vitro* translatability of *GAL1* mRNA was compared to enolase mRNA and GPDH mRNA by immunoprecipitating *in vitro* synthesized *GAL1* polypeptides from aliquots of the same *in vitro* translation as was used to examine enolase and GPDH polypeptides. Autoradiographs were again quantitated by densitometry, and the concentration of *GAL1* polypeptide from each fraction normalized to enolase polypeptide. The data presented in Fig. 8 show that *GAL1* RNA translatability only partially reflected its physical concentration during the cell cycle. *GAL1* mRNA appeared to be significantly more translatable during early G_1 prior to the step in galactokinase enzyme activity.

D. Rate of *GAL1* Protein Synthesis during the Cell Cycle

We would expect the rate of *GAL1* polypeptide synthesis to change extensively during the cell cycle if the step in *GAL1* enzyme activity was regulated through translation controls. To examine this possibility, we grew strain JH7 (kindly provided by James E. Hopper) in minimal vitamin (MVA) medium (commonly used for pulse labeling) containing 10% galactose. This strain, unlike Y185, grew with a generation time of 2.5 hr in both YEP-galactose or MVA–galactokinase medium. Ten minutes prior to harvesting the cells in preparation for zonal rotor analysis, the cells were pulse labeled with 15 mCi of [^{35}S]methionine for 8 min. Incorporation of [^{35}S]methionine into polypeptides was inhibited by the addition of 1 mM unlabeled methionine (final concentration), followed by the addition of cyclohexamide to 100 mM. As shown in Fig. 9, although *GAL1* enzyme activity again stepped during the cell cycle, the rate of *GAL1* polypeptide synthesis was directly proportional to the physical concentration of *GAL1* RNA.

E. Effects of Growth Conditions on *GAL1* Enzyme Accumulation

In several early attempts to examine the rate of *GAL1* protein synthesis during the cell cycle, we grew strain Y185 in MVA–galatose medium. The generation time for growth of Y185 in MVA medium is 4 hr (as compared to 2.5 hr for growth in YEP–galactose). In Fig. 10 we show the cell cycle control assays performed on zonal rotor-fractionated cells after growth in MVA medium. The increase in the percentage of budded cells was normal, and the mean cell volume increased linearly (approximately threefold) during the cell cycle. However, the

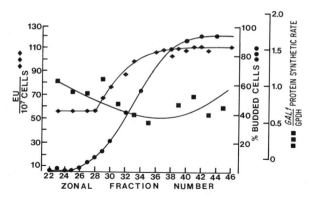

Fig. 9.   Rate of *GAL1* polypeptide synthesis during the cell cycle. Cells of strain JH4 were pulse labeled with [35S]methionine in MVA medium (Lorincz *et al.*, 1982) during asynchronous growth and then fractionated by zonal rotor centrifugation. Cell extracts were immunoprecipitated with *GAL1* antibodies and immunoprecipitated polypeptides analyzed by SDS–polyacrylamide gel electrophoresis, and quantitated by densitometry of gel autoradiograms: % budded cells (●), *GAL1* activity/cell(◆), and *GAL1*/GPDH protein synthetic rate (■).

distribution of cells across the zonal rotor fractions showed an atypically large concentration of cells with large mean cell volumes (Fig. 10). *GAL1* enzyme activity again began to increase in mid-G_1, prior to bud emergence. However, the step in *GAL1* enzyme activity was now spread over a larger fraction of the cell population.

Tyson *et al.* (1979) have shown that the mean cell volume of cells has a tendency to increase up to 2.5-fold when the cellular growth rate increased from 75 to 450 min. Consequently, there is no logical basis for comparing cell cycle assays between cells grown in YEP–galactose medium (Fig. 3) and cells grown in MVA medium (Fig. 10). Empirically, one could compare equivalent increases (midpoints) in enzyme synthesis and bud formation. We thus compared the mean cell volume for cells that were at the 50% level for increase in enzyme activity and the mean cell volume for cells that showed 50% bud emergence. At the 50% level for increase in enzyme activity, cells grown in MVA medium showed a mean cell volume of 111 μm^3 (Fig. 10) as compared to a mean cell volume of 78 μm^3 for cells grown in YEP–galactose medium (Fig. 3). Accordingly, at 50% bud emergence, cells grown in MVA medium showed a mean cell volume of 130 μm^3 as compared to a mean cell volume of 104 μm^3 for cells grown in YEP–galactose medium. An alternative approach to comparing the two cultures was to determine the percentage of the population involved in enzyme synthesis or bud formation. The step in *GAL1* enzyme activity represented 66% of the cell population from cells grown in MVA medium as compared to 29% of the cell population from cells grown in YEP medium. Thus, *GAL1* enzyme synthesis, a G_1 event, is now spread over a larger fraction of the population in the slower-growing cells. The periods of the cell cycle occupied by *GAL1* enzyme synthesis and bud formation can be more easily seen by plotting them directly as a function of mean cell volume (see Fig. 11).

F. Overall Regulation of Galactokinase Expression

There are two areas of concern when cells are size fractionated with a zonal rotor. First, the small, unbudded cells at the top of the gradient may represent only damaged, nonviable cells which would distort the analysis of the beginning of the cell cycle. Second, disruption of the large double cells at the bottom of the gradient may dissociate, causing a distortion in the analysis at the end of the zonal rotor fraction. We showed that the small, unbudded cells in our experiments were approximately 100% viable. In addition, the average mean cell volume of the small cells increased 2.6-fold prior to cell division. These data clearly demonstrated the requirement for small cells to undergo a period of cell growth to attain a critical size (Pringle and Hartwell, 1982) prior to entering the cell cycle.

Care was taken to limit the extent of any mechanical disruption of the double cells at the bottom of the gradient. The mean cell volume of our cells continued to increase linearly to nearly the end of the gradient. Only a minority of the population of cells at the extreme end of the gradient (2%) showed a decrease in mean cell volume representative of cell dissociation.

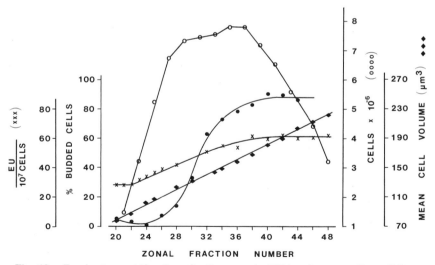

Fig. 10. Zonal rotor centrifugation of cells grown in MVA–galactose medium. Cells were fractionated by gradient centrifugation and assayed as in Fig. 3.

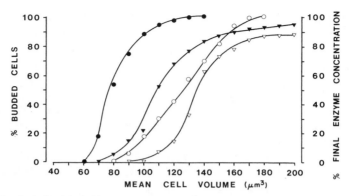

Fig. 11. Periods of the cell cycle occupied by galactokinase enzyme synthesis and bud formation. Data were compiled by plotting percentage of buds and of final galactokinase enzyme concentration versus mean cell volume of cells throughout the zonal rotor fractions. Data show comparison between MVA–galactose medium taken from Fig. 10 [% buds (▽), % final enzyme (○)] and
YEP–galactose medium taken from Fig. 3 [% buds (▼), % final enzyme (●)].

During our initial attempts to examine the rate of *GAL1* polypeptide synthesis
during the cell cycle by pulse-labeling experiments, we were unable to achieve
unperturbed cell cycle fractionations. While examining this problem, we were
subsequently able to derive a possible explanation for the dilemma between
protein analysis versus enzyme analysis (linear versus periodic accumulation).
The perturbation apparently resulted from growing strain Y185 to mid-log phase
in MVA medium. Growth of this strain in the nutrient-poor MVA medium
caused an approximately twofold increase in the generation time with a corresponding lengthening of the G_1 phase of the cell cycle. *GAL1* enzyme synthesis,
a G_1 event, was now spread over a larger fraction of the cell cycle. The result
was that *GAL1* enzyme activity now showed a lengthened synthetic period to the
extent that it would be possible to misinterpret the periodic increase as an exponential increase in enzyme activity. These data are important when one considers
that previous papers showing a general lack of periodic enzyme accumulation
were based on data derived from cells growing with prolonged generation times
in nutrient-poor medium. For example, Elliott and McLaughlin (1978, 1979)
grew cultures of *S. cerevisiae* in yeast nitrogen base supplemented with amino
acids, adenine, and uracil for pulse labeling and two-dimensional gel analysis of
the extracted proteins. Lorincz *et al.* (1982) grew cells in MVA medium prior to
pulse labeling and two-dimensional gel analysis of the extracted proteins. In
addition, J. M. Mitchison (personal communication) grew cultures of *Sch.
pombe* in EMM2 medium and *S. cerevisiae* in yeast extract medium. All of these
poor growth media lengthened cell generation times approximately twofold over
growth in nutrient-rich medium. In every case, there was an inability to detect

periodic synthesis of abundant proteins. We suggest that in some instances this perturbation may be removed either by growing cells in nutrient-rich medium or by selecting a yeast strain that is capable of rapid growth under nutrient-poor growth conditions.

During balanced growth, the components of a cell can double during each cell cycle (1) by increasing once per cycle in a periodic burst, (2) through an exponential rate of synthesis, or (3) at a constant rate of synthesis which doubles once per cell cycle. We have presented data concerning the molecular mechanisms that regulate *GAL1* enzyme synthesis in the yeast *S. cerevisiae*. Data presented elsewhere have suggested that *GAL1* enzyme synthesis is under posttranscriptional control (Yarger and Halvorson, 1983) in addition to the absolute requirement for transcriptional control from the *GAL4* polypeptide (Hopper *et al.*, 1978; Matsumoto *et al.*, 1978). Under our conditions of growth and zonal rotor centrifugation, the *GAL1* enzyme synthesis was periodic during the cell cycle. The high concentration of physical *GAL1* RNA throughout the zonal rotor fractions suggested that the periodicity observed for *GAL1* enzyme synthesis was not controlled through *GAL4* protein at the level of transcription but rather by some posttranscriptional event. In this regard, it is difficult to explain the twofold oscillation in the concentration of physical *GAL1* RNA in these fractions. This oscillation could arise from a regular oscillating pattern of catabolite repression during the cell cycle, since it is known that catabolite repression profoundly affects the amount of *GAL1* enzyme activity (Adams, 1972).

The *in vitro* translability of *GAL1* RNA was significantly higher during early G_1, as compared to the remainder of the cell cycle. These data alone suggest either that the periodicity associated with *GAL1* enzyme activity is due to translational control of *GAL1* RNA, or that RNA translational capacity *in vitro* does not accurately reflect its translatability *in vivo*. The latter now appears to be true. When the rate of *GAL1* polypeptide synthesis was examined, the rate was found to slowly oscillate twofold during the cell cycle in a manner very similar to that observed for the physical concentration of *GAL1* RNA. These data are consistent with our earlier data suggesting that periodic control of *GAL1* enzyme synthesis is not due to transcriptional control. However, the rate data are inconsistent with data obtained from the *in vitro* translation studies and suggest that *in vitro* translational capacity does not always reflect the *in vivo* translational capacity. The rate of *GAL1* polypeptide synthesis observed could not occur in the presence of translational control. In contrast, the rate data suggested that *GAL1* enzyme accumulation occurred through a mechanism involving enzyme stabilization versus enzyme turnover.

We presented data elsewhere (Yarger and Halvorson, 1983) suggesting that control of periodicity could be at the level of *GAL1* enzyme stability. During a short induction period with galactose, followed by a rapid deinduction and continued cell growth in the presence of glucose, *GAL1* enzyme activity was approximately 50% unstable (Yarger and Halvorson, 1983). Therefore, some cell cycle, stage-specific event could occur to stablize the enzyme activity.

It is difficult trying to correlate a twofold increase in *GAL1* enzyme activity with a greater than threefold increase in total cellular protein during the cell cycle. One possible explanation for this paradox is that new *GAL1* enzyme accumulation (or stabilization) may be restricted to the emerging bud. In this regard, it is interesting to note that the acid phosphatase (*PHO5*) polypeptide synthesis has been shown to be restricted to the emerging bud (Field and Schekman, 1980). There are a number of examples of enzyme accumulation regulated at the level of enzyme stability. For example, enzyme stabilization has recently been suggested in the alga *Chlorella sorokiniana* (Turner *et al.*, 1981). The mRNA for the inducible glutamate dehydrogenase (GDH) enzyme is present in both induced and noninduced cells. Pulse-chase experiments demonstrated that GDH polypeptide is rapidly synthesized in the noninduced state, but that it is rapidly, covalently modified and then degraded in the absence of inducer.

Furthermore, it is interesting to note that periodic enzyme synthesis controlled at the level of enzyme stability would not be detected in pulse-labeling experiments in which the polypeptides are analyzed solely by two-demensional gel analysis since the rate of enzyme synthesis would indeed be constant. In light of these new data, two-dimensional gel data alone do not necessarily rule out the possibility of periodic enzyme accumulation in yeast.

III. REGULATION OF PHOSPHORUS METABOLISM

A. Enzymes and Physiology of Phosphorus Metabolism

Phosphorus metabolism in *S. cerevisiae* involves five principal enzymes for the acquisition and metabolic integration of inorganic phosphate (P_i): an exocellular phosphohydrolase (acid phosphatase; APase), catalyzing the hydrolysis of a broad spectrum of phosphoester substrates (Schmidt *et al.*, 1963); a phosphate permease, participating in the facilitated transport of P_i (Roomans and Borst-Pauwels, 1979); polyphosphate kinase, converting intracellular P_i, via ATP, into vacuolar deposits of polyphosphates (polyP) (Harold, 1966; Solimene *et al.*, 1980; Urech *et al.*, 1978; Dawes and Senior, 1973), which are the most abundant phosphorus compounds in yeast; and two enzymes, alkaline phosphatase (AlkPase) and polyphosphatase, located in the yeast vacuole, which hydrolyze polyP (Kaneko *et al.*, 1982; Liss and Langen, 1960; Katchman and Fetty, 1955). These enzymes regulate intracellular concentrations of P_i by a cyclic pathway of polyP synthesis and degradation (Liss and Langen, 1960, 1962), thought to be an important factor in the regulation of cellular homeostasis (Dawes and Senior, 1973; Erecenska *et al.*, 1977).

All of these enzymes are regulated by external growth concentrations of P_i (Harold, 1966; Dawes and Senior, 1973). Under nonlimiting conditions they are present in basal amounts. However, when cells are starved for P_i, these enzymes are derepressed (Toh-e *et al.*, 1973; Schmidt *et al.*, 1963), and exocellular APase may constitute as much as 1% of total yeast protein. When P_i is replenished, polyP is rapidly synthesized and the enzymes are repressed (Harold, 1966; Dawes and Senior, 1973), perhaps by a coordinate mechanism. Little is known about the mechanism or metabolic effectors of this regulation, although the phosphatases have been shown to be regulated at the transcriptional level (Rogers *et al.*, 1982; Kramer and Anderson, 1980) by a complex but genetically well-defined system (see Section B).

B. Genetics and Molecular Biology of Phosphatase Regulation

The phosphatase genetic system in *S. cerevisiae* is a dispersed gene system consisting of numerous structural and regulatory genes involved in the expression of both acid and alkaline phosphatases (Toh-e *et al.*, 1975, 1976; Ueda *et al.*, 1975). These nonspecific phosphatases exist both as constitutive enzymes and enzymes repressible by P_i (Tohe *et al.*, 1973, 1976). The constitutive and repressible APases, *PHO3* and *PHO5* respectively, form a cluster on chromosome II, whereas the repressible AlkPase (*PHO8*) is on chromosome IV (Toh-e *et al.*, 1973; Kaneko *et al.*, 1982). Another structural gene, *PHO84*, involved in P_i transport, may encode a phosphate permease (Ueda and Oshima, 1975). The expression of these genes is mediated by numerous regulatory genes scattered throughout the genome (Toh-e *et al.*, 1975, 1975, 1976; Ueda *et al.*, 1975; Lange and Hansche, 1980). There are seven unlinked regulatory genes: *PHO4*, *PHO81*, *PHO80*, *PHO85*, *PHO2*, *PHO6*, and *PHO7*. Recessive mutations in *PHO4* and *PHO81* block the derepression of both repressible APase and Alk-Pase, in addition to P_i uptake (Ueda *et al.*, 1975; Toh-e *et al.*, 1973). Recessive mutations in *PHO2* do not show the same pleiotrophy, but rather a phenotype lacking only the repressible APase (Ueda *et al.*, 1975; Toh-3 *et al.*, 1973). Recessive mutations occurring in the *PHO80* or PHO85 loci result in the constitutive synthesis of repressible APase and AlkPase. *PHO6* and *PHO7* are involved in expression of the constitutive APase (Toh-e et al., 1975, 1976).

Based on the genetic analysis of various double mutants and the discovery of a cis dominant constitutive mutant (*PHO82*) contiguous to *PHO4* (Ueda *et al.*, 1975), Ueda and colleagues originally proposed a role for these genes involving transcriptional regulation via the sequential functioning of their products (Ueda *et al.*, 1975), reminiscent of control of phage development by repressor and antirepressor proteins (Susskind and Botstein, 1975). Accordingly, *PHO4* is a positive regulatory gene encoding a protein essential for structural gene transcription. Both *PHO80* and *PHO85* code for polypeptides, which together form a repressor under the negative control of *PHO81*. In the presence of aporepressor (P_i or a metabolite thereof), repressor blocks transcription of *PHO4* by binding to an adjacent site *pho82*. The absence of aporepressor inactivates repressor binding at *pho82*, allowing synthesis of the *PHO4* product, which, in turn, binds to activation sites adjacent to the structural genes, *PHO5* and *PHO8*, and activates transcription. For APase, this occurs in concert with *PHO2*, at the *pho83* locus adjacent to *PHO5*.

Several recent observations, however, are inconsistent with this model (Toh-e *et al.*, 1978, 1981). First, fine structure meiotic mapping situates the *pho82* site with a narrow region inside the *PHO4* locus (Toh-e *et al.*, 1981). The presence of two temperature-sensitive mutants and a mutant suppressible by a nonsense suppressor argues strongly that the *PHO4* locus encodes a protein (Toh-e *et al.*, 1981). The fact that the *pho8* site is flanked by two *PHO4* sites which apparently encode a protein argues against a model defining the *pho82* site as an operator of the *PHO4* gene, and *PHO4* as a gene regulated at the transcriptional level. Moreover, APase activity shown by *PHO82 PHO4/pho92* (wild-type) *pho4* diploids grown under repressed conditions varied depending upon the combination of *PHO82* and *pho4* alleles, unlike the *PHO82* homozygous diploids (Toh-e *et al.*, 1981).

These findings led Toh-e and co-workers to propose a new regulatory model involving the simultaneous functioning of the regulatory factors by direct molecular interaction as shown in Fig. 12 (Toh-e *et al.*, 1978, 1981). This model states that a few molecules of the regulatory factors are produced constitutively. Under repressed conditions the *PHO4* gene product aggregates with a complex of the *PHO80* and *PHO85* gene products at a protein domain identified by *pho82*

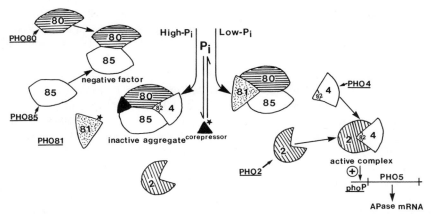

Fig. 12. Current regulatory model for APase regulation in *S. cerevisiae*, as proposed by Toh-e *et al.* (1981). ⋆, Alternatively, corepressor could function by binding to *PHO81* to mediate binding of *PHO81* to the inactive aggregate.

and is unable to activate transcription of *PHO5*. Under depressed conditions, the *PHO81* product binds to the *PHO80, PHO85, PHO4* aggregate, and the *PHO4* product is released. The complex binds, in combination with the *PHO2* protein, to an operator of the structural gene and activates transcription (see the following paragraphs).

Support for a transcriptional regulation model has come from biochemical and molecular studies (Rogers *et al.*, 1982; Kramer and Anderson, 1980; Bostian *et al.*, 1980, 1982; Yarger *et al.*, 1982). In our initial studies of this regulatory system, we developed an *in vitro* assay for the characterization of repressible APase mRNA and identified three mRNAs (p60, p58, and p56) that code for distinct APase polypeptides (Bostian *et al.*, 1980). We demonstrated that depression is the result of a cellular increase in APase mRNA and subsequent *de novo* enzyme synthesis (Bostian *et al.*, 1980), supporting the transcriptional control model.

By analyzing several yeast genes under P_i control, isolated by Kramer and Anderson (1980), we identified the p60 mRNA as a transcript of the genetically defined *PHO5* gene, and the p56 mRNA as the product of a different genetic locus (Rogers, *et al.*, 1982). These P_i-regulated genes were not expressed in *PHO2, PHO4*, or *PHO81* mutants grown under derepressed conditions and were constitutively expressed in *PHO80* or *PHO85* mutants.

More recently, we generated biochemical evidence showing that the *PHO4* gene is expressed constitutively (Yarger *et al.*, 1982), the major tenet of the revised genetic regulatory model. We demonstrated that derepression of APase mRNA occurs normally during continuous exposure and preincubation with a concentration of cycloheximide that effectively blocks *in vivo* protein synthesis, including the appearance of APase. Temperature-sensitive mutants used in temperature-shift experiments in the presence or absence of cycloheximide established that *PHO80* does not regulate *PHO4* transcriptionally. Thus, these biochemical data support the notion that the *PHO4* protein acts at the trancriptional level in controlling *PHO5* and *PHO8* expression. This would occur prior to transcription initiation, or as an antiterminator of transcription, rather than as some pre- or posttranslational event. Nothing, however, is known about the detailed mechanism of action of these regulatory genes.

C. Measurement of Gene Transcripts during Derepression

In addressing the regulation of APase gene expression at the molecular level we have used two quantitative techniques (translation–immunoprecipitation and mRNA hybridization) to measure cellular transcript concentration in different cell populations. These have been applied in a series of experiments with asynchronous cultures to characterize the physiological transitions under which APase is derepressed and to determine to what extent these changing physiological events influence the regulation of APase gene expression. At the same time, these assays have been applied to the analysis of size-fractionated cells generated by zonal centrifugation. The details of these assays are summarized elsewhere (Bostian *et al.*, 1983).

First, we measured total functional mRNA concentrations by *in vitro* translation of extracted RNA in a cell-free wheat germ system, as in our studies of the galactose system (Section II,C). *In vitro* synthesized APase was then measured by immunoprecipitation of the total translation products as shown in Fig. 13. Data were calculated by quantitating APase polypeptide gel autoradiograms by densitometry and standardizing them to a control RNA product (enolase). This control was used in all subsequent experiments. In these experiments, we demonstrated that enolase messenger activity remained at a constant proportion of the total cellular RNA during changing cellular phosphate concentrations. *In vitro* mRNA activities were expressed as a ratio to the maximal level of APase mRNA activity for RNA isolated from a fully derepressed culture of P28-24C. From the translation data the yield of *in vitro* synthesized protein for the three APase mRNAs, as determined by autoradiographic densitometry, was proportional to their specific concentration over a 10-fold range.

Second, we measured the physical concentrations of gene transcripts by an RNA blot hybridization procedure employing the cloned p60 APase gene as a probe (Fig. 13B). The amount of RNA hybridized to the p60 gene as determined by autoradiographic densitometry was expressed in relative density units (as a ratio to the amount of hybridizable RNA in total RNA isolated from a fully derepressed culture of P28-24C). Again, a linearity of hybridization signal with respect to APase mRNA concentration was observed. The limitations of this procedure are presented elsewhere (Bostian *et al.*, 1983).

D. Autoregulation of APase Gene Expression

Using the above experimental approach, physical and functional mRNA levels were measured during derepression. For these experiments yeast cells were grown in a low-P_i-supplemented minimal medium (SMD). In this medium, P_i was absent, but metabolizable organic phosphate in a P_i-free yeast extract/bacto-peptone supplement was supplied at a concentration of 0.17 μmole/ml. Phosphorus was thus available either by consumption of internal reserves, through facilitated transport of the exogenous organic phosphates, or through their degradation by APase, which increased in this medium during P_i-limited growth. Upon continuous growth of cells in this medium, the external phosphorus supply was exhausted. This was accompanied by a period of phosphate-limited growth, following which cells arrested in the G_0 state due to P_i starvation. As shown in Fig. 14, the onset of APase derepression coincided with the time at which mean cell volume began to decrease. This preceded the departure from exponential

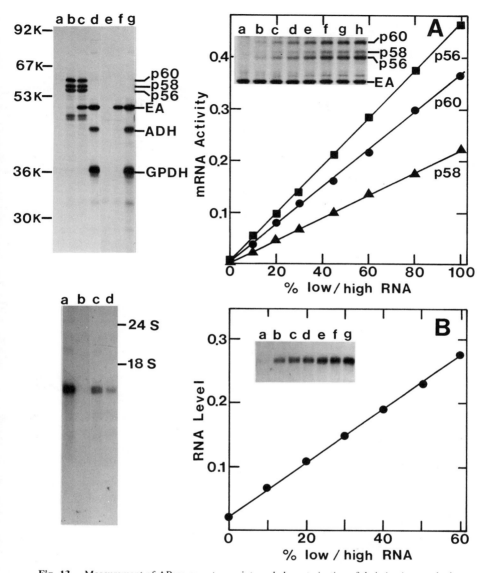

Fig. 13. Measurement of APase gene transcripts and characterization of their *in vitro* synthetic activity. RNA was isolated from cells of strain P28-24C grown in High-P_i or low-P_i SMD medium to an O.D. 660 of 1.0. These total cellular RNA preparations were analyzed by two methods: (A) Immunoprecipitation of their cell-free translation products using APase IgG and other control IgGs. Shown to the left are the translation products of RNA isolated from low-P_i-grown cells immuno-precipitated with (a) preimmune IgG, (b) APase IgG, (c) a mixture of Apase and enolase IgGs, and (d) a mixture of enolase, ADH, and GPDH IgGs; and high-P_i-grown cells (e) immunoprecipitated with preimmune IgG, (f) a mixture of APase and enolase IgGs, and (g) a mixture of ADH, GPDH, and enolase IgGs. (B) RNA blot hybridization to a p60 gene probe. Shown in the autoradiogram to the left: (a,b) P28-24C RNA, and (c, d) P142-44 RNA, isolated from cells grown in high-P_i or low-P_i SMD medium, respectively. (Reprinted with permission from Bostian *et al.*, 1983).

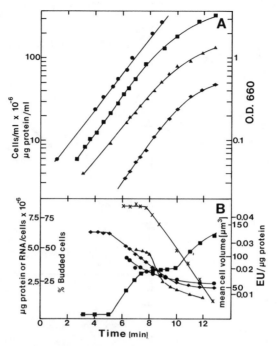

Fig. 14. Effect of phosphate depletion on cellular properties of strain Y185. Cells of strain Y185 were grown overnight at 30°C in high-P_i SMD medium and inoculated at a cell density of 5×10^5 cells/ml into low-P_i SMD medium. (A) Growth and protein concentration were determined: (■) optical density at 660 nm (O.D. 600); (♦) cell/ml determined with Coulter counter, (▲) μg protein/ml, and (●) O.D. 600 control culture grown in high-P_i SMD medium. (B) Changes in cell parameters were measured during growth of the cells in low-P_i medium: (■) APase activity/μg protein, (♦) mean cell volume, (▲) RNA/cell, (●) protein/cell, and (x) percentage of budded cells.

growth, which corresponded with a dramatic decrease in the percentage of un-budded cells in the population, and with a decrease in cellular protein and RNA content. Cells undergo approximately three generations of growth from derepression of APase to G_1 arrest.

To follow the kinetics of APase derepression at the RNA level, cells were initially grown under repressed conditions in high-P_i SMD medium to mid-log phase and then derepressed by transfer to low-P_i SMD medium. In a control culture, cells were transferred into high-P_i SMD in the same fashion. Initial growth rates for both cultures were approximately the same for one generation (Fig. 15A), and enzyme activity in the control (high-P_i) culture remained at a basal level (Fig. 15B). However, cells growing in the absence of exogenous P_i showed an elevation in enzyme levels within the first generation of growth before any noticeable alteration in growth rate (within the first 20 min). Following this initial burst in enzyme accumulation (20–40 min) was a period of reduced accumulation (40–60 min), which preceded a steady state rate of accumulation of enzyme for most of the duration of the experiment. This was reflected by dramatic changes in mRNA levels within the first hr. Both p60 transcript and p60 mRNA activity increased rapidly to maxima at 30 min, followed by a reduction in both mRNA levels, and then a hyperbolic increase to a steady state level. Rates of enzyme synthesis calculated from the slopes of the enzyme plot of Fig. 15B agree well with the functional mRNA data, at least to 240 min.

The oscillations in level of p60 mRNA are consistent with two interpretations. Transcription of the p60 gene may begin within the first 10 min of growth in low-P_i medium and increase hyperbolically with time to a steady state level. The

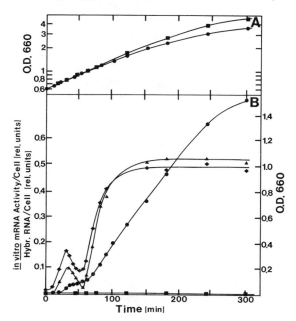

Fig. 15. APase mRNA accumulation during derepression in low-P_i SMD medium. Cells of strain P28-24C were grown in high-P_i SMD medium to an O.D. 600 of 0.6. The culture was divided into two portions and harvested by Millipore filtration. Cells were washed, resuspended, and grown with aeration at 30°C in either high-P_i or low-P_i SMD medium. (A) Growth was monitored by following the O.D. 600 of the culture: (■) high-P_i SMD culture and (●) low-P_i SMD culture. (B) At intervals during growth, samples were removed for the measurement of enzyme activity and for the preparation of RNA. The p60 mRNA activity and hybridizable RNA levels were determined by the procedures used in the experiments of Fig. 14: from the low-P_i SMD culture for measurement of (●) enzyme activity (EU/O.D.660), (◆) p60 transcript (RNA level/cell), and (▲) p60 mRNA level (*in vitro* mRNA activity/cell); from the high-P_i SMD culture for the measurement of enzyme activity (■). The maximal amount of p60 mRNA activity or hybridizable RNA in this experiment was about 50% of that found for the low-P_i RNA. (Reprinted with permission from Bostian *et al.*, 1983.)

oscillatory period would then be explained by a period of accelerated turnover to p60 mRNA. The differences observed in RNA translatability might also be explained by a differential rate of decay of mRNA to an inactive form. Alternatively, the transcription rate could vary with the accumulated level of p60 transcript. Thus, at about 30 min after removal of P_i, the transcription rate would approach zero, and the decrease in RNA accumulation would be explained by the turnover of p60 RNA. The second explanation is consistent with our half-life approximation of p60 mRNA (Bostian *et al.*, 1983). The half-life for p60 transcript is 12.6 min, and for translatable p60 mRNA 8 min, whereas for translatable p58 and p56 mRNA, half-lives are shorter (4.5 and 5.1 min, respectively).

In the derepression experiment of Fig. 15 there are two potential sources of corepressor for the repression of the APase genes: from endogenous polyP reserves and from hydrolysis of the organic phosphate present in the medium. Both methods of generating corepressor could account for the observed autoregulation of mRNA and transcript levels according to the second alternative proposed in the preceding paragraph. Since the regulatory gene products controlling the expression of the p60 gene are expressed constitutively at low- and high-P_i, the generation of corepressor could have a direct effect on the rate of transcription of the p60 gene (Yarger *et al.*, 1982). We have tested this possibility by performing derepression experiments with various concentrations of endogenous and exogenous P_i reserves. By eliminating the external source of P_i in the derepression medium, we observed that enzyme levels and p60 transcript levels accumulated at lower cell densities in low-P_i defined minimal medium than in the low-P_i SMD culture and that the transient decrease in transcript level was reduced (Bostian *et al.*, 1983). Thus, the majority of the transient reduction in transcript level in the low-P_i SMD culture was probably due to feedback repression of transcription by the generation of corepressor from from the early hydrolytic activity of APase.

We have also examined the influence of endogenous reserves on accumulation kinetics of the p60 transcript and mRNA activity during derepression by pregrowing cells under repressed conditions at different P_i concentrations (Bostian *et al.*, 1983) prior to the onset of derepression. Lowering the growth concentrations of P_i in the medium lowered the size of the polyP pool, caused a reduction in the amount of early p60 transcript accumulation, and slightly delayed the time of maximal increase in early p60 transcript levels (Bostian *et al.*, 1983). Since the amount of external organic phosphate was the same for all cultures, this suggests that the differences in levels of mRNA must be due to differences in the utilization or sequestration of intracellular P_i. Regardless of the actual mechanisms that autoregulate p60 mRNA levels, it is clearly influenced by, and perhaps determined by, the metabolic balance of the polyP cycle. The precise nature of the controls governing the utilization of the endogenous and exogenous reserves, however, remains to be determined.

E. Overall Regulation of Phosphorus Metabolism in *S. cerevisiae*

The data presented here and elsewhere support the scheme for phosphorus metabolism and phosphatase regulation in *S. cerevisiae* shown in Fig. 16. Cells growing under nonlimiting conditions of P_i changed polyP levels with changing physiological requirements (Katchman and Fetty, 1955). Under these conditions, polyP synthesis was strongly inhibited by ADP and antagonistic to nucleic acid and phospholipid synthesis. During exponential growth, nucleic acid synthesis inhibits polyP deposition and stimulates its degradation, so that little polyP accumulates. As growth rate declines, degradation is inhibited, and polyP accumulates. Thus, cycling of polyP has the effect of balancing cellular levels of P_i with the energy charge of the cell, as defined by Atkinson (1965), and with the regeneration of ATP for biosynthesis.

Phosphate starvation leads to polyP degradation (Gillies *et al.*, 1981), a decrease in low-density vesicles, sterol esters and triacyl glycerols (Ramsay and Douglas, 1979), and an apparent derepression of polyP, polyphosphatase, and alkaline phosphatase levels. Degradation of polyP is catalyzed by several enzymes, including polyP kinase. However, the direct interconversion of polyP to ATP by polyP kinase does not occur *in vivo*. Rather, polyP is sequentially hydrolyzed to smaller, chain-length molecules by polyphosphatase and by the vacuolar enzyme, alkaline phosphatase (Dawes and Senior, 1973; Katchman and Fetty, 1955; Liss and Langen, 1960).

The derepression of exocellular APase during phosphate starvation results in

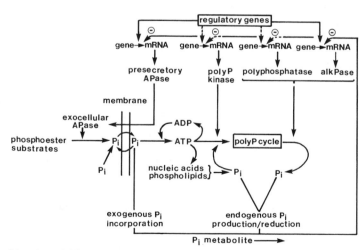

Fig. 16. A model for phosphorus metabolism and phosphatase regulation in *S. cerevisiae*.

the generation of P_i from available external phosphoester substrates (Schmidt *et al.*, 1963; Toh-e and Oshima, 1973), followed by its intracellular incorporation. Depression of the other enzymes results in polyP degradation (Wiame, 1947; Gillies *et al.*, 1981), followed by utilization of the liberated P_i in the biosynthesis of nucleic acids and phospholipids (Harold, 1966). Thus, two potential precursor pools of P_i are available upon derepression. The molecular mechanisms whereby P_i regulates the expression of these enzymes, however, are unknown. For APase, repression by high-P_i has been shown to operate at the level of transcription and to be governed by a genetic system (Toh-e *et al.*, 1981; Toh-e and Oshima, 1973), which also participates in the regulation of alkaline phosphatase (Kaneko *et al.*, 1982) and perhaps in the expression of other phosphorus-metabolizing enzymes.

The autoregulation of APase expression that occurs during derepression may also use this genetic system at the transcriptional level. Support for this comes from measurements of cellular levels of P_i, acid-soluble polyP (mostly PP_i), and acid-insoluble polyP during derepression. Upon derepression, high molecular weight polyP levels rapidly declined just prior to the initial derepression of APase and a surge in acid-soluble polyP. Both the acid-soluble and insoluble polyP fluxed during the lag in APase accumulation prior to exhaustive derepression. Most interesting were the changes in P_i levels. An initial rise coincided with the decrease in polyP reserves, and secondarily with the first derepression of APase. Throughout the remainder of the experiment P_i levels remained high, even during the second phase of APase derepression, and thus may not serve as corepressor for the phosphatase genes. The inverse correlation of APase levels with acid-soluble polyP (primarily PP_i) suggests that PP_i may serve as a regulatory function for APase. Observations that PP_i can serve as an energy source for certain microorganisms and that small molecular weight polyP can regulate RNA synthesis in fungi and the polysaccharide content of the cell well in yeast (for review, see Kulaev, 1975) lend support to the potential role of PP_i as a regulatory signal.

F. Phosphorus Metabolism and APase Expression during the Cell Cycle

Changes in phosphate flow and utilization have also been observed in synchronous cultures of *S. cerevisiae* (Gillies, 1981; Gillies *et al.*, 1981). As in asynchronous culture, synchronously dividing cells consumed polyP only in the absence of an adequate supply of exogenous P_i. In the period during initiation of DNA synthesis, external P_i rapidly decreased simultaneously with an increase in polyP. In the following period of DNA synthesis (S period), if the external P_i level is low, polyP is consumed, presumably by acting as a substitute source of phosphate in place of external P_i (Gillies, 1981; Gillies *et al.*, 1981). During this same period of the cell cycle, APase has been shown to have a stepwise increase in activity (Matur and Berry, 1978), associated with the morphological changes occurring during bud emergence (Field and Schekman, 1980; Linnemans *et al.*, 1977). We observed similar periodic increases in APase activity in our own cell cycle studies, concomitant with bud formation, and have correlated these changes with changes in APase mRNA levels (Yarger *et al.*, 1982). Functional mRNA levels were examined directly after fractionating exponentially growing derepressed cells on a zonal rotor. The mRNA activities for both APase and enolase were measured, and the changes in the ratio (Fig. 13B) were consistent with a continuous rise in enolase mRNA throughout the cell cycle and a high activity and then decline in APase mRNA during the step in APase activity. These findings suggested that under these conditions a periodic transcription, processing the transport of functional APase mRNA, occurred during the cell cycle.

The changes in concentrations of polyP and intracellular P_i or small molecular weight polyP occurring during this period of the cell cycle may play an important role in the periodic expression of APase. Moreover, a study of these phenomena

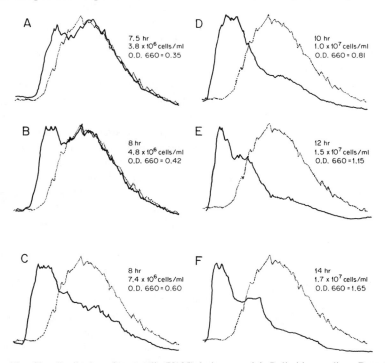

Fig. 17. Size distribution of yeast cells (Y185) during growth in P_i-limiting medium. Dotted line represents control high-P_i culture.

may reveal mechanisms whereby P_i and/or polyP regulate the expression of APase.

To approach intrinsic problems in cell cycle analysis we have included both a number of controls (DNA per cell, percentage of budded cells, and mean cell volume) and internal controls for specific constitutive enzymes. In addition, we have analyzed the physiological transitions that accompany the derepression of APase and have assessed the influence of the cell's changing physiology on regulation of the APase genes (Bostian *et al.*, 1983). From these studies it has become apparent that cells undergoing phosphate starvation do not represent cell cycle growth as in a normally exponentially growing culture. Autoregulation of the APase gene is accompanied by depletion of intracellular polyP pools and by a decrease in cell volume, prior to G_0 arrest. Figure 17 shows the size distribution for cells in culture during different periods of P_i-limited growth. Our cell cycle analyses have been performed on cells from asynchronous cultures grown in an optical density at 660 nm (O.D. 660) of 0.42–0.5. As represented in Fig. 17B, APase is derepressed approximately 20% of its maximal level at this culture density. This just precedes the dramatic shift in cell size distribution the culture undergoes upon further growth. It is thus not clear exactly to what extent cell volume is correlated with cell cycle position in this shifting culture. However, size fractionation of these cells has resulted in zonal rotor fractionation patterns similar to those derived in the galactose system studies for cell cycle control assays: mean cell volume increase, DNA/cell, protein and RNA/cell, and percentage of budded cells. Moreover, the rate of protein synthesis increases across the zonal rotor fractions. The rate of protein synthesis continually increased across the cell cycle. Employing immunoprecipitation of a specific antibody against enolase, we have shown in the same pulse-labeled fractions that show a step in APase that the rate of enolase synthesis rose continuously. Similar data were obtained for a glyceraldehyde-3-phosphate dehydrogenase. Thus, these two enzymes were continuously synthesized and served as internal controls to APase.

To further examine the cell cycle regulation of APase mRNA we have applied the RNA blot hybridization assay for *PHO5* RNA (p60) (Fig. 13B) to the analysis of zonal rotor size-fractionated cells. Shown in Fig. 18A is the cell size distribution, mean cell volume, and DNA content per cell for the asynchronously grown, derepressed culture of Y185. Figure 18B shows the enzyme levels and the specific physical concentrations of *PHO5* RNA. The step in APase activity precedes both the step in DNA concentration as well as the percentage of unbudded cells (data not shown). The *PHO5* RNA data represent the relative amounts of *PHO5* transcript to total cell RNA. The amounts of *PHO5* transcript undergo a twofold net relative increase from the early zonal rotor fractions, followed by a net decrease approaching the initial concentration of the largest cells of the population. These data are consistent with the variation observed in translatable p60 mRNA reported earlier (Yarger *et al.*, 1982). The higher levels of APase

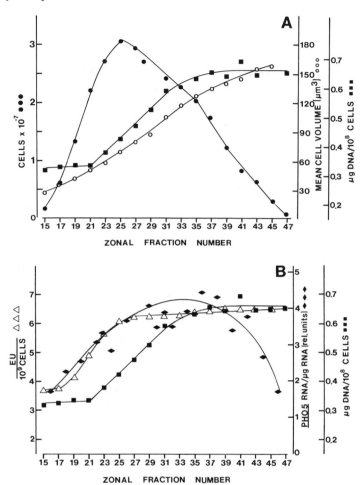

Fig. 18. Analysis of *PH05* (p60) gene transcripts in cells fractionated by zonal rotor centrifugation. Strain Y185 was grown asynchronously to O.D. 660 of 0.45 in low-P$_i$ medium, and the harvested cells fractionated on a zonal rotor as in Fig. 3. Fractions were analyzed for: (A) (●) cell number, (○) mean cell volume, (■) DNA/cell; and (B) (△) APase activity/cell, (◆) PH05 RNA/µg total RNA, (■) DNA/cell.

mRNA present in the mature-sized cells (late zonal fractions) in this and the functional mRNA experiment suggest that posttranscriptional events must regulate APase protein synthesis in these cells if the pattern of enzyme accumulation reflects *de novo* protein synthesis. Furthermore, analysis of these findings will require a clarification of the validity of size-fractionated derepressed cells for cell cycle comparisons.

IV. CONCLUDING REMARKS

We have analyzed the problem of enzyme accumulation in two separate genetic systems in yeast under different growth conditions. Galactokinase enzyme accumulation was studied during steady state growth. In contrast, APase enzyme accumulation was analyzed during transitional growth phases. Both enzymes demonstrated discontinuities with respect to total cellular macromolecular synthesis, and the accumulation of each enzyme appears to be controlled by different mechanisms. APase synthesis was found to be autoregulated; enzyme activity controls the transcription of the APase structural gene. Both transcriptional and posttranscriptional controls appear to regulate the step in APase activity. The periodic accumulation of galactokinase enzyme activity appears to be regulated at the level of enzyme stability. Therefore, the mechanism of regulation of any given enzyme during the cell cycle or varying conditions of growth must be separately analyzed.

We have presented data showing that lengthening the growth rates has profound effects upon the cell cycle pattern of enzyme accumulation, measured by gradient centrifugation. The controversy surrounding step versus linear patterns of enzyme accumulation may subsequently be partially due to the use of nutrient-limiting growth media (with slower growth rates) for experiments in which linear increases are observed, as compared to nutrient-rich growth media (with increased growth rates), from which periodic enzyme accumulations have been observed. Furthermore, we suggest that the data reflecting rates of polypeptide synthesis acquired from two-dimensional gel studies of individual polypeptides cannot be used to answer questions concerning the patterns of enzyme accumulations. Such data can only be properly interpreted when mechanisms controlling the accumulation of each particular enzyme are known.

ACKNOWLEDGMENTS

This work was supported in part by a U.S. Public Health Service Grant AI 10610 (H.O.H.), a National Science Foundation grant PCM-802217 (H.O.H.), and by American Cancer Society Postdoctoral Fellowships (to K.A.B. and to J.G.Y.). We thank Ms. Joan Lemire for technical assistance, and Drs. Akio Toh-e, Yasuji Oshima, Richard Kramer, and David Rogers for their generous provision of information, strain materials, and for unpublished information.

REFERENCES

Adams, B. E. (1972). Induction of galactokinase in *Saccharomyces cerevisiae* – kinetics of induction and glucose effects. *J. Bacteriol.* **111**, 308–315.

Atkinson, D. E. (1965). Biological feedback control at the molecular level. *Science* **150**, 851–857.

Bevan, P., and Douglas, H. C. (1969). Genetic control of phosphoglucomutase varients in *Saccharomyces cerevisiae*. *J. Bacteriol.* **98**, 532–535.

Bostian, K. A., Lemire, J. M., Cannon, L. E., and Halvorson, H. O. (1980). *In vitro* synthesis of repressible yeast acid phosphatase. Identification of multiple mRNAs and products. *Proc. Natl. Acad. Sci. U.S.A.* **77**, 4504–4508.

Bostian, K. A., Lemire, J. M., and Halvorson, H. O. (1982). Synthesis of repressible acid phosphatase in *Saccharomyces cerevisiae* under conditions of enzyme instability. *Mol. Cell Biol.* **2**, 1–10.

Bostian, K. A., Lemire, J. M., and Halvorson, H. O. (1983). Physiological control repressible acid phosphatase gene transcription in *Saccharomyces cerevisiae*. *Mol. Cell. Biol.* **33**, 839–853.

Bullock, J. G., and Coakley, W. T. (1976). The effect of heat on the viability of *Schizosaccharomyces pombe* 972h⁻ growing in synchronous culture. *Exp. Cell Res.* **103**, 447–449.

Cabib, E. (1975). Molecular aspects of yeast morphogenesis. *Annu. Rev. Microbiol.* **29**, 191–214.

Dawes, E. A., and Senior, P. J. (1973). The role and regulation of energy reserve polymers in microorganisms. *Adv. Microb. Physiol.* **10**, 135–266.

Douglas, H. C. (1961). A mutant in *Saccharomyces* that affects phosphoglucomutase activity and galactose utilization. *Biochim. Biophys. Acta* **52**, 209–211.

Douglas, H. C., and Hawthorne, D. C. (1964). Enzymatic expression and genetic linkage of genes controlling galactose utilization in *Saccharomyces*. *Genetics* **49**, 837–844.

Elliott, S. G., and McLaughlin, C. S. (1978). The rate of macromolecular synthesis through the cell cycle of the yeast *Saccharomyces cerevisiae*. *Proc. Natl. Acad. Sci. U.S.A.* **75**, 4384–4388.

Elliott, S. G., and McLaughlin, C. S. (1979). Synthesis and modification of proteins during the cell cycle of the yeast *Saccharomyces cerevisiae*. *J. Bacteriol.* **137**, 1185–1190.

Elliott, S. G., and McLaughlin, C. S. (1983). The yeast cell cycle—coordination of growth and division rates. *Prog. Nucleic Acid Res. Mol. Biol.* (in press).

Erecenska, M., Stubbs, M., Miyata, Y., Ditrey, C. M., and Wilson, D. F. (1977). Regulation of cellular metabolism by intracellular phosphate. *Biochim. Biophys. Acta* **462**, 20–35.

Fantes, P. A. (1976). Control of cell size and cycle time in *Schizosaccharomyces pombe*. *J. Cell Sci.* **24**, 51–67.

Field, C., and Schekma, R. (1980). Localized secretion of acid phosphatase reflects the pattern of cell surface growth in *Saccharomyces cerevisiae*. *J. Cell Biol.* **86**, 123–128.

Fraser, R. S. S., and Moreno, F. J. (1976). Rates of synthesis of polyadenylated messenger RNA and ribosomal RNA during the cell cycle of *Schizosaccharomyces pombe*. *J. Cell Sci.* **21**, 497–521.

Gillies, R. J. (1981). Intracellular pH and growth control in eukaryotic cells. *In* "The Transformed Cell" (I. L. Cameron and T. B. Pool, eds.), pp. 347–395. Academic Press, New York.

Gillies, R. J., Ugurbil, K., Den Hollander, J. A., and Shulman, R. G. (1981). ³¹P NMR studies of intracellular pH and phosphate metabolism during the division cycle of *Saccharomyces cerevisiae*. *Proc. Natl. Acad. Sci. U.S.A.* **78**, 2125–2129.

Gorman, H., Tauro, P., Berge, M., and Halvorson, H. O. (1964). Timing of enzyme synthesis during synchronous division in yeast. *Biochem. Biophys. Res. Commun.* **15**, 43–49.

Halvorson, H. O., Gorman, J., Tauro, P., Epstein, R., and LeBerge, M. (1964). Control of enzyme synthesis in synchronous cultures of yeast. *Fed. Proc. Fed. Am. Soc. Exp. Biol.* **23,** 1002–1008.

Halvorson, H. O., Carter, B. L. A., and Tauro, P. (1971). Synthesis of enzymes during the cell cycle. *Adv. Microb. Physiol.* **6,** 47–106.

Harold, F. M. (1966). Inorganic polyphosphates in biology: Structure, metabolism and function. *Bacteriol. Rev.* **30,** 772–794.

Hartwell, L. H., and Unger, M. W. (1977). Unequal division in *Saccharomyces cerevisiae* and its implications for the control of cell division. *J. Cell Biol.* **75,** 422–435.

Hereford, L. M., Osley, M. A., Ludwig, J. R., and McLaughlin, C. S. (1981). Cell-cycle regulation of yeast histone mRNA. *Cell* **24,** 367–375.

Hopper, J. E., Broach, J. R., and Rowe, L. B. (1978). Regulation of the galactose pathway in *Saccharomyces cerevisiae:* Induction of uridyl transferase mRNA and dependency on *GAL4* function. *Proc. Natl. Acad. Sci. U.S.A.* **75,** 2878–2882.

Kalckar, H. W., Branganca, B., and Munch-Peterson, A. (1953). Uridyl transferase and the formation of uridine diphosphogalactose. *Nature (London)* **172,** 1038–1045.

Kaneko, Y., Toh-e, A., and Oshima, Y. (1982). Identification of the genetic locus for the structural gene and a new regulatory gene for the synthesis of repressible alkaline phosphatase in *Saccharomyces cerevisiae. Mol. Cell. Biol.* **2,** 127–137.

Katchman, B. J., and Fetty, W. O. (1955). Phosphorous metabolism in growing cultures of *Saccharomyces cerevisiae. J. Bacteriol.* **69,** 607–615.

Kosterlitz, H. W. (1943). The fermentation of galactose and galactose-1-phosphate. *Biochem. J.* **37,** 322–326.

Kramer, R. A., and Anderson, N. (1980). Isolation of yeast genes with mRNA levels controlled by phosphate concentration. *Proc. Natl. Acad. Sci. U.S.A.* **77,** 6541–6545.

Kulaev, I. S. (1975). Biochemistry of inorganic polyphosphates. *Rev. Physiol., Biochem. Pharmacol.* **73,** 131–158.

Lange, P., and Hansche, P. E. (1980). Mapping of a centrome-linked gene responsible for constitutive acid phosphatase synthesis in yeast. *Mol. Gen. Genet.* **180,** 605–607.

Leloir, L. F. (1951). The enzymatic transformation of uridine diphosphate glucose into a galactose derivative. *Arch. Biochem.* **33,** 186–190.

Linnemans, W. A. M., Boer, P., and Elbers, P. F. (1977). Localization of acid phosphatase in *Saccharomyces cerevisiae:* A clue to cell wall formation. *J. Bacteriol.* **131,** 638–644.

Liss, E., and Langen, P. (1960). Uber ein hochmolekulares polyphosphat der hefe. *Biochem. Z.* **333,** 193–201.

Liss, E., and Langen, P. (1962). Verouche zur polyphosphat-uberkompensation in hafezzellan nach phosphatverarmung. *Arch. Mikrobiol.* **41,** 383–392.

Lloyd, D., Poole, R. K., and Edwards, S. W. (1982). "The Cell Division Cycle." Academic Press, New York.

Lorincz, A. T., Miller, M., Xuong, N.-H., and Geiduschek, E. P. (1982). Identification of proteins whose synthesis is modulated during the cell cycle of *Saccharomyces cerevisiae. Mol. Cell. Biol.* **2,** 1532–1549.

Lugwig, J. R., II, Foy, J. J., Elliott, S. G., and McLaughlin, C. S. (1982). Synthesis of specific identified, phosphorylated, heat shock and heat stroke proteins through the cell cycle of *Saccharomyces cerevisiae. Mol. Cell. Biol.* **2,** 117–126.

Matsumoto, K., Toh-e, A., and Oshima, Y. (1978). Genetic control of galactokinase synthesis in *Saccharomyces cerevisiae:* Evidence for constitutive expression of the positive regulatory gene *GAL4. J. Bacteriol.* **134,** 446–457.

Matsumoto, K., Adachi, Y., Toh-e, A., and Oshima, Y. (1980). Function of positive regulatory gene *GAL4* in the synthesis of galactose pathway enzymes in *Saccharomyces cerevisiae:*

Evidence that the *GAL81* region codes for part of the *GAL4* protein. *J. Bacteriol.* **141**, 508–527.

Matur, A., and Berry, D. (1978). The use of step enzymes as markers during meiosis and ascospore formation in *Saccharomyces cerevisiae*. *J. Gen. Microbiol.* **109**, 205–213.

Mitchison, J. M. (1971). "The Biology of the Cell Cycle." Cambridge Univ. Press, London and New York.

Mitchison, J. M. (1977). Enzyme synthesis during the cell cycle. *Cell Differ. Microorg., Plants Anim., Int. Symp., 1976* pp. 377–401.

Mitchison, J. M., and Carter, B. L. A. (1975). Cell cycle analysis. *Methods Cell Biol.* **9**, 201–219.

Mortimer, R. K., and Hawthorne, D. C. (1966). Genetic mapping in *Saccharomyces*. *Genetics* **53**, 165–173.

Newlon, C. S., Petes, T. D., Hereford, L. M., and Fangman, W.L. (1974). Replication of yeast chromosomal DNA. *Nature (London)* **247**, 32–35.

Perlman, D., and Hopper, J. E. (1979). Constitutive synthesis of the *GAL4* protein, a galactose pathway regulator in *Saccharomyces cerevisiae*. *Cell* **16**, 89–95.

Polanshek, M. (1977). Effect of heat shock and cyclohexamide on growth and division of the yeast *Schizosaccharomyces pombe*. *J. Cell Sci.* **23**, 1–23.

Pringle, J. R., and Hartwell, L. H. (1982). The *Saccharomyces cerevisiae* life cycle. *In* "The Molecular Biology of the Yeast Saccharomyces: Life Cycle and Inheritance" (J. Strathern, E. Jones, and J. Broach, eds.), pp. 97–142. Cold Spring Harbor Lab., Cold Spring Harbor, New York.

Ramsay, A. M., and Douglas, L. J. (1979). Effects of phosphate limitation of growth and on the cell-wall and lipid composition of *Saccharomyces cerevisiae*. *J. Gen. Microbiol.* **110**, 125–191.

Rogers, D. T., Lemire, J. M., and Bostian, K. A. (1982). Acid phosphatase polypeptides in *Saccharomyces cerevisiae* are encoded by a differentially regulated multigene family. *Proc. Natl. Acad. Sci. U.S.A.* **79**, 2157–2161.

Roomans, G. M., and Borst-Pauwels, G. W. F. H. (1979). Interaction of cations with phosphate uptake by *Saccharomyces cerevisiae*. *Biochem. J.* **178**, 521–527.

St. John, T. P., and Davis, R. W. (1979). Isolation of galactose-inducible DNA sequences by differential plaque filter hybridization. *Cell* **16**, 443–450.

St. John, T. P., and Davis, R. W. (1981). The organization and transcription of the galactose gene cluster of *Saccharomyces*. *J. Mol. Biol.* **152**, 285–316.

Saunders, C. A., Sogin, S. J., Kaback, D. B., and Halvorson, H. O. (1975). Regulation of transcription in yeast. *In* "Control Mechanisms in Development" (R. H. Meints and E. Davies, eds.), pp. 21–34. Plenum, New York.

Schmidt, G., Bartsch, G., Laumont, M. C., Herman, T., and Liss, M. (1963). Acid phosphatase of Bakers yeast: An enzyme of the external cell surface. *Biochemistry* **2**, 126–131.

Sebastian, J., Carter, B. L. A., and Halvorson, H. O. (1971). Use of yeast populations fractionated by zonal centrifugation to study the cell cycle. *J. Bacteriol.* **108**, 1045–1050.

Shilo, B., Riddle, V. G. H., and Pardee, A. B. (1979). Protein turnover and cell cycle initiation in yeast. *Exp. Cell Res.* **123**, 221–227.

Solimene, R., Guerrini, A. M., and Donini, P. (1980). Levels of acid-soluble polyphosphate in growing cultures of *Saccharomcyes cerevisiae*. *J. Bacteriol.* **143**, 710–714.

Slater, M. L. (1973). Effect of reversible inhibition of deoxyribonucleic acid synthesis on the yeast cell cycle. *J. Bacteriol.* **113**, 263–270.

Slater, M. L. (1974). Recovery of yeast from transient inhibition of DNA synthesis. *Nature (London)* **247**, 275–276.

Susskind, M., and Botstein, D. (1975). Mechanism of action of *Salmonella* phage P22 anti-repressor. *J. Mol. Biol.* **98**, 413–424.

Tauro, P., and Halvorson, H. O. (1966). Effect of gene position on the timing of enzyme synthesis and synchronous cultures of yeast. *J. Bacteriol.* **92,** 652–661.

Tauro, P., Halvorson, H. O., and Epstein, R. L. (1968). Time of gene expression in relation to centromere distance during the cell cycle of *S. cerevisiae. Proc. Natl. Acad. Sci. U.S.A.* **59,** 227–284.

Toh-e, A., Ueda, Y., Kakimoto, S., Oshima, Y. Toh-e, A., A., and Oshima, Y. (1973). Isolation and characterization of acid phosphatase mutants in *Saccharomyces cerevisiae. J. Bacteriol.* **113,** 727–738.

Toh-e, A., Kakimoto, S., and Oshima, Y. (1975). Genes coding for the structure of the acid phosphatases in *Saccharomyces cerevisiae. Mol. Gen. Genet.* **141,** 81–83.

Toh-e, A., Nakamura, H., and Oshima, Y. (1976). A gene controlling the synthesis of non specific alkaline phosphatase in *Saccharomyces cerevisiae. Biochim. Biophys. Acta* **428,** 182–192.

Toh-e, A., Kobayashi, S., and Oshima, Y. (1978). Disturbance of the machinery for the gene expression by acidic pH in the repressible acid phosphatase system of *Saccharomyces cerevisiae. Mol. Gen. Genet.* **162,** 139–149.

Toh-e, A., Inouye, S., and Oshima, Y. (1981). Structure and function of the *PHO82-pho4* locus controlling the synthesis of repressible acid phosphatase of *Saccharomyces cerevisiae. J. Bacteriol.* **145,** 221–232.

Toh-e, A., and Oshima, Y. (1973). Isolation and characterization of acid phosphatase mutants in *Saccharomyces cerevisiae. J. Bacteriol.* **113,** 727–738.

Tsuyuma, S., and Adams, B. G. (1973). Population analysis of the deinduction kinetics of galactose long-term adaption mutants in yeast. *Proc. Natl. Acad. Sci. U.S.A.* **70,** 919–923.

Turner, K., Bascomb, F., Lynch, J., Molin, W., Thurston, C., and Schmidt, R. (1981). Evidence for messenger ribonucleic acid of an ammonium-inducible glutamate dehydrogenase and synthesis, covalent modification, and degradation of enzyme subunits in uninduced *Chlorella sorokiniana* cells. *J. Bacteriol.* **146,** 578–589.

Tyson, C. C., Lord, P. G., and Wheals, A. E. (1979). Dependency of size of *Saccharomyces cerevisiae* cells on growth rate. *J. Bacteriol.* **138,** 92–98.

Ueda, Y., and Oshima, Y. (1975). A constitutive mutation, *phoT,* of repressible acid phosphatase synthesis with inability to transport inorganic phosphate in *Saccharomyces cerevisae. Mol. Gen. Genet.* **136,** 255–259.

Ueda, Y., Toh-e, A., and Oshima, Y. (1975). Isolation and characterization of recessive, constitutive mutations for repressible acid phosphatase. *J. Bacteriol.* **122,** 911–922.

Urech, K., Duerr, M., Boller, T. H., and Wiemken, A. (1978). Localization of phosphate in vacuoles of *Saccharomyces cerevisiae. Arch. Microbiol.* **116,** 275–278.

Wain, W. H., and Staatz, W. D. (1973). Rates of synthesis of ribosomal protein and total ribonucleic acid through the cell cycle of the fission yeast *Schizosaccharomyces pombe. Exp. Cell Res.* **81,** 269–278.

Wiame, J. M. (1947). Etude d'une substance polyphosphoresis, basophile et mitochromatique chez les levures. *Biochim. Biophys. Acta* **1** 234–255.

Williamson, D. H., and Scopes, A. W. (1960). The behavior of nucleic acids in dividing cultures of *Saccharomyces cerevisiae. Exp. Cell. Res.* **20,** 338–349.

Yarger, J. G., and Halvorson, H. O. (1983). Regulation of galactokinase enzyme activity in *Saccharomyces cerevisiae. J. Mol. Cell. Biochem.* (in press).

Yarger, J. G., Bostian, K. A., and Halvorson, H. O. (1982). Developmental regulation of enzyme synthesis in *Saccharomyces cerevisiae. In* ''Cell Growth'' (C. Nicolini, ed.), pp. 271–304. Plenum, New York.

Yashphe, J., and Halvorson, H. O. (1976). ψ-Galactosidase activity in single cells during cell cycle of *Saccharomyces lactis. Science* **191,** 1283–1284.

4

Gene Expression during the Cell Cycle of *Chlamydomonas reinhardtii*

DAVID HERRIN AND ALLAN MICHAELS

Department of Biology
University of South Florida
Tampa, Florida

I. INTRODUCTION: *CHLAMYDOMONAS* AS AN EXPERIMENTAL ORGANISM

Chlamydomonas reinhardtii is a eukaryotic, unicellar green alga. It exhibits synchronous growth in nature and has vegetative and sexual aspects in its life cycle. Axenic, vegetative cultures of *Chlamydomonas* can be synchronized in the laboratory under physiological conditions by a repeating light–dark regimen (Bernstein, 1960). Mutants of this organism can be generated, and the patterns of inheritance determined by genetic crosses (Sager, 1972). An extensive collection of mutants including cell wall, cell cycle, photosynthetic, flagellar, antibiotic resistance, and others has been obtained, and a *Chlamydomonas* Genetics Center has been established at Duke University. These characteristics have made *C. reinhardtii* a useful organism for investigations of the eukaryotic cell cycle.

87

RECOMBINANT DNA AND
CELL PROLIFERATION

C. reinhardtii contains and utilizes three major genetic systems: nuclear, chloroplast, and mitochondrial. Investigations to date have characterized protein and rRNA synthesis in the nuclear–cytoplasmic and chloroplast systems. However, recent reports on the isolation and construction of a restriction map for mitochondrial DNA from *Chlamydomonas* represent a first step toward the analysis of gene expression in this system (Grant and Chiang, 1980).

Evidence obtained so far has shown that regulation of expression of nuclear and chloroplast genes occurs in this organism. In fact, stage-specific synthesis of the major cellular proteins seems to be the rule rather than the exception as it is in yeast and mammalian cells (Ludwig *et al.*, 1982; Bravo and Cellis, 1980). Synthesis of cytoplasmic and chloroplast rRNAs occurs during the light portion of the cell cycle (mid- to late G_1 phase) (Wilson and Chiang, 1977). Ribulose-1,5-bisphosphate carboxylase (Rubisco, an abundant soluble protein) and the major photosynthetic membrane proteins are also synthesized/assembled during the light period when the bulk of macromolecular synthesis takes place (Iwanij *et al.*, 1975; Beck and Levine, 1974). Individual thylakoid membrane polypeptides show different patterns of synthesis within the light period, resulting in stage-specific assembly of the membrane. In contrast to the chloroplast components, α- and β-tubulin are synthesized in the dark period during and after cell division when flagellar regeneration occurs (Weeks and Collis, 1979). Cell cycle, stage-specific synthesis of the major cellular proteins makes *Chlamydomonas* an accessible system for determining the levels of gene regulation as cells progress through the cell cycle.

The advent of recombinant DNA technology has enabled the isolation and investigation of several chloroplast and nuclear genes from this organism (von Wettstein, 1981; Silflow and Rosenbaum, 1981; Minami *et al.*, 1981). Also, a recent report of the transformation of *C. reinhardtii* with cloned yeast DNA has great potential for isolating genes from clonal libraries by complementation of mutants (Rochaix and van Dillewijn, 1982). This technique may enable the isolation of genes whose expression regulates the progression of cells through the cell cycle by complementation of cell cycle mutants.

In this chapter we summarize what is presently known about the expression of genes for rRNAs, α- and β-tubulin, and for chloroplast proteins. These cellular components form the bulk of what is understood about the expression of specific genes during the cell cycle of *Chlamydomonas*.

II. CHARACTERISTICS OF SYNCHRONOUS CULTURES OF *C. REINHARDTII*

Cultures of *C. reinhardtii* can be synchronized by a repeating light–dark regimen of 12 hr of light followed by 12 hr of darkness. By convention, 0–12 hr is the light period and 12–24 hr, the dark period. Both mixotrophic and autotrophic cultures of the alga can be synchronized by light. Complete synchrony is achieved after three light–dark cycles (Bernstein, 1960). The cells divide once each 24 hr into two, four, six, or eight daughter cells with the average depending on the culture media and light intensity. Figure 1 shows several growth parameters of a synchronous culture of a wild-type strain of *C. reinhardtii* 137 mt[+] grown autotrophically (Ohad *et al.*, 1967) under a light intensity of 2000 lux. No increase in cell number is observed during the 12-hr light period. Cell division begins approximately 2 hr into the dark period and is complete by 6 hr into the dark period. Under these conditions there is a twofold increase in cell number.

The cells increase in size during the light period, showing a twofold increase in protein and chlorophyll (Fig. 1). Photosynthetic membrane protein also increases during the light period, comprising approximately 40% of total cellular protein at the beginning of the light period and 80% by the end of the light period (Fig. 1). The levels of these constituents remain constant during the dark period. These cells commit a great deal of energy and protein-synthetic machinery to synthesis of thylakoid membranes. Pulse-labeling experiments, to be discussed later in this chapter (Section VI), have confirmed that the light period is when maximal synthesis of total cellular and photosynthetic membrane protein occurs.

The accumulation of cellular RNA in a synchronous culture of *Chlamydomonas* is similar to protein and chlorophyll (Wilson and Chiang, 1977). Howev-

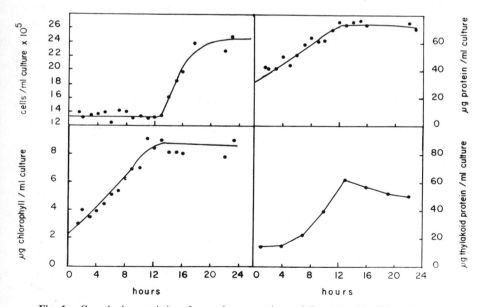

Fig. 1. Growth characteristics of a synchronous culture of *C. reinhardtii*. Cell number, chlorophyll, protein, and thylakoid protein content were determined on a synchronous culture of *Chlamydomonas*. Thylakoid protein content was estimated by multiplying chlorophyll content (μg/ml culture) times the protein-to-chlorophyll ratios (wt/wt) of isolated thylakoid membranes.

er, accretion of cellular DNA shows a two-step increase, one occurring at 3–5 hr of the cycle, and a much greater increase occurring at the light–dark transition (12 hr). The smaller increase at 3–5 hr results from chloroplast DNA replication and the larger increase represents nuclear DNA replication (Chiang and Sueoka, 1967). Mitosis is followed almost immediately by cell division. Thus, the cell cycle of *C. reinhardtii* is characterized by a long G_1 period (18 hr) and a very short G_2 period (0.5 hr).

Spudich and Sager (1980) have recently investigated the regulation of the cell cycle of *Chlamydomonas* by light and dark. They have determined two critical points within the 24-hr cell cycle: the arrest point (A) and the transition point (T). The A point occurs in the early G_1 phase (late dark/early light) when further progression through the cell cycle is light dependent. Cells maintained in darkness or shifted to dark conditions during the first 6 hr of the light period do not proceed through the cell cycle but remain metabolically active. After 6 hr into the light period (T point), further progression through the cell cycle leading to cell division becomes light independent. Further experiments by Spudich and Sager (1980) indicated that electron transport through Photosystem II was essential for cells to proceed to the T point even though they contained sufficient starch reserves to generate energy without photosynthesis. The results suggest a block in starch utilization until after the T point, wherein photosynthesis could be inhibited and cells would still progress through the cycle, degrading starch for energy. Whether this transition from light dependence to light independence requires synthesis of starch-degrading enzymes or activation of existing enzymes is not known.

III. EXPRESSION OF rRNA GENES DURING THE CELL CYCLE

Wild-type cells of *Chlamydomonas reinhardtii* contain four major, high-molecular-weight species of RNA (Hoober and Blobel, 1969). They have sedimentation coefficients of 25, 23, 18, and 16 in nondenaturing sucrose gradients. The 25 and 18 S species are components of cytoplasmic ribosomes, and the 23 and 16 S RNAs are components of chloroplast ribosomes. The cytoplasmic species are approximately twice as abundant as the chloroplast rRNAs.

Rochaix (1978) and his colleagues (Rochaix and Malnoe, 1978) have cloned the majority of the chloroplast genome into the bacterial plasmids pCR1 and pBR313. They have constructed a restriction map of this relatively large (126 megadalton), circular molecule using the restriction endonucleases *Bam*HI, *Eco*-RI, and *Bgl*II (Fig. 2). The genes for the chloroplast rRNAs have been localized on the restriction map (Fig. 2) by Southern hybridization. Genes for the 23, 16, 7, 5, and 3 S RNAs are found *en bloc* in two copies on the genome, in inverted orientation. The 7 and 3 S RNAs are components of the large subunit of chlo-

Fig. 2. Physical map of the chloroplast DNA of *C. reinhardtii*. The *Eco*RI and *Bam*HI restriction fragments are denoted by R and Ba, respectively. The portions of the genome that have been cloned by J.-D. Rochaix are indicated by a thick line on the inner circle. The two ribosomal gene units are indicated on the outside, and only the 16, 23, and 5 S genes are identified. The arrows indicate fragments that contain 4 S RNA genes, and the fragments labeled with large letters hybridize to 4 S RNA. The position of the genes encoding the large subunit of Rubisco (denoted by *LS*), and membrane polypeptides D-1 (homologous to the 32-kilodalton protein of higher plants) and D-2 are shown. The tentative location of the genes for the α and β subunits of coupling factor CF_1, and for apoproteins of chlorophyll–protein complexes III and IV are also indicated (Chl_a-AP_2 and Chl_a-AP_3).

Not shown in the figure is the location of the gene coding for elongation factor, *Tu*, which has been shown by heterologous hybridization with a cloned *E. coli Tu* gene to span the *Eco*RI site between fragments R26 and R03. (Adapted from von Wettstein, 1981; Watson and Surzycki, 1982.)

roplast ribosomes. The gene order in the direction of transcription is 16, 7, 3, 23, and 5 S. The 5, 23, 3 and 7 S genes are transcribed on the same strand (Rochaix and Malnoe, 1978).

Electron microscopy of rRNA–DNA hybrids revealed a strech of DNA [950 base pairs (bp)] within the coding region of the 23 S gene that did not have a complementary sequence in the mature 23 S species (Rochaix and Malnoe, 1978). Therefore, this chloroplast rRNA gene shows a property characteristic of many eukaryotic genes; it contains an intervening sequence which is removed during maturation. The presence of an intervening sequence in the 23 S rRNA gene suggests that chloroplasts contain the necessary processing enzymes for splicing precursor transcripts to their mature form. The presence of such a processing function also provides another potential regulatory site besides transcription for determining the levels of 23 S RNA.

Expression of the major chloroplast (23 and 16 S) and cytoplasmic (25 and 18 S) rRNA genes during the cell cycle has been investigated by pulse labeling synchronous cells with $H_3{}^{32}PO_4$ and analyzing radioactivity incorporated into these species (Wilson and Chiang, 1977; Cattolico *et al.*, 1973). The relative amounts of the rRNAs occurring in a synchronous culture were also measured optically after separation by gel electrophoresis. Figure 3 shows the pattern of rRNA accumulation during the cell cycle and the synthetic pattern measured by ^{32}P incorporation. Both chloroplast and cytoplasmic rRNA species accumulate throughout the light period (G_1 phase). Pulse labeling with ^{32}P also shows this to be the period when 90% of synthesis takes place. During the dark period no further accumulation of rRNA is seen, and synthesis of both chloroplast and cytoplasmic species is depressed. ^{32}P incorporation into ribosomes during the cell cycle was also measured and showed a pattern similar to the rRNAs, indicating that the newly synthesized rRNA is immediately assembled into ribosomes (Wilson and Chiang, 1977).

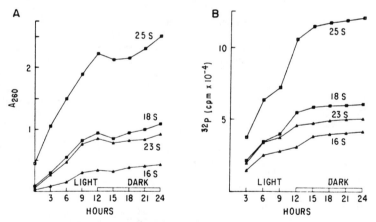

Fig. 3. The rRNA synthetic pattern during the cell cycle of *C. reinhardtii* measured optically (A) and by pulse labeling with $H_3{}^{32}PO_4$ (B). (A) Equivalent amounts of S-10 samples from synchronous cells were electrophoresed and the gels scanned at 260 nm. The heights of the four major peaks are plotted versus time over one cell cycle. (B) The pattern of ^{32}P incorporation into the four major rRNA species during a cell cycle. Aliquots of a synchronous cell culture were taken at 3-hr intervals, pulse labeled for 3 hr, and the RNA extracted, electrophoresed, and counted. The data are plotted as the total accumulative incorporation of ^{32}P into the four rRNA species up to a given point in the cell cycle. (Adapted from Wilson and Chiang, 1977.)

Transcription of the region of the chloroplast genome that contains the rRNA genes has also been examined by pulse labeling synchronous cells with $H_3{}^{32}PO_4$ and hybridizing the radiolabeled RNA to Southern blots of *Eco*RI-restricted chloroplast DNA (Matsuda and Surzycki, 1980). The extent of hybridization to rDNA restriction fragments increased during the light period, peaking at the midlight stage and declining to a low level in the dark period. These results enforce those of Wilson and Chiang (1977) and demonstrate that transcription of the chloroplast rRNA genes is restricted to the light period (G_1 phase) of the cell cycle.

A more accurate and thorough analysis of expression of the chloroplast rRNA genes is now possible due to the availability of cloned rRNA genes (Rochaix and Malnoe, 1978). The clones genes will allow accurate measurements of rates of transcription of specific rRNA genes *in vivo*. Unfortunately, it is not possible to isolate intact chloroplasts from *Chlamydomonas* to measure *in vitro* rates of transcription. However, it is possible to purify chloroplast RNA polymerase and to transcribe *in vitro* chloroplast DNA (Jolly *et al.*, 1981; Briat *et al.*, (1982). Investigations of homologous *in vitro* transcription systems from chloroplasts should yield useful information for understanding the expression of chloroplast genes.

DNA sequence analysis (Maxam and Gilbert, 1977) of the genes and S1 nuclease mapping (Berk and Sharp, 1977) of the rRNA on the cloned gene will determine the origin of transcription. Further sequencing of the flanking regions will also aid in determining what sequence-specific signals may operate in influencing the rates of transcription of these genes.

IV. EXPRESSION OF TUBULIN GENES DURING THE CELL CYCLE

Chlamydomonas reinhardtii possesses two flagella. These flagella are used for movement and in pairing of gametic cells during mating. The most abundant flagellar protein is tubulin, of which there are two basic types, α and β.

Tubulin shows a stage-specific pattern of synthesis during the cell cycle. Synthesis is initiated during the late light period, 1.5–2 hr prior to cytokinesis. Synthesis continues through cytokinesis, lasting until the late dark period (20 hr), when fully flagellated daughter cells are released (Weeks and Collis, 1979).

A more rapid and extensive induction of tubulin synthesis can be obtained by deflagellation of vegetative cells, and it is this effect that has been most studied (Silflow and Rosenbaum, 1981; Minami *et al.*, 1981). RNA obtained from deflagellated cells was used as a substrate for reverse transcriptase and the resulting cDNA cloned into plasmid pBR322. Clones for both α- and β-tubulins were obtained and have been used to determine the number of tubulin genes and their expression following deflagellation. Southern blot analysis of genomic DNA cut with several different restriction enzymes has revealed the existence of four tubulin genes in *C. reinhardtii,* two α and two β. RNA–DNA hybridizations have revealed that all four of these genes are expressed coordinately during flagellar regeneration following excision and during cell division in synchronous cells (Brunke *et al.,* 1982a). Therefore, it appears that tubulin synthesis is regulated at the level of transcription and/or mRNA stability.

Ares and Howell (1982) used heterologous probes (cloned chicken α- and β-tubulin genes) (Fig. 4) and *in vitro* translation to confirm that mRNA levels regulate the synthesis of tubulin during the cell cycle. They also found that tubulin synthesis was independent of the illumination conditions of the culture. This was done by analyzing cells placed into continuous light during the fourth light–dark cycle. The cells complete a normal synchronous cell division, only they are in light during the normal dark period. Although tubulin gene expression seems to be independent of illumination conditions and dependent only on the cell cycle stage, it seems unlikely that all genes, especially those of photosynthetic proteins, will show the same characteristic (see Section VI).

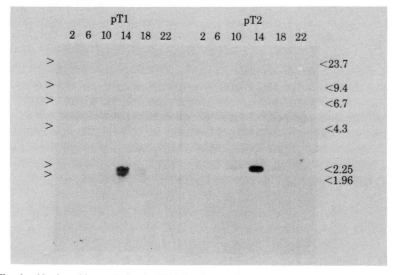

Fig. 4.   Northern blot analysis of mRNA levels encoding α- (left) and β- (right) tubulin during the cell cycle. Poly(A)$^+$ RNA (1 μg), isolated from synchronous cells at the times indicated at the top of the lanes was fractionated on 0.8% agarose gels containing methylmercurihydroxide and transferred to nitrocellulose. The blots were then hybridized to cloned, ^{32}P-labeled, chicken α-tubulin DNA (pT1) and β-tubulin DNA (pT2), respectively. *Hind*III fragments of λ-phage DNA were used as molecular weight markers (in kilobases). (Adapted from Ares and Howell, 1982.)

The cloned tubulin cDNAs have also been used to isolate the corresponding genomic genes from a phage recombinant library (Brunke *et al.*, 1982b). Sequence analysis of the 5′ flanking regions of the genes reveals a homologous region of dyad symmetry just upstream from the TATA box (which is involved in the initiation of transcription). *In vitro* mutagenesis of this region (Shorthe *et al.*, 1981) and subsequent reintroduction of the cloned, mutated genes into cells may determine whether this region of dyad symmetry is important in the coordinate regulation of the four genes.

V. RIBULOSE-1,5-BISPHOSPHATE CARBOXYLASE SYNTHESIS DURING THE CELL CYCLE

Rubisco is a chloroplast enzyme that catalyzes the initial step in the fixation of CO_2 via the Calvin cycle. It is an abundant, soluble protein composed of two nonidentical subunits with molecular weights of 55,000 and 20,000, respectively (Iwanij *et al.*, 1975). The large subunit is synthesized in chloroplasts, and the small subunit is synthesized in the cytoplasm and posttranslationally imported into chloroplasts (Chua and Schmidt, 1978).

The synthesis of Rubisco and its subunits during the cell cycle of *Chlamydomonas* has been investigated by Iwanij *et al.* (1975). They pulse labeled synchronous cells with [³H]arginine and then isolated the holoenzyme on sucrose gradients for radioactivity measurements. They found an increasing rate of synthesis of the holoenzyme and its subunits (after separation by SDS gel electrophoresis) during the light period, reaching a maximum value at the end of the photoperiod. Little synthesis of the enzyme or its subunits was detected in the dark period. They also found slow turnover occurring during the dark period. Their measurements of Rubisco activity in synchronous cultures correlated with synthesis and indicated that the biogenesis of the enzyme during the light period was responsible for the increased Rubisco activity in the culture. Close coordination of the synthesis of the two subunits was observed in synchronous, control cells. Also, when synthesis of one of the subunits was blocked by inhibitors, synthesis of the holoenzyme could not be detected, suggesting further that synthesis of the subunits is tightly coupled. However, some recent work suggests that posttranslational proteolysis may regulate levels of the enzyme in some cases (Mishkind *et al.*, 1982). When synthesis of the large subunit is blocked by chloramphenicol, the small subunit is still synthesized but rapidly degraded. Because Iwanij *et al.* (1975) analyzed only the holoenzyme they would not have detected synthesis and rapid degradation of subunits without assembly.

We have analyzed the synthesis of soluble and thylakoid membrane proteins during the cell cycle of *Chlamydomonas*. Figure 5 shows a fluorograph of a 7.5–15% acrylamide gel of soluble proteins pulse labeled (for 1 hr) with [³H]

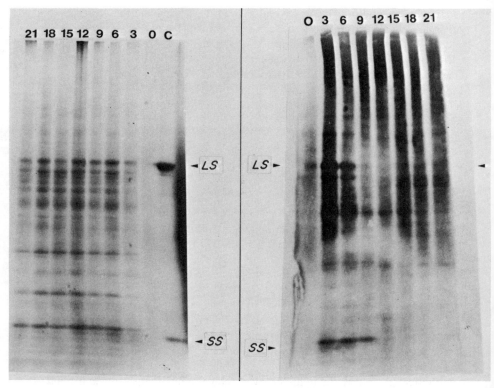

Fig. 5. Synthesis of Rubisco subunits during the cell cycle of *C. reinhardtii*. Equal-sized aliquots of a synchronous culture were removed at 3-hr intervals, pulse labeled for 1 hr, and 100,000 *g* supernatants of cell homogenate were precipitated with trichloroacetic acid (TCA) and applied to a 7.5–15% gradient gel. Purified Rubisco (lane C) was electrophoresed alongside the 100,000 *g* supernatants to localize the large subunit (*LS*) and small subunit (*SS*) on the fluorograph (right). The numbers at the top of each lane indicate the time of the cell cycle that the cells were removed and labeled with [³H]arginine (1μCi/ml).

arginine at the indicated times of the cell cycle. Total soluble proteins from equal aliquots of the culture were applied to the gel along with purified Rubisco. Synthesis of Rubisco subunits was confined to the light period, exhibiting a maximum at 3–4 hr in the light period. These results are similar to Iwanij *et al.* (1975) in that synthesis of Rubisco occurs during the light period. However, we observe the peak of synthesis of Rubisco subunits (assembled and free) occurring considerably earlier (3–5 hr) in the light period than they found for subunits assembled into the holoenzyme. This difference may reflect a lag in assembly of the subunits into the holoenzyme and could provide an interesting system for investigating the regulation of assembly of a multimeric protein. Large subunits have been shown to occur in the chloroplast in unassembled and aggregate form and appear to be normal intermediates in the assembly pathway (Roy *et al.*, 1982).

The gene for the large subunit (*LS*) of Rubisco from *Chlamydomonas* has been cloned, identified, and localized on the restriction map of chloroplast DNA (Malnoe *et al.*, 1979). There appears to be but one copy of the gene per chloroplast DNA molecule as determined by Southern hybridization of the cloned gene to restriction digests of chloroplast DNA. The location of the gene was initially determined by *in vitro* transcription–translation of cloned chloroplast DNA and immunoprecipitation of the products with antibodies to *LS*. The mRNA for this polypeptide (and apparently for most chloroplast-coded polypeptides) is not polyadenylated. The major species of total, cellular A⁻ RNA that hybridizes to the *LS* gene is 1.5×10^3 bases in length. Faint hybridization signals to other, larger RNA species were also observed and may represent precursor transcripts.

The regulation of Rubisco gene expression during the cell cycle provides an interesting problem because there appears to be but one copy of the nuclear-encoded gene for the small subunit (*SS*) (Cashmore, 1979), but the multiple copies of chloroplast DNA that exist provide for many copies of the *LS* gene. Nevertheless, the synthesis of the two subunits does appear to be coordinated during the normal cell cycle. The cloned, identified *LS* gene can be used to probe synchronous cells for both the level and synthesis of its corresponding mRNA. The *SS* gene from a higher plant has been cloned, and it may be possible to use it as a heterologous probe to detect *Chlamydomonas SS* mRNA or to isolate the homologous gene from a clonal library. Using these techniques investigators could determine whether regulation is at the transcriptional or translational level. If regulation occurs at the transcriptional level, it will also be necessary to rule out rapid turnover of the mRNAs.

The regulation of Rubisco gene expression has been studied in higher plants and cloned DNA probes obtained for the *LS* and *SS* proteins (Bedbrook *et al.*, 1979, 1980). Link *et al.* (1978) demonstrated cell-specific expression of the *LS* gene in the C_4 plant *Zea mays*. Bundle sheath cells were found to contain Rubisco and *LS* mRNA whereas mesophyll cells did not contain Rubisco, nor could *LS* mRNA be detected.

The expression of Rubisco genes during light-induced chloroplast development has also been investigated using cloned DNA probes (Smith and Ellis, 1981). Light was shown to stimulate accumulation of both *SS* and *LS* transcripts compared to dark-grown plants. These investigators have not ruled out RNA turnover or translational regulation, but it appears that regulation of transcription contributes to the accumulation of Rubisco during development.

Some insight into the expression of the *LS* gene has been obtained by Jolly *et al.* (1981). They find that in the presence of S factor (a 27-kilodalton polypeptide found in maize chloroplasts), maize chloroplast RNA polymerase will preferentially transcribe the *LS* gene over vehicle sequences or another chloroplast gene (the β subunit of CF_1) contained on the same supercoiled plasmid. This suggests that the *LS* gene has an inherently stronger promoter function than the β gene. However, transcripts of the β subunit gene can be found in mesophyll and bundle sheath cells of maize whereas *LS* transcripts are found only in bundle sheath cells. As suggested by these investigators, S factor may recognize promoter strength, but other factors may be required to suppress transcription, which appears to occur for the *LS* gene in mesophyll cells. These investigators also have not ruled out instability of the *LS* transcripts in mesophyll cells.

VI. SYNTHESIS OF PHOTOSYNTHETIC MEMBRANE PROTEINS DURING THE CELL CYCLE

Chlamydomonas reinhardtii contains large amounts of photosynthetic (thylakoid) membranes which have been characterized biochemically and ultrastructurally (Chua *et al.*, 1975; Chua and Bennoun, 1975; Delepelaire and Chua, 1979; Wollman *et al.*, 1980). The sites of synthesis, either cytoplasmic or chloroplastic, of the major thylakoid polypeptides have been investigated using specific inhibitors of protein synthesis *in vivo* (Chua and Gillham, 1977). Evidence has also been presented that the thylakoid polypeptides synthesized within chloroplasts are synthesized on membrane-bound ribosomes (Herrin *et al.*, 1980; Herrin and Michaels, 1982a,b).

During the light portion of the cell cycle, *Chlamydomonas* generates its photosynthetic apparatus (Fig. 1) in preparation for cell division. We have investigated the cell cycle program of photosynthetic protein synthesis and assembly by pulse labeling synchronous cells with [³H]arginine (Michaels and Herrin, 1983). [³H]arginine is taken up by the cells and used for protein synthesis throughout the cell cycle. The cells were pulse labeled (for 1 hr) at 3-hr intervals and thylakoid membranes isolated. The specific radioactivity of the labeled membranes was determined and expressed as a function of protein and chlorophyll. The results showed an increasing rate of synthesis of thylakoid protein during the light, peaking near the midlight stage, and then declining to a low level in the dark.

To analyze the pattern of synthesis of individual polypeptides, thylakoid membranes were dissolved with SDS and analyzed by electrophoresis on 7.5–15% gradient SDS gels. This gel system has been used in identifying the function of a number of polypeptides. A 10–18% gradient SDS gel containing 2 *M* urea was also used to obtain better separation of low-molecular-weight polypeptides. A fluorograph of the 7.5–15% gel is shown in Fig. 6. The gel shows the relative rate of synthesis of a given polypeptide at a particular point in the cell cycle. To compare different time points the fluorographs were quantitated by densitometry, and the portion of total radioactivity (per lane of the gel) found in a given

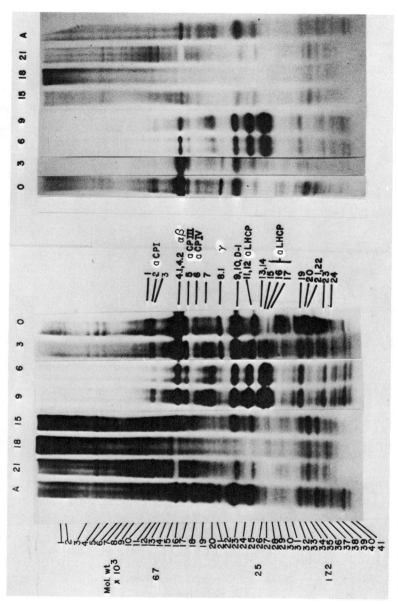

Fig. 6. Thylakoid membrane polypeptide synthesis/assembly during the cell cycle. Synchronous cells were pulse labeled with [³H]arginine (1 μCi/ml) for 1 hr at the indicated times of the cell cycle. Thylakoid membranes were isolated, radioactivity determined, and equal amounts of radioactivity applied to the 7.5–15% acrylamide gel. The polypeptides are labeled according to the nomenclature of Chua and Gillham (1977), and to the left we have simply numbered all the polypeptides that appear any time in the cell cycle. Also, the apoproteins of chlorophyll–protein complexes I, III, and IV (aCPI, aCPIII and aCPIV), the major apoproteins of the light-harvesting chlorophyll a/b protein complex (aLHCP), and the α, β, and γ subunits of coupling factor CF₁ are noted on the far right. Polypeptide D-1 is homologous to the 32-kilodalton protein of higher plants. The figure shows two different exposures of the same gel.

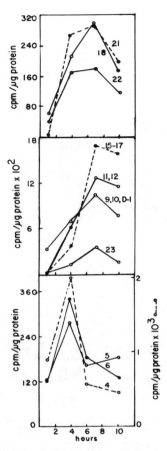

Fig. 7. Patterns of synthesis of some thylakoid polypeptides during the light period of the cell cycle. Synthesis was quantitated as described in the text. Data for polypeptides 18, 21, 22, and 23 were obtained from a 10–18% gradient urea gel (not shown). Data for the synthesis of the other polypeptides were obtained from different exposures of the gel shown in Fig. 6. The line labeled 4 represents polypeptides 4.1, 4.2, and 8.1.

polypeptide band was multiplied by the specific radioactivity (cpm/μg protein) of the membrane preparations. This provides a more absolute measure of the rate of synthesis and assembly of the different polypeptides. The result of this analysis is shown in Fig. 7.

Analysis of individual polypeptides reveals a complex, stage-specific program for the synthesis and integration of membrane polypeptides (Figs. 6 and 7). The α, β, and γ subunits of coupling factor CF_1 (polypeptides 4.1, 4.2, and 8.1) (Piccioni *et al.*, 1981) showed peak rates of synthesis and integration early in the light period but were synthesized at significant rates throughout the photoperiod. Polypeptides 5 and 6, the apoproteins of chlorophyll–protein complexes III and

IV (Delepelaire and Chua, 1979), also showed rates of synthesis and integration that peaked in the early light period. Similarly, polypeptide 2, the apoprotein of chlorophyll–protein complex I (Chua *et al.*, 1975), was synthesized throughout the photoperiod with maximum synthesis occurring early in the light period (not shown in Fig. 7).

The major apoproteins of the light-harvesting chlorophyll *a*/*b* protein (LHCP) complex (polypeptides 11, 15, 16, and 17) exhibited a different pattern of synthesis and assembly. Synthesis and integration of these polypeptides was confined almost exclusively to the last half of the light period. Incorporation of these polypeptides into thylakoids could not be detected during the first hr of the light period and, like the other major thylakoid proteins, assembly was not detected during the middark period (cell division).

Microheterogeneity in the synthesis and integration of thylakoid polypeptides during the 1-hr pulse periods could be detected with successive 15-min pulses. However, the 1-hr pulses provided good resolution of the temporal program of polypeptide synthesis and insertion during the 24-hr cell cycle (Michaels and Herrin, 1983).

The stability of the major polypeptides synthesized and inserted into membranes during the midlight period was examined by pulse-chase analysis (Michaels and Herrin, 1983). The polypeptides were stable through the ensuing dark period, and differential turnover of specific membrane polypeptides did not occur. These results confirm and extend the earlier findings of Beck and Levine (1974). Thylakoid membrane assembly during the cell cycle is a multistep process where synthesis and integration of CF_1 and the chlorophyll proteins of the photochemical reaction centers is initiated several hr before synthesis of the LHCP complex. The delayed synthesis of the LHCP complex until the latter half of the period of membrane biogenesis may be related to its proposed association with the Photosystem II reaction center proteins (polypeptides 5 and 6) and its role in membrane stacking (Michaels and Herrin, 1983).

Shepherd *et al.* have investigated the regulation of LHCP synthesis during the cell cycle (Shepherd *et al.*, 1983). Utilizing *in vitro* translation of poly(A)$^+$ RNA and immunoprecipitation with antisera to the light-harvesting complex from duckweed (*Lemna*), they find levels of mRNA for what appear to be precursors to polypeptides 11, 16, and 17 to be elevated during the last half of the light period. Translatable mRNA for LHCP was not detected during the early light period or in the dark portion of the cell cycle. In contrast to tubulin synthesis, the levels of LHCP mRNAs were affected by illumination. Light, during the normal light portion of the cell cycle, was required for induction of the mRNAs. However, the mRNAs disappeared (as they do under normal conditions) even when the lights were left on during the ensuing dark period. This suggests that different mechanisms operate in the light-dependent induction phase versus the light-independent disappearance of LHCP mRNA during the cell cycle.

These investigators have also made cDNA libraries to poly(A)$^+$ RNA from synchronous cells of *C. reinhardtii* (Shepherd *et al.*, 1983). They have isolated several clones which preferentially hybridize to RNA isolated from cells during the light portion of the cell cycle. At least one clone, pHS16, appears to code for one of the 32-kilodalton precursors to the LHCP apoprotein. Cloned DNA containing coding sequences for these polypeptides will enable further analysis of mechanisms regulating LHCP mRNA during the cell cycle.

Approximately 10–12 thylakoid polypeptides are synthesized within the chloroplast in *Chlamydomonas* (including the α and β subunits of CF_1, the apoproteins of the Photosystem II reaction center, and the 32-kilodalton, herbicide-binding protein). The genes encoding these polypeptides appear to reside on chloroplast DNA (Fig. 2). In order to obtain DNA probes for these polypeptides we have cloned the chloroplast DNA of *C. reinhardtii* into the *Eco*RI site of plasmid pBR325 (Herrin and Michaels, 1982a). Approximately 65% of the genome has been cloned, including many fragments for which there exists at least tentative evidence for the presence of functional genes.

Previously, we had found that thylakoid-bound polysomes isolated during the light but not the dark portion of the cell cycle synthesize a 29-kilodalton thylakoid polypeptide, D-2 (function unknown) (Herrin *et al.*, 1980). Pulse labeling synchronous cells showed that D-2 was synthesized *in vivo* during the light but not the dark period of the cell cycle. In collaboration with J. D. Rochaix we have begun analyzing synchronous cells for levels of D-2 mRNA. Preliminary results using RNA isolated from cells at 3 hr and 18 hr of the cell cycle suggest that D-2 synthesis is regulated in part by mRNA levels. Little hybridization of 18-hr RNA to a cloned DNA probe specific for D-2 was obtained whereas 3-hr RNA gave a strong signal in a Dot hybridization assay (Kafatos *et al.*, 1979) (Michaels and Herrin, unpublished results). Recently, we have also found that the α and β subunits of CF_1 and the 32-kilodalton, herbicide-binding protein are synthesized by thylakoid-bound polysomes isolated during the light but not dark portion of the cell cycle (Herrin and Michaels, 1982a,b). One of the goals of the present research is to determine at what level the expression of chloroplast-encoded membrane protein genes is regulated during the cell cycle, utilizing cloned DNA probes for these polypeptides.

VII. CONCLUSION

It is clear from the results presented and discussed here that gene expression in both the nuclear–cytoplasmic and chloroplast systems is highly regulated during the cell cycle of *Chlamydomonas*. Although much work remains to be done the results obtained so far suggest that gene regulation occurs primarily by controlling the levels of mRNA for a given polypeptide. It will be necessary to rule out translational regulation more thoroughly and to determine whether mRNA levels are regulated through transcription and/or RNA turnover.

The isolation of transcriptionally active nuclei from a cell wall-deficient strain of *Chlamydomonas* has been reported (Keller *et al.*, 1981). This will enable rates of transcription of nuclear genes to be measured *in vitro*. The isolation of intact chloroplasts from *Chlamydomonas* is not presently possible, and rates of transcription will therefore have to be measured *in vivo*. *In vitro* systems capable of faithful transcription of chloroplast DNA may be useful in determining the inherent promoter strength of particular genes (Jolly *et al.*, 1981). Although chloroplast genetic systems have a great deal of prokaryotic character, operons have not been found and the regulation of gene transcription may prove to be considerably more complicated than one might have expected.

Compared to the nucleocytoplasmic system, however, the factors determining gene expression in chloroplasts may be relatively simple. Chloroplast protein and rRNA synthesis *in vivo* are greatly depressed in the dark and elevated in the light. ATP levels within the chloroplast may be the major factor regulating gene expression. Some recent work has shown that ATP levels remain relatively constant in the cytoplasm while fluctuating in the chloroplast and mitochondria of plant cells under different conditions of illumination. This arrangement would allow protein synthesis in the cytoplasm to proceed in the light or dark whereas chloroplast and mitochondrial protein synthesis would be highly dependent on illumination conditions. In contrast, gene expression in the nucleocytoplasmic system may have many contributing factors. Tubulin gene expression is light independent whereas that of the LHCP is light dependent and cell cycle dependent.

Another important problem is how the expression of genes encoding proteins that have subunits synthesized in the cytoplasm and in chloroplasts (such as Rubisco and CF_1) is coordinated. Do the subunits produced in the cytoplasm regulate chloroplast gene expression or vice versa? The analysis of mutants may be helpful in answering this question. Also, in systems other than *Chlamydomonas*, it may be possible to isolate chloroplasts whose pattern of gene expression *in vitro* can be modified by the addition of proteins (such as subunit precursors of multimeric chloroplast proteins) or factors from the cytoplasm.

Finally, the isolation and identification of genes whose expression regulates (or is required for) progression of cells through the cell cycle may be possible by complementation of cell cycle mutants through DNA-mediated transformation (Rochaix and van Dillewijn, 1982). More effective use of the ability to mutate *Chlamydomonas* and the development of an effecient transformation system should provide answers to some of these questions.

ACKNOWLEDGMENTS

We would like to thank Dr. Steven Howell for communicating results prior to publication. Work in this laboratory has been supported by grants to A.M. from the National Science Foundation. D.H. is also supported by a predoctoral fellowship from the University of South Florida.

REFERENCES

Ares, M., and Howell, S. H. (1982). Cell cycle stage-specific accumulation of messenger RNAs encoding tubulin and other polypeptides in *Chlamydomonas*. *Proc. Natl. Acad. Sci. U.S.A.* **79**, 5577–5581.

Beck, D. P., and Levine, R. P. (1974). Synthesis of chloroplast membrane polypeptides during synchronous growth of *Chlamydomonas reinhardtii*. *J. Cell Biol.* **63**, 759–772.

Bedbrook, J. R., Coen, D. M., Beaton, A. R., Bogorad, L., and Rich, A. (1979). Location of the single gene for the large subunit of ribulose bisphosphate carboxylase on the maize chloroplast chromosome. *J. Biol. Chem.* **254**, 905–910.

Bedbrook, J. R., Smith, S., and Ellis, R. J. (1980). Molecular cloning and sequencing of cDNA encoding the prescursor to the small subunit of chloroplast ribulose-1,5-bisphosphate carboxylase. *Nature (London)* **287**, 692–697.

Berk, A. J., and Sharp, P. A. (1977). Sizing and mapping of early Adenovirus mRNA's by gel electrophoresis of S1 endonuclease-digested hybrids. *Cell* **12**, 721–732.

Bernstein, E. (1960). Synchronous division in *Chlamydomonas moewusii*. *Science* **131**, 1528–1529.

Bravo, R., and Cellis, J. E. (1980). A search for differential polypeptide synthesis throughout the cell cycle of Hela cells. *J. Cell Biol.* **84**, 795–802.

Briat, J.-F., Dron, M., Loiseaux, S., and Mache, R. (1982). Structure and transcription of the spinach chloroplast rDNA leader region. *Nucleic Acids Res.* **10**, 6865–6878.

Brunke, K. J., Young, E. E., Buchbinder, B. U., and Weeks, D. P. (1982a). Coordinate regulation of the four tubulin genes of *Chlamydomonas reinhardi*. *Nucleic Acids Res.* **10**, 1295–1310.

Brunke, K. J., Buchbinder, B. U., Sternberg, E. J., Anthony, J. G., and Weeks, D. P. (1982b). Sequency homology in the 5′ flanking regions of the coordinately-regulated tubulin genes from *Chlamydomonas reinhardi*. *J. Cell Biol.* **95**, 216a.

Cashmore, A. R. (1979). Reiteration frequency of the gene coding for the small subunit of ribulose-1,5-bisphosphate carboxylase. *Cell* **17**, 383–388.

Cattalico, R. A., Senner, J. W., and Jones, R. F. (1973). Changes in cytoplasmic and chloroplast ribosomal ribonucleic acid during the cell cycle of *Chalmydomonas reinhardtii*. *Arch. Biochem. Biophys.* **156**, 58–65.

Chiang, K.-S., and Sueoka, N. (1967). Replication of chloroplast DNA in *Chlamydomonas reinhardi* during vegetative cell cycle: Its mode and regulation. *Proc. Natl. Acad. Sci. U.S.A.* **57**, 1506–1513.

Chua, N.-H., and Bennoun, P. (1975). Thylakoid membrane polypeptides of *Chlamydomonas reinhardtii:* Wild-type and mutant strains deficient in Photosystem II reaction center. *Proc. Natl. Acad. Sci. U.S.A.* **72**, 2175–2179.

Chua, N.-H., and Gillham, N. (1977). The sites of synthesis of the principal thylakoid membrane polypeptides in *Chlamydomonas reinhardtii*. *J. Cell Biol.* **74**, 441–452.

Chua, N.-H., and Schmidt, G. W. (1978). Post-translational transport into intact chloroplasts of a precursor to the small subunit of ribulose-1,5-bisphosphate carboxylase. *Proc. Natl. Acad. Sci. U.S.A.* **75**, 6110–6114.

Chua, H.-H., Matlin, K., and Bennoun, P. (1975). A chlorophyll-protein complex lacking in Photosystem I mutants of *Chlamydomonas reinhardtii*. *J. Cell Biol.* **67**, 361–377.

Delepelaire, P., and Chua, N.-H. (1979). Lithium dodecyl sulfate polyacrylamide gel electrophoresis of thylakoid membranes at 4°C: Characterizations of two additional chlorophyll a-protein complexes. *Proc. Natl. Acad. Sci. U.S.A.* **76**, 111–115.

Grant, D., and Chiang, K.-S. (1980). Physical mapping and characterization of *Chlamydomonas* mitochondrial DNA molecules: Their unique ends, sequence homogeneity, and conservation. *Plasmid* **4**, 82–96.

Herrin, D., and Michaels, A. (1982a). *In vitro* synthesis of the 32 kd membrane protein by thylakoid-bound ribosomes and cloning of its presumptive gene. *J. Cell Biol.* **95,** 279a.

Herrin, D., and Michaels, A. (1982b). The peripheral subunits of chloroplast coupling factor CF_1 (alpha and beta) are synthesized on membrane-bound ribosomes: A model for CF_1 synthesis and assembly. *J. Cell Biol.* **95,** 408a.

Herrin, D., Michaels, A., and Hickey, E. (1980). Synthesis of a chloroplast membrane polypeptide on thylakoid-bound ribosomes during the cell cycle of *Chlamydomonas reinhardtii* 137+. *Biochim. Biophys. Acta* **655,** 136–145.

Hoober, J. K., and Blobel, G. (1969). Characterization of the chloroplastic and cytoplasmic ribosomes of *Chlamydomonas reinhardi. J. Mol. Biol.* **41,** 121–138.

Iwanij, V., Chua, N.-H., and Siekevitz, P. (1974). The purification and some properties of ribulose bisphosphate carboxylase and of its subunits from the green alga *Chlamydomonas reinhardtii. Biochim. Biophys. Acta* **358,** 329–340.

Iwanij, V., Chua, N.-H., and Siekevitz, P. (1975). Synthesis and turnover of ribulose bisphosphate carboxylase and of its subunits during the cell cycle of *Chlamydomonas reinhardtii. J. Cell Biol.* **64,** 572–585.

Jolly, S. O., McIntosh, L., Link, G., and Bogorad, L. (1981). Differential transcription *in vivo* and *in vitro* of two adjacent maize chloroplast genes: The large subunit of ribulosebisphosphate carboxylase and the 2.2-kilobase gene. *Proc. Natl. Acad. Sci. U.S.A.* **78,** 6821–6825.

Kafatos, F. C., Jones, C. W., and Efstatiadis, A. (1979). Determination of nucleic acid sequence homologies and relative concentrations by a Dot hybridization procedure. *Nucleic Acids Res.* **7,** 1541–1552.

Keller, L. R., Silflow, C. D., and Rosenbaum, J. L. (1981). Transcription *in vitro* of alpha and beta tubulin genes in isolated *Chlamydomonas reinhardii* nuclei. *J. Cell Biol.* **91,** 375a.

Link, G., Coen, D. M., and Bogorad, L. (1978). Differential expression of the gene for the large subunit of ribulose biosphophate carboxylase in maize leaf cell types. *Cell* **15,** 725–731.

Ludwig, J. R., Foy, J. J., Elliott, S. G., and McLaughlin, C. S. (1982). Synthesis of specific identified, phosphorylated, heat shock, and heat stroke proteins through the cell cycle of *Saccharomyces cerevisiae. Mol. Cell. Biol.* **2,** 117–126.

Malnoe, P., Rochaix, J.-D., Chua, N.-H., and Spahr, P.-F. (1979). Characterization of the gene and messenger RNA of the large subunit of ribulose-1,5-diphosphate carboxylase in *Chlamydomonas reinhardii. J. Mol. Biol.* **133,** 417–434.

Matsuda, Y., and Surzycki, S. J. (1980). Chloroplast gene expression in *Chlamydomonas reinhardi. Mol. Gen. Genet.* **180,** 463–474.

Maxam, A. M., and Gilbert, W. (1977). A new method for sequencing DNA. *Proc. Natl. Acad. Sci. U.S.A.* **74,** 560–564.

Minami, S. A., Collis, P. S., Young, E. E., and Weeks, D. P. (1981). Tubulin induction in *C. reinhardtii:* Requirement for tubulin mRNA synthesis. *Cell* **24,** 89–95.

Michaels, A., and Herrin, D. (1983) In preparation.

Mishkind, M., Jensen, J. H., Plumley, F. G., and Schmidt, G. W. (1982). Coordinated accumulation of Rubisco subunits is achieved by proteolysis. *J. Cell Biol.* **95,** 276a.

Ohad, I., Siekevitz, P., and Palade, G. E. (1967). Biogenesis of chloroplast membranes. II. Plastid differentiation during greening of a dark-grown algal mutant (*Chlamydomonas reinhardi*). *J. Cell Biol.* **35,** 553–584.

Piccioni, R. G., Bennoun, P., and Chua, H.-H. (1981). A nuclear mutant of *Chlamydomonas reinhardtii* defective in photosynthetic photophosphorylation. Characterization of the algal coupling factor ATPase. *Eur. J. Biochem.* **117,** 93–102.

Rochaix, J.-D. (1978). Restriction endonuclease map of the chloroplast DNA of *Chlamydomonas reinhardtii. J. Mol. Biol.* **126,** 597–617.

Rochaix, J.-D., and Malnoe, P. (1978). Anatomy of the chloroplast ribosomal DNA of *Chlamydomonas reinhardii. Cell* **15,** 661–670.

Rochaix, J.-D., and van Dillewijn, J. (1982). Transformation of the green alga *Chlamydomonas reinhardii* with yeast DNA. *Nature (London)* **296,** 70–72.

Roy, H., Bloom, M., Milos, R., and Monroe, M. (1982). Studies on the assembly of large subunits of ribulose bisphosphate carboxylase in isolated pea chloroplasts. *J. Cell Biol.* **94,** 20–27.

Sager, R. (1972). "Cytoplasmic Genes and Organelles." Academic Press, New York.

Shepherd, H. S. Ledoigt, G. and Howell, S. (1983). Submitted for publication.

Shortle, D., Diamio, D., and Nathans, D. (1981). Directed mutagenesis. *Annu. Rev. Genet.* **15,** 265–294.

Silflow, D. C., and Rosenbaum, J. (1981). Multiple alpha and beta tubulin genes in *Chlamydomonas* and regulation of tubulin mRNA levels after deflagellation. *Cell* **24,** 81–88.

Smith, S. M., and Ellis, R. J. (1981). Light-stimulated accumulation of transcripts of nuclear and chloroplast genes for ribulosebisphosphate carboxylase. *J. Mol. Appl. Genet.* **1,** 127–137.

Spudich, J. L., and Sager, R. (1980). Regulation of *Chlamydomonas* cell cycle by light and dark. *J. Cell Biol.* **85,** 136–145.

von Wettstein, D. (1981). Chloroplast and nucleus: concerted interplay between genomes of different cell organelles. *In* "International Cell Biology 1980–1981" (H. G. Schweiger, ed.), pp. 250–272. Springer-Verlag, Berlin and New York.

Watson, J. C., and Surzycki, S. J. (1982). Extensive sequence homology in the DNA coding for elongation factor Tu from *Escherichia coli* and the *Chlamydomonas reinhardii* chloroplast. *Proc. Natl. Acad. Sci. U.S.A.* **79,** 2264–2267.

Weeks, D. P., and Collis, P. S. (1979). Induction and synthesis of tubulin during the cell cycle and life cycle of *Chlamydomonas reinhardi. Dev Biol.* **69,** 400–407.

Wilson, R., and Chiang, K.-S. (1977). Temporal programming of chloroplast and cytoplasmic ribosomal RNA transcription in the synchronous cell cycle of *Chlamydomonas reinhardtii. J. Cell Biol.* **72,** 470–481.

Wollman, F.-A., Olive, J., Bennoun, P., and Recouvreur, M. (1980). Organization of the Photosystem II centers and their associated antennae in the thylakoid membranes: A comparative ultrastructural, biochemical and biophysical study of *Chlamydomonas* wild type and mutants lacking in Photosystem II reaction centers. *J. Cell Biol.* **87,** 728–735.

5

Expression of Histone Genes during the Cell Cycle in Human Cells

G. S. STEIN,[1] M. A. PLUMB,[1] J. L. STEIN,[2] F. F. MARASHI,[1] L. F. SIERRA,[1,3] AND L. L. BAUMBACH[1]

I. INTRODUCTION

This chapter summarizes the progress made during the past several years toward understanding the structure, organization, and expression of human histone genes. By way of introduction, we would like to share our rationale for pursuing this problem: First, an understanding of the molecular mechanisms by which specific genetic sequences are differentially expressed during the cell cycle may provide insight into regulation of the complex and interdependent biochemical events required for cell proliferation. Second, histone proteins are essential for packaging DNA and are therefore involved with DNA replication,

[1]Department of Biochemistry and Molecular Biology, University of Florida College of Medicine, Gainesville, Florida.

[2]Department of Immunology and Medical Microbiology, University of Florida College of Medicine, Gainesville, Florida.

[3]Present address: Swiss Institute for Experimental Cancer Research, CH-1066 Epalinges, Switzerland.

107

RECOMBINANT DNA AND
CELL PROLIFERATION

RNA transcription, and the transcriptional potential of genomic domains. Third, but equally important, histone genes appear to be organized and regulated somewhat differently from most single-copy genes.

It has been within this context that we have been directing our efforts toward examining the human histone genes and their control. This chapter focuses on three points: (1) the structure and organization of human histone genes, primarily because of the functional interrelationships among structure, organization, and regulation, but also because specific regions of genomic histone sequences have been used as probes for analysis of human histone gene organization and expression; (2) the levels of control of histone gene expression and variations in control during the cell cycle; and (3) the relationship between human histone gene expression and DNA replication, with emphasis on both the temporal and putative functional coupling of these biochemical processes.

In this chapter we consider only the general features of human histone gene structure and organization, with emphasis on those aspects most directly related to expression of the human histone genes in conjunction with cell proliferation. We have presented a more in-depth description of the structural features of human histone genes elsewhere (Stein *et al.*, 1984; Sierra *et al.*, 1982a).

II. STRUCTURE AND ORGANIZATION OF HUMAN HISTONE GENES

We have isolated a series of genomic clones containing human histone-coding sequences and their flanking regions (Sierra *et al.*, 1982a) from a λCharon 4A human gene library constructed by Lawn *et al.* (1978). These clones were initially characterized by restriction mapping and were then analyzed with respect to the representation and location of histone-coding regions by hybridization using both homologous and heterologous histone gene probes. Assignment of histone-coding sequences was confirmed by hybrid selection–*in vitro* translation (see Fig. 1) and by partial nucleotide sequencing (Stein *et al.*, 1984). Restriction maps illustrating the organization of four types of human histone gene clusters are shown in Fig. 2. Despite the apparent clustering of these human histone genes, they do not conform to the tandem repeat organization observed for the *Drosophila* histone genes and for the "early" sea urchin histone genes. Rather, the human histone gene clusters we have isolated are polymorphic and exhibit at least four types of arrangements with respect to restriction sites and the order of coding sequences. Although most of the cloned human histone gene clusters we have isolated contain various arrangements and representations of the sequences encoding the core histones (H2A, H2B, H3, and H4), λHHG415 contains sequences encoding H1 as well as the four core histones (Carozzi *et al.*, 1984). Heintz *et al.* (1981) and Clark (1982) have isolated clones exhibiting additional

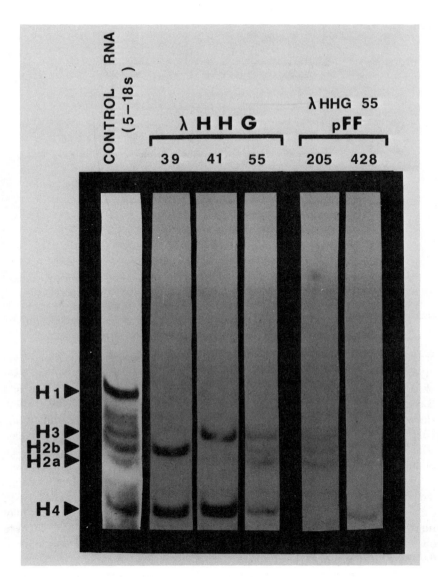

Fig. 1. Hybrid selection–*in vitro* translation analysis of DNAs from λHHG recombinant phage containing human histone gene sequences and of recombinant pBR322 plasmids containing specific regions of the genomic human histone sequences subcloned from λHHG phage. Phage or plasmid DNA was immobilized on nitrocellulose filters and hybridized with HeLa S-phase polysomal RNA. Hybridized RNAs were eluted, translated in a wheat germ cell-free protein-synthesizing system, and the *in vitro*-translated polypeptides were fractionated electrophoretically in acetic acid–urea poly-acrylamide gels. Analysis was by autoradiography. The left lane shows *in vitro* translation products (all five histone polypeptides) translated from the RNA preparation used for hybrid selection (control).

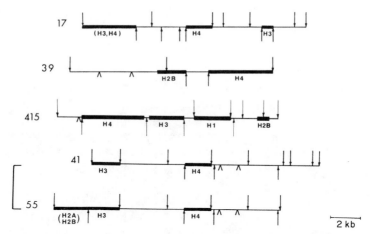

Fig. 2. Restriction map of five representative λHHG clones, indicating the relative positions of human histone-coding regions. Clones 41 and 55 are bracketed to emphasize their similarity with respect to restriction enzyme recognition sites, as well as histone-coding regions. The location of the H2A gene in λHHG 415 has not been definitively resolved and is therefore not indicated (↓) *Eco*RI; (↑) *Hind*III; (∧) *Bam*HI. Scale on figure is in kilobases (kb).

arrangements of the core histone genes which provide a further indication of the diversity of human histone gene organization. Similar polymorphism has been observed in the organization of mouse (Sittman *et al.*, 1981; Seiler-Tuyns and Birnstiel, 1981), chicken (Harvey and Wells, 1979; Crawford *et al.*, 1979; Harvey *et al.*, 1981, 1982; Engel and Dodgson, 1981), *Xenopus* (Zernik *et al.*, 1980; Van Dongen *et al.*, 1981; Old *et al.*, 1982; Ruberti *et al.*, 1982), yeast (Hereford *et al.*, 1979; Wallis *et al.*, 1980; Choe *et al.*, 1982), and late sea urchin (Childs *et al.*, 1982) histone genes.

The cloned human histone genes, like most of the human genome, are interspersed with highly reiterated Alu-like DNA sequences (Duncan *et al.*, 1979; Calabretta *et al.*, 1981). It has been suggested that Alu family sequences may function as genomic origins of replication (Jelinek and Schmid, 1982). These observations, together with those identifying *ars* sequences in close proximity to the yeast histone genes (Hereford *et al.*, 1982), suggest the possibility that those genomic sequences that are under cell cycle regulation may be localized near specific genomic origins of replication.

The location of histone-coding regions in λHHG phage has been confirmed in several cases by direct DNA sequencing. For example, the nucleotide sequence of the H4 gene present in λHHG41 (pFO108A) is shown in Fig. 3. Partial sequences are also available for the H2B genes in λHHG39 (pTN521) and λHHG55 (pFF435B), and for the H3 genes present in λHHG41 (pFO535) and λHHG17 (pST519) (refer to Fig. 6). These sequences have been presented elsewhere (Stein *et al.*, 1984).

It is worthy of note that amino acid sequence analysis of human H3 and H4 histone proteins has not yet been reported; hence, the nucleotide sequences of the histone genes described here have provided the first direct information pertaining to the primary structure of these human histone proteins. With the exception of the H2B gene present in λHHG55 (pFF435B), which appears to be nonfunctional (pseudogene) (Marashi *et al.*, 1984), the rest of the human histone gene sequences that have been analyzed (Stein *et al.*, 1984) illustrate the evolutionary conservation of histone amino acid sequences, when compared with mouse (Seiler-Tuyns and Birnstiel, 1981), chicken (Harvey *et al.*, 1981; Engel and

```
-240        -230         -220        -210         -200         -190
AATTC TCCCG GGGAC CGTTG CGTAG GCGTT AAAAA AAAAA AAGAG TGAGA GAGGG ACTGA

-180        -170         -160        -150         -140         -130
GCAGA GTGGA GGAGG AGGGA GAGGA AAACA GAAAA GAAAT GACGA AATGT CGAGA GGGCG

-120        -110         -100        -90          -80          -70
GGGAC AATTG AGAAC GCTTC CCGCC GGCGC GCTTT CGGTT TtCAA TCTGG TCCGA TAtCt

-60         -50          -40         -30          -20          -10
CtGTA TATtA CGGGG AAGaC GGtGa CGCtC CGatC GaNcN Nctat CGGGC TCCtG CGGTC


ATG TCC GGC tGt GGa aAG GGC GGA AAG GGC TTA GGC AAA GGT GGC GCT AAG CGC
Met Ser Gly Arg Gly Lys Gly Gly Lys Gly Leu Gly Lys Gly Gly Ala Lys Arg
   0   1                5                     10                   15

CAC CGC AAG GTC TTG AGA GAC AAC ATT CaG GGC ATC ACC aAG CCT GCC aTT CGG
His Arg Lys Val Leu Arg Asp Asn Ile Gln Gly Ile Thr Lys Pro Ala Ile Arg
          20                  25                  30                   35

CGT NTA GCT CGG CGT GGC GGC GTT AAG CGG ATC TCT GGC CTC ATT TAC GAG GAG
Arg Leu Ala Arg Arg Gly Gly Val Lys Arg Ile Ser Gly Leu Ile Tyr Glu Glu
                 40               45                  50

ACC CGC GGT GTG CTG AAa GTG TTC TTG GAG AAT GTG ATT CGG GAC GCA GTC ACC
Thr Arg Gly Val Leu Lys Val Phe Leu Glu Asn Val Ile Arg Asp Ala Val Thr
       55               60                  65                   70

TAC ACC GAG CAC GCC AAG CGC AAG ACC GTC ACA GCC ATG GAT GTG GTG TAC GCG
Tyr Thr Glu His Ala Lys Arg Lys Thr Val Thr Ala Met Asp Val Val Tyr Ala
             75               80                  85

CTC AAG CGN CAG GGG AGN aCC CtC TAC GGC TTC GGA GGC TAG GCCGC CGCTC
Leu Lys Arg Gln Gly Arg Thr Leu Tyr Gly Phe Gly Gly Stop
     90               95                  100       102
                                                        mRNA 3' end
CAGCT TTGCA CGTTT CGATC CCAAA GGCCC TTTTT GGGCC GACCA CTTGC TCAtC CTGAG

GAGTT GGACA CTTGA CTGCG TAAAG TGCAA CAGTA ACGAT GTTGG AAGGT AACTT TGGCA

GTGGG GCGAC AATCG GATCT GAAGT TAACG GAAAG acata accgc
```

Fig. 3. Complete nucleotide sequence of the human H4 histone gene (mRNA coding and flanking sequences) cloned in λHHG41.

Dodgson, 1981; Krieg *et al.*, 1982; Ruiz-Vesquez and Ruiz-Carillo, 1982; Clark, 1982), or yeast (Wallis *et al.*, 1980). Conservative nucleotide substitutions in the human histone-coding regions have been observed, indicating that these genes have evolved independently. The H4 gene from λHHG41 has no intervening sequences (Fig. 3), a common feature of histone genes (Hentschel and Birnstiel, 1981), with the exception of one chicken H3 gene (Engel *et al.*, 1982).

Although functions for the sequences flanking mRNA-coding regions of eukaryotic genes have yet to be definitively established, putative regulatory regions appear to be conserved among sequences upstream (5′) from several genes transcribed by RNA polymerase II. These sequences include a TATA box, located about 30 nucleotides upstream from the mRNA start site (Goldberg, 1979; Efstratiadis *et al.*, 1980). As shown in Fig. 3, this consensus sequence is also present in the human histone H4 gene and has been identified in numerous other histone genes isolated from a variety of species (Hentschel and Birnstiel, 1981). The CAAT box, located approximately 70 to 80 nucleotides upstream from the mRNA start site (Efstratiadis *et al.*, 1980; Benoist *et al.*, 1980), has been found flanking histone H2A, H2B, and H3 genes (Hentschel and Birnstiel, 1981). Although the CAAT box has not previously been found upstream from H1 or H4 histone genes, analysis of the human H4 histone gene shown in Fig. 3 indicates the presence of one clearly defined CAAT box and a second somewhat modified version of this consensus sequence. Further upstream, several short direct repeats and a large purine-rich region have been identified. The purine-rich sequence may have some functional significance for histone genes, as they have been found in the spacer regions separating histone genes in several species of sea urchins (Schaffner *et al.*, 1978; Sures *et al.*, 1978), in *Drosophila* (Goldberg, 1979), and in mouse (Seiler-Tuyns and Birnstiel, 1981).

The 3′ region of this human histone H4 gene also shares sequence similarities with those of mouse (Seiler-Tuyns and Birnstiel, 1981), chicken (Clark, 1982), and sea urchin (Schaffner *et al.*, 1978; Grunstein *et al.*, 1981; Busslinger *et al.*, 1980). For example, at the 3′ end of the H4 mRNA coding sequence, there is a region exhibiting hyphenated dyad symmetry followed by an ACCA termination sequence. This symmetrical sequence has the potential for forming a hairpin loop and may have some functional significance (e.g., as a protein-binding site). The GA box several nucleotides downstream from the putative mRNA termination site is another sequence shared in common with other histone genes (Hentschel and Birnstiel, 1981).

The last point we will consider with regard to human histone gene organization is some of the evidence for sequence differences in the multiple copies of the H4 genes, despite the stringent conservation of the amino acid sequences of the H4 histone proteins. Our first indication of the polymorphic nature of human H4 histone genes preceded the availability of cloned genomic human histone sequences and was based on direct analysis of HeLa H4 histone mRNAs.

The primary observation that permitted the analysis of HeLa H4 mRNAs was that polysomal histone H4 mRNA subspecies could be resolved by using a combination of gel electrophoretic systems (Lichtler *et al.*, 1977, 1980, 1982). By alternating the use of denaturing and nondenaturing electrophoretic systems, and by characterizing the mRNA species thus resolved by both *in vitro* translation (Lichtler *et al.*, 1977, 1980) and two-dimensional "fingerprinting" after T_1 ribonuclease digestion (Lichtler *et al.*, 1982), sequence differences in at least seven H4 mRNA species were identified. We then assigned specific histone H4 mRNA species to individual cloned H4 histone genes using a modification of the Berk and Sharp procedure (Lichtler *et al.*, 1982). As shown in Fig. 4, when *in vivo* labeled polysomal histone H4 mRNAs were collectively annealed to individual, cloned genomic histone H4 sequences, each of the four cloned probes used formed an S1 nuclease-resistant hybrid with a distinct H4 mRNA. These results provide evidence that some, if not all, of the HeLa histone H4 mRNAs identified are not a consequence of posttranscriptional processing of a lesser number of distinct mRNAs, but are rather the transcriptional products of different histone H4 genes. It is worthy of note that where two histone H4 genes reside in the same cluster (λHHG17, Fig. 2), two different H4 histone mRNAs are protected from S1 nuclease digestion after solution hybridization, i.e., the adjacent H4 genes encode different H4 mRNAs.

The obvious question that arises is the biological significance of multiple forms of H4 histone mRNAs. Multiple histone mRNAs have been shown to be differentially expressed during sea urchin development (Kunkel *et al.*, 1975; Hieter *et al.*, 1979), and it is generally accepted that variant H2A, H2B, and H3 histone proteins are differentially expressed during development (Carroli and Ozaki, 1979; Trostle-Weige *et al.*, 1982; Newrock *et al.*, 1978) and during the cell cycle (Wu and Bonner, 1981; Wu *et al.*, 1982). However, the identification of multiple forms of H4 histone mRNAs, which serve as templates for the synthesis of apparently identical polypeptides, requires a more subtle explanation than one in which distinct histone amino acid sequences are required under defined biological circumstances. Although we can only speculate, our current thinking encompasses a working model with the following components: (1) Groups of histone genes are coordinately regulated at both the transcriptional and posttranscriptional levels, and their expression is modulated by common regulatory sequences and/or regulatory molecules. (2) Different histone genes are preferentially expressed in different cell types and/or as a response to changing biological circumstances. (3) The expression of specific groups of histone genes is largely determined either by the cell's requirement for variant histone H2A, H2B, H3, or H1 proteins, or by the proximity of the genes to other genomic

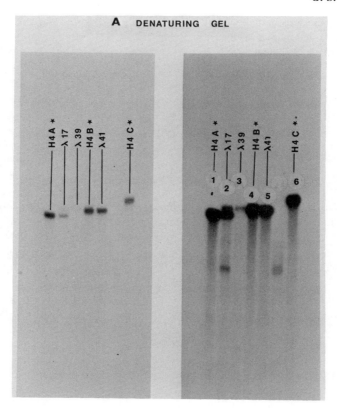

Fig. 4. Identification of H4 histone mRNA species which are S1 nuclease-resistant after hybridization to histone DNA clones. Ten to twenty micrograms of λHHG recombinant DNAs were mixed with 500–2000 cpm of ^{32}P-labeled total H4 histone mRNA isolated from a denaturing gel, and nucleic acids were precipitated with ethanol in a 1.5-ml microfuge tube. The pellet was drained thoroughly and dissolved in 20 μl of redistilled but not deionized formamide (BRL). Four microliters of 2 M NaCl, 0.4 M PIPES (pH 6.5), and 5 mM EDTA was added and the reaction was incubated at 80°C for 5 min. Hybridization was at 50°C for 5 hr. After hybridization 225 μl of ice-cold 0.25 M NaCl, 25 mM sodium acetate (pH 4.4), and 0.45 mM ZnSO$_4$ was added, and the sample was mixed thoroughly and immediately placed on ice. Then, 110 units of S1 nuclease (Sigma) in 5 μl were added, and the sample incubated at 37°C for 30 min. Twenty microliters of 10% (w/v) SDS, 20 μl 0.2 M EDTA (pH 8.0), 20 μg tRNA, 200 μl phenol, and 200 μl chloroform/isoamyl alcohol [24:1(v/v)] were added, and the mixture was vortexed, centrifuged, and the supernatant was ethanol-precipitated. The dried pellet was dissolved in 8.3 M urea–5 mM EDTA, heated to 100°C for 2 min, and electrophoresed in an acrylamide gel. S1-resistant samples were electrophoresed in an 8.3-M urea gel (A) or in a nondenaturing gel (B). Results from two different autoradiographic exposure times are shown. Starred lanes H4A, H4B, and H4C correspond to marker RNAs.

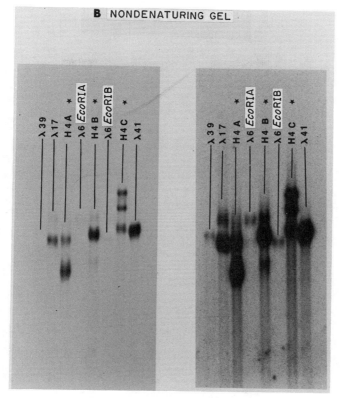

Fig. 4. (*Continued*)

sequences, for example, specific origins of replication, which by definition must also be under cell cycle control.

III. LEVELS OF CONTROL OF HUMAN HISTONE GENE EXPRESSION

There are at least five levels at which gene expression may be regulated: (1) transcription; (2) transcript processing (splicing, 5'-capping, and 3'-polyadenylation); (3) transport from the nucleus to the cytoplasm; (4) translation; and (5) posttranslational modifications. To understand the level at which regulation of human histone gene expression resides, it is necessary to establish whether or not histone proteins and histone mRNAs are synthesized in a cell cycle stage-specific manner. Equally important is an understanding of the relationship be-

Fig. 5. (a) Two-dimensional electrophoretic analysis of [^{35}S]methionine-labeled total cellular proteins. A, S phase; B, G$_1$ phase; C, cytosine– arabinoside-treated S phase. HeLa cells in suspension culture were synchronized by double thymidine block (2 mM). After release from the second thymidine block, cells were maintained at a density of 5×10^5/ml for 1 hr. S-phase cells were incubated for 2 additional hr at 37°C either in the presence or in the absence of the DNA synthesis inhibitor, cytosine–arabinoside (40 µg/ml). Ten milliliters of either S-phase cells or cytosine–arabinoside-treated S-phase cells was centrifuged, washed in spinner salts solution containing 5 µM methionine (37°C), and resuspended in 2.8 ml of low (5 µM) methionine medium

NEPHGE

SDS

(b)

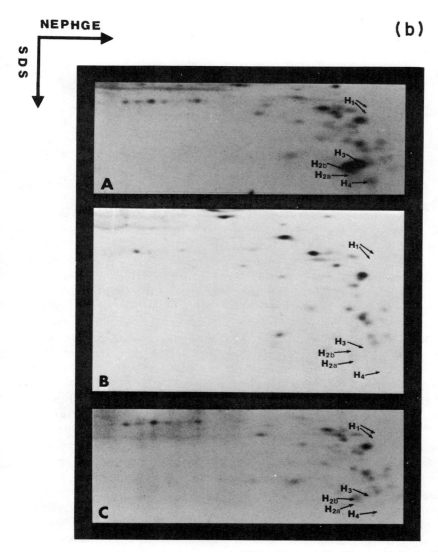

(containing cytosine–arabinoside when appropriate). [^{35}S]methionine (140 μCi) was added and cells were pulse labeled for 45 min at 37°C. G$_1$-phase HeLa cells were obtained by mitotic selective detachment and labeled as described earlier for S-phase cells. Cells were harvested by centrifugation at 1500 g for 5 min, rinsed with ether, and dried before solubilizing in 9.5 M urea, 2% (w/v) NP-40, 2% (w/v) ampholine, 5% (w/v) mercaptoethanol, 0.3 M NaCl, and 1.0 mg/ml protamine sulfate. ^{35}S-labeled peptides were fractionated in a two-dimensional electrophoretic system in which the first dimension was nonequilibrium pH gradient electrophoresis (NEPHGE) and the second dimension was an SDS-containing, 15% (w/v) acrylamide gel. (b) Two-dimensional NEPHGE/SDS electrophoresis of acid-extracted nuclear proteins of (A) S-phase, (B) G$_1$-, and (C) cytosine–arabinoside-treated S-phase cells. [^{35}S]methionine-labeled cells were lysed in 10 mM KCl, 10 mM Tris, and 1.3 mM MgCl$_2$ (pH 7.4) containing 0.65% (v/v) Triton X-100, and nuclei were pelleted by centrifugation at 800 g. Nuclear pellets were extracted with 0.4 M H$_2$SO$_4$ for 30 min and the acid-soluble nuclear proteins were precipitated from the supernatant at −20°C overnight after addition of three volumes of 95% ethanol.

tween histone protein synthesis, histone gene transcription, and the abundance of cellular histone mRNA sequences.

A. Histone Protein Synthesis during the Cell Cycle

It has been well documented that in most eukaryotic cells, the bulk synthesis of histones (that which is required during DNA replication) occurs in the S phase of the cell cycle (Marashi *et al.*, 1982; Spalding *et al.*, 1966; Robbins and Borun, 1967; Delegeane and Lee, 1982; Moll and Wintersberger, 1976; Stein and Borun, 1972). Exceptions include the recent identification of "basal levels" of histone protein synthesis in G_1 and G_{10} in several mammalian cell lines; these basal levels correspond to approximately 8% of the rate of synthesis observed in S phase (Wu and Bonner, 1981; Wu *et al.*, 1982). Similarly, histone H1 synthesis may occur outside cell cycle control under special circumstances (Tarnowka *et al.*, 1978), and a constant rate of histone synthesis during the cell cycle has been reported in S49 mouse lymphoma cells (Groppi and Coffino, 1980).

Early studies with synchronized HeLa cells and with nondividing human diploid fibroblasts stimulated to proliferate indicated that the specific activity of nuclear histones was highest when cells were pulse labeled with ^3H- or ^{14}C-labeled amino acids during S phase (Spalding *et al.*, 1966; Robbins and Borun, 1967; Park *et al.*, 1976). To eliminate the possibility of cell cycle-specific compartmentalization of newly synthesized histone proteins, we recently pulse labeled G_1- and S-phase HeLa cells with [^{35}S]methionine, and analyzed total cellular proteins (Fig. 5a) or 0.4 M H_2SO_4-soluble cellular proteins (Fig. 5b) for the precence of radiolabeled histones. As shown in Fig. 5, S-phase total cellular histone proteins have a considerably higher specific activity than those isolated from G_1 cells, an observation consistent with S-phase-specific histone protein synthesis (Marashi *et al.*, 1982). The synthesis of histone proteins primarily during S phase is also suggested by the very low levels of both histone protein synthesis and cytoplasmic histone mRNAs at nonpermissive temperatures in temperature-sensitive cell cycle mutants (Delegeane and Lee, 1982; R. Hirschhorn, F. Marashi, R. Baserga, J. Stein, and G. Stein, submitted for publication).

B. Representation and Synthesis of Histone mRNA Sequences during the Cell Cycle

To gain further understanding of the regulation of histone gene expression during the cell cycle, we have used cloned genomic human histone sequences to examine the representation and synthesis of histone mRNAs in synchronized, continuously dividing HeLa S3 cells (Rickles *et al.*, 1982; Plumb *et al.*, 1983) and in the quiescent human diploid fibroblasts stimulated to proliferate (Green *et* al., 1983). This problem is a complex one that requires the analysis of heterologous cellular histone mRNAs. The complexity stems from the reiterated nature of the human histone genes, approximatley 40 copies of each histone-coding sequence per haploid genome (i.e., 160 genes coding for four distinct proteins and their variants), and their sequence and structural polymorphism (discussed previously). We therefore employed an experimental approach that permits the resolution of more than 15 different histone gene transcripts and used the technique to obtain results that are consistent with the cellular abundance of human histone mRNAs being coordinately regulated at both the transcriptional and posttranscriptional levels during the cell cycle. In agreement with earlier observations from this laboratory and with findings of others (Delegeane and Lee, 1982; Stein et al., 1977b; Detke et al., 1979; Parker and Fitschen, 1980; Chiu et al., 1979; G. Stein et al., J. L. Stein et al., 1975) in which histone mRNAs were analyzed by RNA excess hybridization with homologous histone cDNA, histone mRNA sequences are predominantly synthesized and present during S phase, when DNA replication and histone protein synthesis occur.

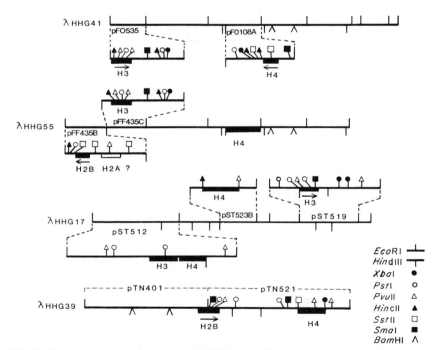

Fig. 6. Restriction endonuclease maps of λCh4A recombinants containing human histone gene clusters isolated and characterized as described (Sierra et al., 1982a) and fine restriction mapping of fragments subcloned into pBR322. Histone coding sequences were identified by hybrid selection–in vitro translation, hybridization to homologous and heterologous probes, and partial sequencing.

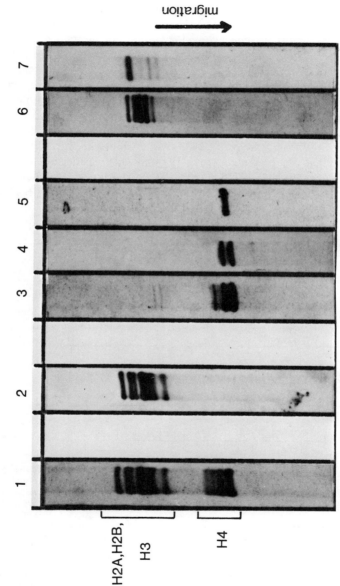

Fig. 7. The effect of temperature on hybrid selection of histone mRNA variants. One hundred fifty micrograms of *in vivo* [³H]uridine pulse-labeled total cellular RNA was hybridized with 50 μg of filter-immobilized cloned human histone sequences. Hybridized RNAs were eluted and resolved by denaturing 6% (w/v) polyacrylamide, 50% (w/v) urea gel electrophoresis, and visualized by fluorography. Lane 1: pFO108A (H4), pFF435B (H2A+H2B), and pFF435C (H3) hybridized at 43°C; lane 2: pFF435B (H2A+H2B) hybridized at 45°C; lanes 3, 4, and 5: pFO108A (H4) hybridized at 43, 45, and 48°C respectively; and lanes 6 and 7: pFF435C hybridized at 45 and 48°C respectively.

1. *Representation of Histone mRNAs during the Cell Cycle*

Given the multiplicity of human core histone mRNAs it was imperative to use several cloned genomic human histone sequences (see Fig. 6) as hybridization probes to ensure that we were detecting as many histone mRNA subspecies as possible in cellular RNAs from the synchronized cells. Hybridization results have suggested considerable variations in sequence homology for the different histone mRNA subspecies. For example, when *in vivo* [³H]uridine-labeled total cellular RNA was hybridized to filter-immobilized plasmid DNAs containing human histone gene sequences and the RNAs were eluted, resolved by denaturing polyacrylamide gel electrophoresis, and visualized by fluorography, several distinct mRNA size classes were selected by each probe (Fig. 7). As shown for the histone H3 and H4 probes, the complexity of the resolved mRNAs is dependent on the temperature of hybridization [43–48°C, in 50% (w/v) formamide]. At an intermediate temperature of hybridization (45°C) we detect over 15 mRNA size classes that code for core histone proteins.

To compensate for the variable sequence homology between different histone mRNAs, we adopted the following approach to analyze total histone mRNA levels during the cell cycle: HeLa cells were synchronized at the G_1–S-phase boundary by two cycles of 2 m*M* thymidine treatment, and total cellular RNA was isolated at various times as the cells progressed through S phase after release. Northern blot analysis of this RNA was then performed with four different cloned human histone gene probes and a heterologous chicken histone (H3 and H4) gene fragment (p2.6H) (Clark, 1982). The temperature of hybridization was varied (45–50°C) to alter the complexity of the histone mRNAs detected. Three Northern blot analyses are shown in Fig. 8a, and the data from three experiments are quantitatively summarized in Fig. 8b.

The relative abundance of the four core histone mRNAs varied only within limits of experimental error, indicating that their relative cellular levels are coordinately regulated. The *in vivo* rates of DNA synthesis were monitored in parallel, and as shown in Fig. 8b, the relative abundance of histone mRNAs is temporally coupled with the relative rate of DNA synthesis.

To eliminate the possibility that this temporal coupling was an artifact of either the synchronization technique or the cell line, two analogous experiments were performed. We found that the relative abundance of histone mRNAs in HeLa cells progressing from G_1 to S (synchronized by mitotic selective detachment) paralleled the observed relative rates of DNA synthesis (Fig. 9). A similar coupling of histone mRNA levels and the rate of DNA synthesis has been observed following the stimulation of quiescent WI38 human diploid fibroblasts to proliferate (Fig. 10) (Green *et al.*, 1984) and in yeast cells progressing into S phase after synchronization in G_1 with the yeast mating pheromone (Hereford *et al.*, 1981). These results indicate that the temporal coupling is not determined by

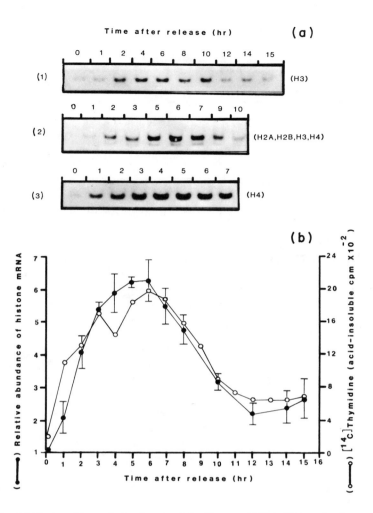

Fig. 8. (a) Northern blot analysis of total cellular histone mRNA in HeLa cells after synchronization by two cycles of 2 m*M* thymidine treatment. At various times after release from the second thymidine block, total cellular RNA was isolated, 50 μg of RNA from each preparation was resolved by 1.5% (w/v) agarose, 6% (w/v) formaldehyde gel electrophoresis, and the RNAs were transferred to nitrocellulose. Filters were hybridized at 48°C in 50% (w/v) formamide to one of three nick-translated probes: (1) pFF435C (H3); (2) pFF435(B+C) (H2A, H2B, and H3) and p2.6H (chicken H3+H4); and (3) pFO108A (H4). Hybrids were visualized by autoradiography. (b) The relative abundance of histone mRNAs (●——●, ± SEM) in cells released into S phase after synchronization by two cycles of thymidine treatment, and assayed by Northern blot analysis. Hybrids were quantitated for nine Northern blots by densitometry and/or liquid scintillation spectrometry, after hybridization with the five histone gene probes as described above and in the text. Values represent the mean relative abundances of histone H4, H3, or H2A+H2B mRNAs. The rate of DNA synthesis (○——○) was monitored by pulse labeling 10⁶ cells with 0.2 μCi of [¹⁴C]thymidine for 20 min and measuring the incorporation of radiolabel into acid-precipitable material.

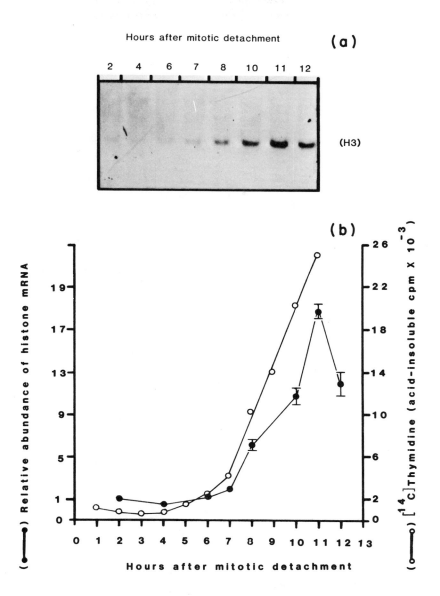

Fig. 9. (a) Northern blot analysis of total cellular histone mRNA in HeLa cells synchronized by selective mitotic detachment. At various times after mitotic detachment, total cellular RNA was isolated, 50 μg of each was resolved by 1.5% (w/v) agarose, 6% formaldehyde gel electrophoresis, and the RNAs transferred to nitrocellulose. The filter was hybridized to nick-translated pST519 (H3) probe and visualized by autoradiography. (b) Histone mRNA accumulation (●——● ± SEM) assayed for three Northern blots probed with pST519 (H3), pFO108A (H4), or pFF435B (H2A + H2B), quantitated as described in Fig. 8b. DNA synthesis rates (○——○) were monitored by measuring the incorporation of [¹⁴C]thymidine into acid-precipitable material in a 20-min pulse.

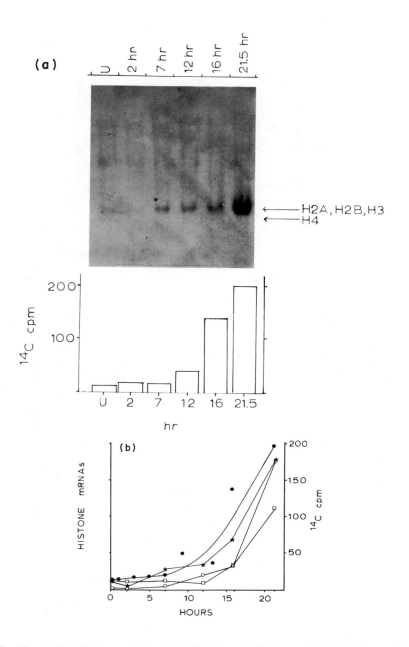

Fig. 10. (a) Hybridization of total cellular RNAs from WI38 human diploid fibroblasts with a cloned human histone DNA sequence (pFF435) encoding histones H2A, H2B, and H3. The signals shown were obtained using 50 μg of electrophoretically fractionated, nitrocellulose-immobilized total cellular RNAs from quiescent WI38 human diploid fibroblasts isolated at various times after serum stimulation (2, 7, 12, 16, and 21.5 hr) and from unstimulated cells (U). The lower bar graph

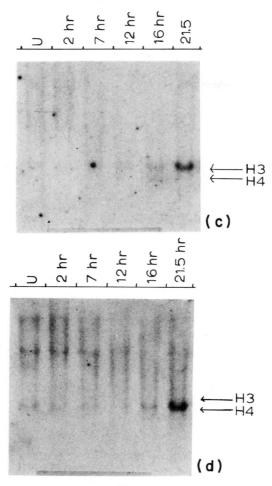

depicts the levels of DNA synthesis at various times after serum stimulation (2, 7, 12, 16, and 21.5 hr) and in unstimulated cells (U). The levels are expressed as ^{14}C cpm incorporated into DNA, determined by pulse labeling duplicate cultures with [^{14}C]thymidine for 30 min followed by assaying trichloroacetic acid (TCA)-precipitable radioactivity. (b) Histone mRNA levels and DNA synthesis at various times after serum stimulation of quiescent WI38 human diploid fibroblasts. The histone mRNA levels were determined by densitometric analysis of hybridization signals obtained when 50 μg of electrophoretically fractionated, nitrocellulose-immobilized WI38 total cellular RNAs was hybridized with cloned human histone DNA sequences: ★——★, H3 (pFO422); □——□, H4 (pFO108A); ●——●, H2A, H2B, H3 (pFF435). The DNA synthesis levels are expressed as ^{14}C cpm incorporated into DNA by pulse labeling with [^{14}C]thymidine for 30 min followed by assaying TCA-precipitable radioactivity, ○——○. (c) and (d) Hybridization of total cellular RNAs from WI38 human diploid fibroblasts to cloned human histone DNA sequences (pFO535 and pFO108A) encoding histones H3 and H4, respectively. The signals shown were obtained using 50 μg of electrophoretically fractionated, nitrocellulose-immobilized total cellular RNA from WI38 cells isolated at various times after stimulation (2, 7, 12, 16, 21.5 hr) and from unstimulated (U) cells (c) H3 and (d) H4.

125

the cell synchronization protocol, nor is it a characteristic of the HeLa genotype or transformed human cells grown in culture.

The maximal levels of HeLa histone mRNAs in S phase represent a six- to sevenfold increase compared with those found in cells blocked at the G_1–S boundary (Fig. 8). It was therefore of interest to determine whether the histone mRNA levels in blocked cells represent a "basal" pool which is not under cell cycle control or whether these histone mRNA levels reflect a "leaky" or incomplete inhibition of DNA synthesis in cells at the G_1–S-phase boundary in the presence of 2 mM thymidine. Wu and Bonner (1981) have defined cellular "basal" levels of histone protein synthesis as those that are insensitive to the DNA synthesis inhibitor hydroxyurea. We therefore used the same approach to determine the origins of histone mRNAs present in thymidine-blocked cells. Synchronized cells at 0, 2, 6, or 11 hr after release from the second thymidine treatment were exposed to 1 mM hydroxyurea for 30 min, total cellular RNA was isolated, and histone mRNA levels were determined by Northern blot analysis. We found that the relative abundance of histone mRNAs decreased to 30–60% of the levels found in cells blocked at the G_1–S boundary with thymidine, irrespective of the levels observed in the appropriate S-phase control cells (Plumb *et al.*, 1983). Thus, only 50% of the total histone mRNAs found in cells arrested at the G_1–S boundary by synchronization with double thymidine blocks represents a "basal" pool which, by these criteria, is not S-phase-specific (~5% of maximal levels).

2. Histone mRNA Synthesis during the HeLa Cell Cycle

To determine whether the temporal coupling between histone mRNA levels and rates of DNA synthesis occurs at the transcriptional level, we measured the incorporation of [^{3}H]uridine into histone mRNAs in 1-hr labeling intervals at various times during the HeLa cell cycle. Although shorter labeling times would have given a more accurate indication of transcription rates, 1-hr labeling intervals yielded RNA preparations of sufficiently high specific activity to permit detection of radiolabel incorporation into distinct histone mRNA size classes and therefore detection of the differential regulation of member(s) of the heterologous histone mRNA population.

Radiolabeled total cellular RNA was hybridized to filter-immobilized plasmid DNAs containing cloned human histone gene sequences, and the eluted RNAs were resolved by denaturing polyacrylamide gel electrophoresis. To maximize the complexity of mRNAs hybridizing to the immobilized probe, the temperature of hybridization was kept at 45°C [in 50% (w/v) formamide, see Fig. 7]. At this temperature we have observed some cross hybridization with ribosomal RNAs, probably due to the high G–C content of both ribosomal RNA and histone mRNAs. This cross hybridization can be eliminated by increasing the temperature of hybridization to above 48°C, but as shown in Fig. 7, this reduces the complexity of detected histone mRNAs. For quantitation therefore, either RNA was hybridized at a higher temperature and the radioactivity bound to a pBR322 control was subtracted, or radioactivity was monitored in specific areas of the polyacrylamide gel after electrophoretic resolution and fluorography of RNAs.

As shown in Fig. 11, there are no obvious qualitative differences in the histone mRNAs synthesized at different times during S phase, either for H4 (pFO108A), H2A + H2B (pFF435B), or H3 (pFF435C). Moreover, the specific activity of any one mRNA size class relative to another in the same sample does not change appreciably as cells traverse S phase. In total, five histone clones [pST512 (H3 + H4), pST519 (H3), pFO108A (H4), pFF435B (H2A + H2B), and pFF435C (H3), see Fig. 6] were used to hybridize total cellular RNA, and the same pattern was observed at two hybridization temperatures (45 and 48°C) and over a three-fold range of RNA concentrations (50–150 μg of total cellular RNA/50 μg of filter-immobilized plasmid DNA).

The results are quantitatively summarized in Fig. 11 along with the observed rates of DNA synthesis and the total histone mRNA levels detected by Northern blot analysis. The maximal incorporation of [^{3}H]uridine into histone mRNAs of all size classes occurs between the first and second hr of S phase. As histone mRNA accumulation and DNA synthesis rates have only reached 50–60% of maximal levels after 2 hr, we conlude that transcription is the predominant regulatory mechanism modulating the concentration of cellular-core histone mRNAs in the first half of S phase. We have observed a two- to threefold stimulation of total cellular transcription in the first 2–3 hr of S phase, but as this is significantly less than the apparent increase in [^{3}H]uridine incorporation into histone mRNAs (Fig. 11), we conclude that there is a preferential stimulation of histone gene transcription as cells enter S phase. Five to six hr into S phase, a transition point is reached at which both DNA synthesis rates and total histone mRNA levels are maximal but the relative abundance of pulse-labeled mRNAs is reduced (Fig. 11). Thereafter, total histone mRNA levels decreased with an apparent half-life of about 2.0 hr. As the apparent incorporation of [^{3}H]uridine into histone mRNAs remains low but relatively constant at this time, we conclude that both the transcription and the mRNA turnover rates are approximately constant in the second half of S phase, but the turnover rate exceeds the rate of synthesis. In the second half of S phase, therefore, the concentration of cellular histone mRNAs is largely regulated at the posttranscriptional level.

It is unclear whether the reduced concentration of pulse-labeled histone mRNAs in the second half of S phase (Fig. 11) represents an S-phase-specific "maintenance level" of transcription, or whether it reflects synchrony decay. We therefore examined [^{3}H]uridine incorporation into histone mRNA over 1-hr labeling intervals in cells progressing from G_1 to S phase after synchronization by selective mitotic detachment. HeLa cells were pulse labeled at various times after mitotic detachment, and total cellular RNA was hybrid-selected for histone

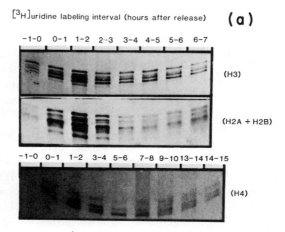

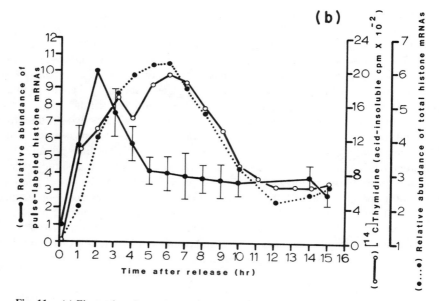

Fig. 11. (a) Electrophoretic resolution of histone mRNAs pulse labeled *in vivo* for 1 hr (for the times indicated) after synchronization of HeLa cells by double thymidine blocks. One hundred fifty micrograms of 1-hr [³H]uridine pulse-labeled total cellular RNA was hybrid selected at 45°C with 50 μg of filter-immobilized plasmid DNA. Eluted RNAs were electrophoretically resolved on denaturing 6% (w/v) polyacrylamide, 50% urea gels, and visualized by fluorography. Histone mRNAs selected by pFF435C (H3), pFF435B (H2A+H2B), and pFO108A (H4). (b) (●———● ± SEM) Quantitation for three hybrid selections using three probes [pFF435B (H2A+H2B), pFF435C (H3), and pFO108A (H4)]. Values are plotted at the end of the 1-hr pulse label and represent the mean relative abundances of pulse-labeled histone H4, H3, or H2A+H2B mRNAs. The DNA synthesis rates (○———○) and the relative abundance of total histone mRNAs (●– –●) are reproduced from Fig. 8b.

mRNAs, which were then electrophoretically resolved as shown in Fig. 12. We observed no obvious qualitative differences in the histone mRNA size classes as cells progressed from G_1 to S phase, nor was there a significant change in the specific activity of one mRNA size class relative to another in the same sample. As these histone mRNAs are qualitatively equivalent to those from cells synchronized by double thymidine treatments (Fig. 11), we conclude that all the detected histone mRNA size classes are coordinately regulated in the HeLa cell cycle.

The *in vivo* incorporation of [^{3}H]uridine during 1-hr labeling intervals into histone mRNAs in cells progressing from G_1 to S phase is quantitatively summarized in Fig. 12. ^{3}H-labeled histone mRNAs are detected in G_1 (0–7 hr after mitosis), but these have a significantly lower radiolabel content than those found in S phase (8–12 hr after mitosis) (Fig. 12). There is therefore a detectable but reduced apparent level of transcription in G_1 cells. It is not yet clear whether this represents a G_1 basal level of transcription, or whether it reflects contamination with a small proportion of S-phase cells after mitotic shake-off. As there is not a significant accumulation of histone mRNAs until 8 hr after mitosis (Fig. 12), we infer that the reduced apparent transcription in G_1 is insufficient to permit accumulation above the putative G_1-specific "basal" histone mRNAs which are translated in the presence of hydroxyurea as described by Wu and Bonner (1981).

There is a sharp increase in the relative abundance of pulse-labeled histone mRNAs 7–8 hr after mitosis, and this level does not change significantly over the next 4 hr (Fig. 12). As maximal histone mRNA levels are not reached until 11 hr after mitosis, we conclude that these data are consistent with maximal histone mRNA synthesis in early S phase as was observed in cells synchronized by double thymidine treatment (Fig. 11). The relatively constant abundance of pulse-labeled histone mRNAs in early and mid-S phase after selective mitotic detachment is attributed to the synchrony decay observed as cells in G_1 enter S phase (Prescott *et al.*, 1980). These cells therefore probably represent a semi-synchronous population that initiates DNA synthesis over a 3–4 hr period.

C. Regulation of Histone Gene Expression

We have shown that over 15 heterologous histone mRNAs are coordinately regulated during the HeLa cell cycle. A temporal coupling between cellular histone mRNA levels and DNA synthesis rates has now been demonstrated for HeLa S3 cells (Fig. 13), WI38 human diploid fibroblasts (Fig. 10), and yeast cells (Hereford *et al.*, 1981). As the relative cellular abundance of histone mRNAs closely parallels both the *in vivo* histone protein synthesis reported for cells under equivalent conditions (Marashi *et al.*, 1982; Spalding *et al.*, 1966; Robbins and Borun, 1967) and the *in vitro* translatability of polysomal histone

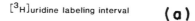

[³H]uridine labeling interval **(a)**

(hours after mitotic detachment)

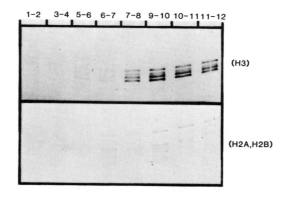

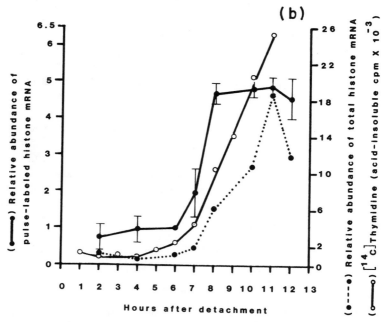

Fig. 12. (a) Electrophoretic resolution of histone mRNAs pulse labeled *in vivo* for 1 hr (at the times indicated) after synchronization of HeLa cells by selective mitotic detachment. One hundred fifty micrograms of [³H]uridine pulse-labeled total cellular RNA was hybridized at 45°C with 50 μg of filter-immobilized plasmid DNA. Bound RNAs were eluted and electrophoretically resolved on denaturing 6% (w/v) polyacrylamide, 50% urea gels and visualized by fluorography. RNA selected by pFF435C (H3) and pFF435B (H2A+H2B). (b) (●———●, ± SEM) Quantitation for three hybrid selections using three probes [pFF435B (H2A+H2B), pFF435C (H3) and pFO108A (H4)]. Values are plotted at the end of the 1-hr pulse label and represent the mean relative abundances of pulse-labeled histone H4, H3, or H2A+H2B mRNAs. The DNA synthesis rates (O———O) and the relative abundance of total histone mRNAs (●— —●) are reproduced from Fig. 9b.

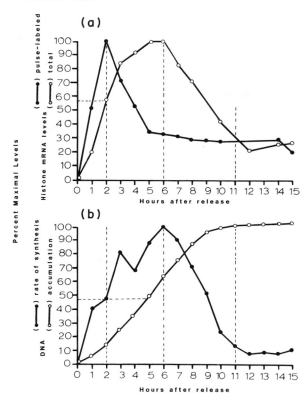

Fig. 13. Summaries for the data obtained for cells synchronized by double thymidine blocks. Data are expressed as the percentage of maximal levels detected above those observed in cells just before release from the second thymidine block. (a) (O———O) The relative abundance of total histone mRNA. (●———●) The relative incorporation of [³H]uridine into histone mRNA in 1-hr labeling intervals, plotted at the end of the pulse. (b) (●———●) Relative rates of DNA synthesis monitored by pulse labeling with [¹⁴C]thymidine. (O———O) Accumulation of DNA calculated from the rates of DNA synthesis.

mRNAs (Borun *et al.*, 1975; Breindl and Gallwitz, 1974a; Jacobs-Lorena *et al.*, 1973), it is not unreasonable to propose that histone gene expression (histone protein synthesis) is regulated primarily by modulating the cellular concentration of histone mRNAs available for translation. Histone mRNA synthesis apparently precedes maximal accumulation of mRNAs in both synchronized HeLa (Fig. 13) and yeast (Hereford *et al.*, 1982) cells, indicating that the cellular abundance of histone mRNAs in early S phase is predominantly under transcriptional control. For the moment, we cannot determine whether the onset of histone gene transcription in HeLa cells precedes the initiation of DNA synthesis, as is apparently the case in yeast (Hereford *et al.*, 1982), although our data (Fig. 13) would suggest that the two occur within 30 min of each other (~6% of the HeLa S phase).

In the second half of the HeLa S phase, cellular histone mRNA levels are regulated predominantly at the posttranscriptional level as suggested by the decay of cellular histone mRNAs with apparent half-lives of 2.0 hr, and by the reduced and relatively constant incorporation of [^{3}H]uridine into histone mRNAs. These data are compatible with constant rates of both synthesis and turnover of histone mRNAs in the second half of S phase, but with the turnover rate exceeding the rate of synthesis. As studies in HeLa and yeast cells (Hereford *et al.*, 1982; Plumb *et al.*, 1983) suggest that histone gene transcription rates vary during S phase, it is now of interest to determine whether the turnover of histone mRNAs varies during the cell cycle in order to determine whether only one level of regulation (regulated transcription) or two (regulated synthesis and turnover) is responsible for the control of histone gene expression.

As a working model, we propose that the turnover of histone mRNAs remains constant throughout the cell cycle of HeLa cells. This model is consistent with Perry and Kelley's observation (1973) that the mean lifetime of polysomal histone mRNAs in exponentially growing mouse L cells is equal to the length of S phase. As will be discussed later, however, histone mRNA destabilization can be induced, but this may reflect a secondary regulatory mechanism which is only applied under extreme biological circumstances. Our working model therefore relies predominantly on the regulation of histone mRNA synthesis. In the first half of S phase an elevated rate of synthesis permits histone mRNA accumulation, and in the second half of S phase a reduced transcription allows the decay of histone mRNA in the presence of a constant turnover rate. Furthermore, the kinetics of synthesis and accumulation of histone mRNAs in early S phase are suggestive of a feedback regulatory mechanism. One possibility currently being investigated is that histone gene replication is temporally and functionally related to transcription of the histone genes.

Other potential regulatory aspects of human histone gene expression are suggested by observations reported for other biological systems. For example, the differential expression of "early" and "late" histone gene clusters in developing sea urchin embryos (Kunkel *et al.*, 1975; Hieter *et al.*, 1979) has stimulated the search for an analogous situation in higher eukaryotes. It is becoming evident that differential translation of variant histone proteins occurs during the cell cycle (Wu and Bonner, 1981; Wu *et al.*, 1982), and differential expression of variant histones during development of higher eukaryotes has also been observed (L. Cohen, personal communication). In lower eukaryotes, histone gene expression immediately prior to and shortly after fertilization is regulated almost exclusively at posttranscriptional levels. During the extremely rapid rate of cell division just after fertilization, it appears that developing *Xenopus* oocytes rely predominantly on a maternal store of histone proteins (Woodland and Adamson, 1977), and sea urchin embryos rely predominantly on a maternal store of histone mRNAs (Skoultchi and Gross, 1973; Woods and Fitschen, 1978; Maxson et al., 1983; Maxson and Wilt, 1982) to provide the histone proteins required for packaging the rapidly replicating DNA. Although the rate of cell division of embryos in higher eukaryotes is significantly slower, it is of some interest to determine whether such posttranscriptional regulation occurs during early development and whether there are equivalents of the "early" and "late" histone genes of lower eukaryotes.

IV. RELATIONSHIP BETWEEN HISTONE GENE EXPRESSION AND DNA REPLICATION

As described previously in this chapter, there is a plethora of information which supports temporal coupling of histone gene expression and DNA replication. Although the molecular basis for the coupling remains to be elucidated, it is becoming clear that the inhibition of cellular DNA synthesis by a variety of unrelated exogenous perturbations results in a parallel decrease in cellular histone mRNA levels and an equivalent decrease in histone protein synthesis. Early studies, for example, established that the inhibition of DNA replication by the metabolic inhibitors hydroxyurea (Wu and Bonner, 1981; Wu et al., 1982; Jacobs-Lorena et al., 1973; Stahl and Gallwitz, 1977; Breindl and Gallwitz, 1974a,b; Gallwitz and Mueller, 1969; Shephard et al., 1982; Butler and Mueller, 1973; Stein et al., 1977a), cytosine–arabinoside (Marashi et al., 1982; Spalding et al., 1966; Jacobs-Lorena et al., 1973; Perry and Kelley, 1973; Borun et al., 1967, 1975; Stein et al., 1977b; Jansing et al., 1976; Kedes and Gross, 1969; Craig et al., 1971), and actinomycin D (Gallwitz and Mueller, 1969; Breindl and Gallwitz 1974b; Butler and Mueller, 1973; Borun et al., 1967) resulted in the rapid reduction of histone protein synthesis and the loss of polysomal histone mRNAs. More recently, we have monitored total cellular levels of histone mRNAs after inhibition of DNA replication by: (1) metabolic inhibitors, (2) infection with adenovirus (Flint, S., Plumb, M., Stein, G., and Stein, J., manuscript submitted), and (3) shifting temperature-sensitive cell cycle mutants to the nonpermissive temperature (Hirschhorn, R., Marashi, F., Baserga, R., Stein, J., and Stein, G., manuscript submitted). In all cases, the perturbation of distinct cellular processes resulted in both the inhibition of DNA replication and an equivalent reduction in total cellular histone mRNA levels.

When DNA replication is completely (>95%) inhibited by compounds that disrupt the metabolism of deoxynucleotide precursors (100 μM cytosine–arabinoside or 1 mM hydroxyurea) or that inhibit DNA polymerase α [2 μg/ml aphidicolin (Huberman, 1981) (see Fig. 14)], the cellular representation of histone mRNAs is reduced to an inhibitor-resistant "basal level" corresponding to approximately 10% of the levels observed in exponentially growing cells (see

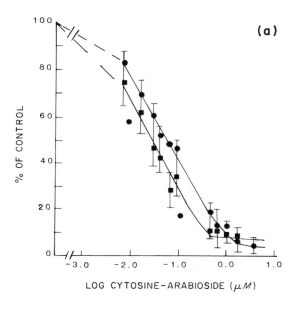

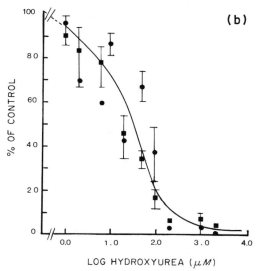

Fig. 14. The effects of (a) cytosine–arabinoside, (b) hydroxyurea, and (c) aphidicolin on histone mRNA levels (■———■) and DNA synthesis (●———●) in exponentially growing HeLa cells. DNA synthesis was monitored by incorporation of [³H]thymidine (2 μCi/ml and 4 × 10⁵ cells/ml) into 10% TCA-precipitable material. Cells were pretreated for 30 min with each of the various inhibitors and then pulse labeled for 30 min. Total cellular RNA was prepared and histone mRNA steady state levels determined as described in Fig. 8. Each value represents the mean (±SEM) of five Northern blots using various histone gene probes: pFO108A (H4), pFF435B (H2A+H2B), pFF435C (H3), pST519D (H3), and pFO535 (H3).

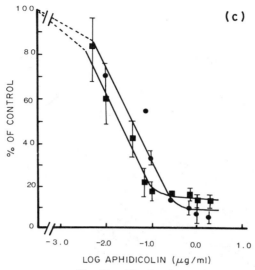

Fig. 14. (*Continued*)

Fig. 14) and to approximately 6% of the peak levels observed in synchronized S-phase HeLa cells. These estimates for "basal levels" of histone mRNAs are quantitatively consistent with the basal levels of histone protein synthesis observed in G_1- and S-phase cells in the presence of 1 mM hydroxyurea (Wu and Bonner, 1981; Wu *et al.*, 1982). Both the rates of inhibition of DNA synthesis and the rates of histone mRNA decay occurred with apparent half-lives of less than 10 min. These kinetics differ significantly from the reported mean lifetime of 9 hr for polysomal histone mRNAs in exponentially growing L cells (Perry and Kelley, 1973), and differ from the estimated apparent histone mRNA half-lives of 2.0 hr described in Section B,2 for the latter part of S phase in synchronized HeLa cells (Plumb *et al.*, 1983) (Fig. 13.) The inhibitors therefore apparently cause the rapid destabilization of histone mRNAs. This posttranscriptional coupling is supported by the observation that after transcriptional inhibition with 5 µg/ml actinomycin D, cellular histone mRNA levels decay in parallel with inhibition of DNA synthesis [95% inhibition within 2 hr (unpublished results, and Borun *et al.*, 1967)] despite complete inhibition of transcription within ~6 min.

We are presently examining the coupling of DNA synthesis and histone mRNA levels under conditions that result in the incomplete inhibition of DNA synthesis, using reduced concentrations of the metabolic inhibitors hydroxyurea (10 µM–1 mM), cytosine–arabinoside (7 nM–100 µM), and aphidicolin (10 ng/liter–2 mg/liter). As shown in Fig. 14, partial inhibition of DNA synthesis results in the stoichiometric reduction of total cellular histone mRNAs irrespec-

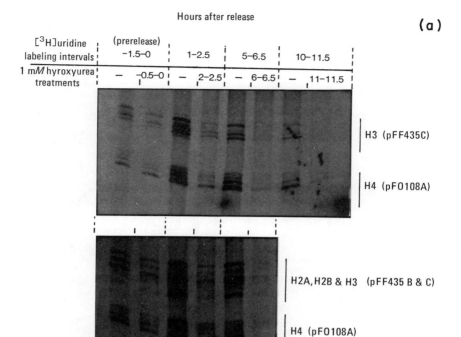

Hours after release

(a)

[³H]uridine labeling intervals	(prerelease) −1.5−0	1−2.5	5−6.5	10−11.5
1 mM hyroxyurea treatments	−	−0.5−0	−	2−2.5	−	6−6.5	−	11−11.5

H3 (pFF435C)

H4 (pFO108A)

H2A, H2B & H3 (pFF435 B & C)

H4 (pFO108A)

Fig. 15. (a) The effect of hydroxyurea on S-phase HeLa histone mRNA species prelabeled with [³H]uridine at various times after release from double thymidine block synchronization. At −1.5, 1, 4.5, and 9.5 hr after release from thymidine, cells were labeled with [³H]uridine (0.1 mCi/ml and 2.5 × 10⁶ cells/ml) for 1 hr and then exposed to 1 mM hydroxyurea for 30 min in the presence of the radiolabel. Control cells (not treated with hydroxyurea) were labeled for 90 min over the same time intervals. Total cellular RNA was prepared and 300 μg hybridized at 43°C to filter-immobilized plasmid DNA as described in Fig. 7. Bound RNAs were eluted, resolved electrophoretically, and visualized by fluorography. (b) The incorporation of [³H]uridine into S-phase HeLa histone mRNA species after 30 min pretreatment with hydroxyurea at various times after release from double thymidine block synchronization. At −1.5, 1, 4.5 and 9.5 hr after release, cells were treated for 30 min with 1 mM hydroxyurea, and then labeled for 1 hr with [³H]uridine (0.1 mCi/ml and 2.5 × 10⁶ cells/ml) in the presence of hydroxyurea. Control cells were labeled for the last hr of the 1.5 hr time interval. Total cellular RNA was prepared and 300 μg hybridized at 43°C to filter-immobilized plasmid DNA as described in Fig. 7. Bound RNAs were eluted, resolved electrophoretically, and visualized by fluorography.

tive of the inhibitor or its concentration. Time course experiments have indicated that both the rate of DNA synthesis and the level of histone mRNAs equilibrate within 30 min of the metabolic perturbation, implying that the primary coupling mechanism occurs at the posttranscriptional level.

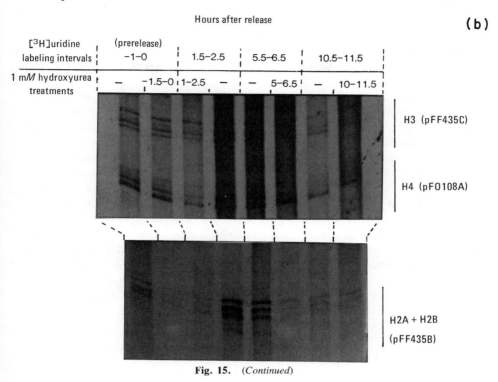

Fig. 15. (*Continued*)

As our studies on the effects of metabolic inhibitors discussed above only yielded a generalized indication of the fate of the core histone mRNAs after complete or partial inhibition of DNA synthesis, it was of interest to determine whether coordinate regulation of the expression of the polymorphic histone gene family was operative. This question was also relevant to the observations that specific histone variant proteins are differentially synthesized in G_1, G_{10}, and S-phase cells exposed to 1 mM hydroxyurea (Wu and Bonner, 1981, Wu *et al.*, 1982). We have initially approached this problem by posing three questions: (1) Are newly synthesized S-phase histone mRNA subspecies differentially turned over in the presence of 1 mM hydroxyurea? (2) Are specific subspecies differentially transcribed in S-phase cells exposed to 1 mM hydroxyurea? And (3) are specific histone mRNAs differentially sensitive to destabilization by 1 mM hydroxyurea at different times in the cell cycle? Using the hybrid selection of [³H]uridine pulse-labeled histone mRNAs to visualize the incorporation of radiolabel into over 15 distinct histone mRNA size classes, we have found the following: (1) that levels of these multiple histone mRNA species are apparently coordinately controlled at the posttranscriptional level when HeLa cells with prelabeled histone mRNAs are exposed to hydroxyurea either before, during, or after S phase (see Fig. 15a); and (2) that the incorporation of [^{3}H]uridine into the multiple histone mRNA species is reduced, but qualitatively and quantitatively similar, after cells at various times in the cell cycle have been pretreated with 1 mM hydroxyurea for 30 min (see Fig. 15b).

Several cautionary provisos must be maintained when interpreting these results. As alluded to earlier (Fig. 7), the detection of [^{3}H]uridine pulse-labeled histone mRNAs is largely governed by the choice of probe and the hybridization conditions used. It is therefore possible that, with the probes we have used, we are not detecting all histone mRNA species and that those not detected may be regulated by a distinct mechanism. Similarly, we have relied to a greater or lesser degree on the assumption that the turnover of all histone mRNA species is sufficiently high to detect incorporation of [^{3}H]uridine in 1-hr pulse periods, but this assumption may not be valid. Finally, it must be emphasized that the interpretation of data obtained from perturbed cells may in fact reflect a secondary (fail-safe) regulatory mechanism which is not normally operative during the G_1–S–G_2 transitions. It is also possible that histone gene expression is functionally coupled to DNA replication by means of the histone gene products themselves, as it is not unreasonable to predict that the constitutive synthesis of histone proteins might severely affect cell viability.

V. CONCLUSION

In this chapter we have restricted our considerations to the regulation of histone genes, a specific set of repetitive genetic sequences that are differentially expressed during the cell cycle. Of broader biological relevance is the relationship between expression of histone genes and other genetic sequences required for execution of the proliferative process. While it is well documented that cell proliferation involves a complex and interdependent series of biochemical events, the molecular basis for the control of these events or for their interrelationships remains to be resolved. An intriguing possibility is that genetic sequences whose expression is related to specific cell cycle processes are coordinately controlled. Within this context it is tempting to postulate that expression of genes functionally related to DNA replication shares common structural and/or regulatory elements. Do histone genes have regulatory sequences similar to those, for example, of genes encoding enzymes required for DNA replication or deoxynucleotide metabolism? Do histone genes and other S phase-specific sequences respond to similar cellular signals or regulatory molecules? These questions are being approached experimentally and will establish the credibility of such speculation.

ACKNOWLEDGMENT

Studies described in this chapter were supported by grants PCM80-18075, PCM81-18951, and PCM83-18177 from the National Science Foundation, GM32010 from the National Institutes of Health, and 1-813 from the March of Dimes Birth Defects Foundation.

REFERENCES

Benoist, C., O'Hare, K., Breathnach, R., and Chambon, P. (1980). The ovalbumin gene–sequence of putative control regions. *Nucleic Acids Res.* **8,** 127–142.

Borun, T. W., Scharff, M., and Robbins, F. (1967). Rapidly labeled, polyribosome-associated RNA having the properties of histone messenger. *Proc. Natl. Acad. Sci. U.S.A.* **58,** 1977–1983.

Borun, T. W., Gabrielli, F., Ajiro, K., Zweidler, A., and Baglioni, C. (1975). Further evidence of transcriptional and translational control of histone messenger RNA during the HeLa S3 cycle. *Cell* **4,** 59–67.

Breindl, M., and Gallwitz, D. (1974a) Effects of cordycepin, hydroxyurea and cycloheximide on histone mRNA synthesis in synchronized HeLa cells. *Mol. Biol. Rep.* **1,** 263–268.

Breindl, M., and Gallwitz, D. (1974b). On the translational control of histone synthesis. *Eur. J. Biochem.* **45,** 91–97.

Busslinger, M., Portmann, R., Irminger, J. C., and Birnstiel, M. L. (1980). Ubiquitous and gene-specific regulatory 5' sequences in a sea urchin histone DNA clone coding for histone protein variants. *Nucleic Acids Res.* **8,** 957–977.

Butler, W. B., and Mueller, G. C. (1973). Control of histone synthesis in HeLa cells. *Biochim. Biophys. Acta* **294,** 481.

Calabretta, B., Robberson, D. L., Maizel, A. L., and Saunders, G. F. (1981). mRNA in human cells contains sequences complementary to the Alu family of repeated DNA. *Proc. Natl. Acad. Sci. U.S.A.* **78,** 6003–6007.

Carozzi, N., Marashi, F., Stein, G., and Stein, J. (1984). Isolation and characterization of a human genomic sequence containing H1 and core (H2A, H2B, H3 and H4) histones. (Submitted for publication.)

Carroll, A. G., and Ozaki, H. (1979). Changes in the histones of the sea urchin *Strongylocentrotus purpuratus* at fertilization. *Exp. Cell Res.* **119,** 307–315.

Childs, G., Nocente-McGrath, C., Lieber, T., Holt, C., and Knowles, J. A. (1982). Sea urchin (*L. pictus*) late stage histone H3 and H4 genes: Characterization and mapping of a clustered but nontandemly linked multigene family. *Cell* **31,** 383–393.

Chiu, I. M., Cooper, D., and Marzluff, W. F. (1979). Unscheduled synthesis of histone H1 in isoleucine starved cells. *Abstr., 2nd Annu. Meet. Am. Cancer Soc.* No. 38.

Choe, J., Kolodrobetz, D., and Grunstein, M. (1982). The two yeast histone H2A genes encode similar protein subtypes. *Proc. Natl. Acad. Sci. U.S.A.* **79,** 1484–1487.

Clark, S. J. (1982). Chicken and human histone genes. Ph.D. Thesis, University of Adelaide.

Craig, N., Kelley, D. E., and Perry, R. P. (1971). Lifetime of the messenger RNAs which code for ribosomal proteins in L-cells *Biochim. Biophys. Acta* **246,** 493–498.

Crawford, R. J., Krieg, P., Harvey, R. P., Hewish, D. A., and Wells, J. R. E. (1979). Histone genes are clustered with a 15-kilobase repeat in the chicken genome. *Nature (London)* **279,** 132–136.

Delegeane, A. M., and Lee, A. L. (1982). Coupling of histone and DNA synthesis in the somatic cell cycle. *Science* **215,** 79–81.

Detke, S., Lichtler, A., Phillips, I., Stein, J. L., and Stein, G. S. (1979). Reassessment of histone gene expression during the cell cycle in human cells by using homologous H4 histone cDNA. *Proc. Natl. Acad. Sci. U.S.A.* **76,** 4995–4999.

Duncan, C., Biro, P. A. Choudary, P. V., Elder, J. T., Wang, R. R. C., Forget, B. G., deRiel, J. K., and Weissman, S. M. (1979). RNA polymerase III transcriptional units are interspersed among human non-α-globin genes. *Proc. Natl. Acad. Sci. U.S.A.* **76,** 5095–5099.

Efstratiadis, A., Posakony, J. W., Maniatis, T., Lawn, R. M., O'Connell, C., Spritz, R. A., deRiel, J. K., Forget, B. G., Weissman, S. M., Slightom, J. L., Blechl, A. E., Smithies, O., Baralle, F. E., Shoulders, C. C., and Proudfoot, N. J. (1980). The structure and evolution of the human β-globin gene family. *Cell* **21,** 653–668.

Engel, J. D., and Dodgson, J. B. (1981). Histone genes are clustered but not tandemly repeated in the chicken genome. *Proc. Natl. Acad. Sci. U.S.A.* **78,** 2856–2860.

Engel, J. A., Sugarman, B. J., and Dodgson, J. B. (1982). A chicken histone H3 gene contains intervening sequences. *Nature (London)* **297,** 434–436.

Gallwitz, D., and Mueller, G. C. (1969). Histone synthesis *in vitro* on HeLa cell microsomes. *J. Biol. Chem.* **244,** 5947–5952.

Goldberg, M. (1979). Ph.D. Thesis, Stanford University, Stanford, California (1979).

Green, L., Stein, G., and Stein, J. (1984). Histone Gene Expression in human diploid fibroblasts: Analysis of histone mRNA levels using cloned human histone genes *Mol. Cell. Biochem.* (in press).

Groppi, V. E., and Coffino, P. (1980). G1 and S phase mammalian cells synthesize histones at equivalent rates. *Cell* **21,** 195–204.

Grunstein, M., Diamond, K. E., Knoppel, E., and Grunstein, J. E. (1981). Comparison of the early histone H4 gene sequence of *Strongylocentrotus purpuratus* with maternal, early, and late histone H4 mRNA sequences. *Biochemistry* **20,** 1216–1223.

Harvey, R. P., and Wells, J. R. E. (1979). Isolation of a genomal clone containing chicken histone genes. *Nucleic Acids Res.* **7,** 1787–1798.

Harvey, R. P., Krieg, P. A. Robins, A. J., Coles, L. S., and Wells, J. R. E. (1981). Non tandem arrangement and divergent transcription of chicken histone genes. *Nature (London)* **294,** 49–52.

Harvey, R. P., Robins, A. J., and Wells, J. R. E. (1982). Independently evolving chicken H2B genes: Identification of a ubiquitous H2B-specific 5′ element. *Nucleic Acids Res.* **10,** 7851–7863.

Heintz, N., Zernik, M., and Roeder, R. G. (1981). The structure of the human histone genes: Clustered but not tandemly repeated. *Cell* **24,** 661–668.

Hentschel, C. C., and Birnstiel, M. L. (1981). The organization and expression of histone gene families. *Cell* **25,** 301–313.

Hereford, L., Bromley, S., and Osley, M. A. (1982). Periodic transcription of yeast histone genes. *Cell* **30,** 305–310.

Hereford, L. M., Fahrner, K., Woolford, J., and Rosbash, M. (1979). Isolation of yeast histone genes H2A and H2B. *Cell* **18,** 1261–1271.

Hereford, L. M., Osley, M. A., Ludwig, J. R., and McLaughlin, C. S. (1981). Cell cycle regulation of yeast histone mRNA. *Cell* **24,** 367–375.

Hieter, P. A., Hendricks, M. B., Hemminki, K., and Weinberg, E. S. (1979). Histone gene switch in the sea urchin embryo: Identification of late embryonic histone mRNAs and the control of their synthesis. *Biochemistry* **18,** 2707–2716.

Huberman, J. (1981). New views of the biochemistry of eucaryotic DNA replication revealed by aphidicolin an unusual inhibitor of DNA polymerase α. *Cell* **23,** 647–648.

Jacobs-Lorena, M., Gabrielli, F., Borun, T. W., and Baglioni, C. (1973). Studies on the translational control of histone synthesis. *Biochim. Biophys. Acta* **324,** 275–281.

Jansing, R. L., Stein, J. L., and Stein, G. S. (1976). Activation of histone gene transcription by nonhistone chromosomal proteins in WI-38 human diploid fibroblasts. *Proc. Natl. Acad. Sci. U.S.A.* **74**, 173–177.

Jelinek, W. R., and Schmid, C. W. (1982). Repetitive sequences in eukaryotic DNA and their expression. *Annu. Rev. Biochem.* **51**, 813–844.

Kedes, L. H., and Gross, P. R. (1969). Identification in cleaving embryos of three RNA species serving as templates for the synthesis of nuclear proteins. *Nature (London)* **223**, 1335–1339.

Krieg, P. A., Robins, A. J., Gait, M. J., Titmas, R. C., and Wells, J. R. E. (1982). Chicken histone H5: Selection of a cDNA recombinant using an extended synthetic primer. *Nucleic Acids Res.* **10**, 1495–1502.

Kunkel, N. S., and Weinberg, E. S. (1978). Histone gene transcripts in the cleavage and mesenchyme blastula embryo of the sea urchin, *S. purpuratus. Cell* **14**, 313–319.

Lawn, R. M., Fritsch, E. F., Parker, R. C., Blake, G., and Maniatis, T. (1978). The isolation and characterization of a linked δ- and β-globin genes from a cloned library of human DNA. *Cell* **15**, 1157–1174.

Lichtler, A. C., Stein, G. S., and Stein, J. L. (1977). Isolation and characterization of two mRNAs from HeLa S3 cells coding for histone H4. *Biochem. Biophys. Res. Commun.* **77**, 845–853.

Lichtler, A. C., Detke, S., Phillips, I. R., Stein, G. S., and Stein, J. L. (1980). Multiple forms of H4 histone mRNA in human cells. *Proc. Natl. Acad. Sci. U.S.A.* **77**, 1942–1946.

Lichtler, A. C., Sierra, F., Clark, S., Wells, J. R. E., Stein, J. L., and Stein, G. S. (1982). Multiple H4 histone mRNAs of HeLa cells are encoded in different genes. *Nature (London)* **298**, 195–198.

Marashi, F., Baumbach, L., Rickles, R., Sierra, F., Stein, J. L., and Stein, G. S. (1982). Histone proteins in HeLa S3 cells are synthesized in a cell cycle stage specific manner. *Science* **215**, 683–685.

Marashi, F., Stein, J. L., and Stein, G. S. (1984). Identification of a human histone pseudogene. *Proc. Natl. Acad. Sci. U.S.A.* (in press).

Maxson, R., and Wilt, F. (1982). Accumulation of the early histone mRNAs during development of *S. purpuratus. Dev. Biol.* **94**, 435–440.

Maxson, R., Mohuer, T., Gameazans, G., Childs, G., and Kedes, L. (1983). Distinct organizations and patterns of expression of early and late histone gene sets in the sea urchin. *Nature (London)* **301**, 120–126.

Moll, R., and Wintersberger, E. (1976). Synthesis of yeast histones in the cell cycle. *Proc. Natl. Acad. Sci. U.S.A.* **73**, 1863–1867.

Newrock, K. M., Cohen, L. H., Hendricks, M. B., Donnelly, R. J., and Weinberg, E. S. (1978). Stage-specific mRNAs coding for subtypes of H2A and H2B histones in the sea urchin embryo. *Cell* **14**, 327–336.

Old, R. W., Woodland, H. R., Ballantine, J. E. M., Aldridge, T. C., Newton, C. A., Bains, W. A., and Turner, P. C. (1982). Organization and expression of cloned histone gene clusters from *Xenopus laevis* and *X. borealis. Nucleic Acids Res.* **10**, 7561–7580.

Park, W. D., Thrall, C. L., Stein, J. L., and Stein, G. S. (1976). Activation of histone gene transcription from chromatin of G1 HeLa cells by S phase nonhistone chromosomal proteins. *FEBS Lett.* **62**, 226–229.

Parker, I., and Fitschen, W. (1980). Histone mRNA metabolism during the mouse fibroblast cell cycle. *Cell Differ.* **9**, 23–30.

Perry, R. P., and Kelley, D. E. (1973). Messenger RNA turnover in mouse L cells. *J. Mol. Biol.* **79**, 681–696.

Plumb, M., Stein, J., and Stein, G. (1983). Coordinate regulation of multiple histone mRNAs during the cell cycle in HeLa cells. *Nucleic Acids Res.* **11**, 2391–2410.

Prescott, D. M., Liskay, R. M., and Stancel, G. M. (1980). The cell life cycle and the G1 period. *NATO Adv. Study Inst. Ser., Ser. A* **38**, 305–314.

Rickles, R., Marashi, F., Sierra, F., Clark, S., Wells, J., Stein, J., and Stein, G. (1982). Analysis of histone gene expression during the cell cycle in HeLa cells by using cloned human histone genes. *Proc. Natl. Acad. Sci. U.S.A.* **79**, 749–753.

Robbins, E., and Borun, T. W. (1967). The cytoplasmic synthesis of histones in HeLa cells and its temporal relationship to DNA replication. *Proc. Natl. Acad. Sci. U.S.A.* **57**, 409–416.

Ruberti, I., Fragapane, P., Pierandrei-Amaldi, P., Beccari, E., Amaldi, F., and Bozzoni, I. (1982) Characterization of histone genes isolated from *Xenopus laevis* and *Xenopus tropicalis* genomic libraries. *Nucleic Acids Res.* **10**, 7543–7559.

Ruiz-Vazquez, R., and Ruiz-Carillo, A. (1982). Construction of chimeric plasmids containing histone H5 cDNA from hen erythrocyte. DNA sequence of a fragment derived from the 5′ region of H5 mRNA. *Nucleic Acids Res.* **10**, 2093–2108.

Schaffner, W., Kunz, G., Daetwyler, H., Telford, J., Smith, H. O., and Birnstiel, M. L. (1978). Gene and spacers of cloned sea urchin histone DNA analyzed by sequencing. *Cell* **14**, 655–671.

Seiler-Tuyns, A., and Birnstiel, M. L. (1981). Structure and expression in L-cells of a cloned H4 histone gene of the mouse. *J. Mol. Biol.* **151**, 607–625.

Shephard, E. A., Phillips, I. R., Davis, J., Stein, J. L., and Stein, G. S. (1982). Evidence for the resumption of DNA replication prior to histone synthesis in HeLa cells after release from treatment with hydroxyurea. *FEBS Lett.* **140**, 189–192.

Sierra, F., Lichtler, A., Marashi, F., Rickles, R., Van Dyke, T., Clark, S., Wells, J. R. E., Stein, G. S., and Stein, J. L. (1982a). Organization of human histone genes. *Proc. Natl. Acad. Sci. U.S.A.* **79**, 1795–1799.

Sierra, F., Leza, A., Marashi, F., Plumb, M., Rickles, R., Van Dyke, T., Clark, S., Wells, J. R. E., Stein, G. S., and Stein, J. L. (1982b). Human histone genes are interspersed with members of the Alu family and with other transcribed sequences. *Biochem. Biophys. Res. Commun.* **104**, 785–792.

Sittman, D. B., Chiu, I. M., Pan, C.-J., Cohn, R. H., Kedes, L. H., and Marzluff, W. F. (1981). Isolation of two clusters of mouse histone genes. *Proc. Natl. Acad. Sci. U.S.A.* **78**, 4078–4082.

Skoultchi, A., and Gross, P. R. (1973). Maternal histone messenger RNA: Detection by molecular hybridization. *Proc. Natl. Acad. Sci. U.S.A.* **70**, 2840–2844.

Spalding, J., Kajiwara, K., and Mueller, G. C. (1966). The metabolism of basic proteins in HeLa cell nuclei. *Proc. Natl. Acad. Sci. U.S.A.* **56**, 1535–1542.

Stahl, H., and Gallwitz, D. (1977). Fate of histone messenger RNAs in synchronized HeLa cells in the absence of initiation of protein synthesis. *Eur. J. Biochem.* **72**, 385–392.

Stein, G. S., and Borun, T. W. (1972). The synthesis of acidic chromosomal proteins during the cell cycle of HeLa S3 cells. *J. Cell Biol.* **52**, 292–307.

Stein, G. S., Park, W. D., Thrall, C. L., Mans, R. J., and Stein, J. L. (1975). Regulation of histone gene transcription during the cell cycle by nonhistone chromosomal proteins. *Nature (London)* **257**, 764–767.

Stein, G. S., Stein, J. L., Shephard, E. A., Park, W. D., and Phillips, I. R. (1977a). Evidence that the coupling of histone gene expression and DNA synthesis in HeLa S3 cells is not mediated at the transcriptional level. *Biochem. Biophys. Res. Commun.* **77**, 245–252.

Stein, G. S., Stein, J. L., Park, W. D., Detke, S., Lichtler, A. C., Shepard, E. A., Jansing, R. L., and Phillips, I. R. (1977b). Regulation of histone gene expression in HeLa S3 cells. *Cold Spring Harbor Symp. Quant. Biol.* **42**, 1107–1120.

Stein, G. S., Sierra, F., Stein, J. L., Plumb, M., Marashi, F., Carozzi, N., Prokopp, K., and Baumbach, L. (1984). Organization and expression of human histone genes. *In* "Histone Genes" (G. S. Stein, J. L. Stein, and W. F. Marzluff, eds.). Wiley, New York.

Stein, J. L., Thrall, C. L., Park, W. D., Mans, R. J., and Stein, G. S. (1975). Hybridization analysis of histone messenger RNA association with polyribosomes during the cell cycle. *Science* **189,** 557–558.

Sures, I., Lowry, J., and Kedes, L. H. (1978). The DNA sequence of sea urchin (*S. purpuratus*) H2A, H2B and H3 histone coding and spacer regions. *Cell* **15,** 1033–1044.

Tarnowka, M. A., Baglioni, C., and Basilico, C. (1978). Synthesis of H1 histones by BHK cells in G1. *Cell* **15,** 163–171.

Trostle-Weige, P. K., Meistrich, M. L., Brock, W. A., Nishioka, K., and Bremer, J. W. (1982). Isolation and characterization of TH2A, a germ cell-specific variant of histone 2A in rat testis. *J. Biol. Chem.* **257,** 5560–5567.

Van Dongen, W., de laaf, L., Zaal, R., Moorman, A., and Destree, O. (1981). The organization of the histone genes in the genome of *Xenopus laevis. Nucleic Acids Res.* **9,** 2297–2311.

Wallis, J. W., Hereford, L., and Grunstein, M. (1980). Histone H2B genes of yeast encode two different proteins. *Cell* **22,** 799–805.

Woodland, H. R., and Adamson, E. D. (1977). The synthesis and storage of histones during the oogenesis of *Xenopus laevis. Dev. Biol.* **57,** 118–135.

Woods, D. E., and Fitschen, W. (1978). The mobilization of maternal histone messenger RNA after fertilization of sea urchin eggs. *Cell Differ.* **7,** 103–114.

Wu, R. S., and Bonner, W. M. (1981). Separation of basal histone synthesis from S phase histone synthesis in dividing cells. *Cell* **27,** 321–330.

Wu, R. S., Tsai, S., and Bonner, W. M. (1982). Patterns of histone variant synthesis can distinguish G0 from G1 cells. *Cell* **31,** 367–374.

Zernik, M., Heintz, N., Boime, I., and Roeder, R. G. (1980). *Xenopus laevis* histone genes: Variant H1 genes are present in different clusters. *Cell* **22,** 807–815.

6

Expression of the α-Fetoprotein Gene during Development, Regeneration, and Carcinogenesis

TAIKI TAMAOKI

Department of Medical Biochemistry
University of Calgary
Calgary, Alberta, Canada

NELSON FAUSTO

Department of Pathology
Division of Biology and Medicine
Brown University
Providence, Rhode Island

145

RECOMBINANT DNA AND
CELL PROLIFERATION

I. INTRODUCTION

α-Fetoprotein (AFP) is of great biological and medical interest because it belongs to a group of proteins classified as "oncodevelopmental," that is, proteins that are present at high levels in embryonic and neoplastic tissues. AFP levels, which are low in the serum and liver of adult animals and humans, are very high in the yolk sac and embryonic liver as well as in the majority of germ cell tumors and hepatomas. Thus, studies of the structure, regulation, and expression of the AFP gene contribute to our understanding of normal and neoplastic growth processes, and at the same time have direct clinical applicability, as AFP is used in the diagnosis of tumors and birth defects.

The number of publications on AFP is large and most of the work has been previously reviewed by Ruoslahti and Seppälä (1979), Hirai (1979), and Stillman and Sell (1979). This chapter discusses studies of AFP gene structure and regulation and those involving measurements of AFP mRNA. Other work is quoted only if directly pertinent to these studies.

Cloning of AFP cDNA and AFP genomic sequences has so far been accomplished with three animal sources: mouse, rat, and human. Characterization of these clones has revealed useful and interesting information on the structure of AFP genes and their molecular evolution. The relationship between the AFP gene (which has been found in all mammalian species examined, in chickens and in sharks) and the albumin gene is an important biological problem because of the evolutionary origin and possible coordinated expression of these two genes during embryonic development. Studies reviewed in this article indicate that the amount of AFP synthesized in liver and yolk sac correlates well with the cellular content of AFP mRNA. The levels of AFP mRNA in the cell appear to be regulated by at least two processes: gene transcription and stabilization of the messenger. Little is known about the regulatory signals that regulate the transcription of the gene, but the extent of methylation of the AFP gene is inversely related to its activity.

II. STRUCTURE OF THE α-FETOPROTEIN GENE

In recent years, rapid progress has been made in the analysis of the structure of the AFP gene with the application of recombinant DNA technology. Table I shows cDNA and genomic clones of AFP and albumin from mice, rats, and humans which have been isolated and characterized. The availability of these clones allows structural comparisons between AFP genes of various origins and intra- and interspecies comparisons between AFP and albumin genes. These studies have revealed useful information on the structure of the AFP gene and evolutionary relationships between AFP and albumin genes.

TABLE I

cDNA and Genomic Clones of AFP and Albumin from Mice, Rats, and Humans

Species	cDNA clone		Genomic clone	
	AFP	Albumin	AFP	Albumin
Mouse	Full-size[a,b]	Full-size[e]	Full-size[e]	Full-size[e]
Rat	Partial[c]	Full-size[f]	—	Full-size[j]
Human	Full-size[d]	Full-size[g,h]	Full-size[i]	—

[a] Law and Dugaiczyk (1981).
[b] Gorin et al. (1981).
[c] Jagodzinski et al. (1981).
[d] Morinaga et al. (1983).
[e] Kioussis et al. (1981).
[f] Sargent et al. (1981a).
[g] Lawn et al. (1981).
[h] Dugaiczyk et al. (1982).
[i] Sakai et al. (1983).
[j] Sargent et al. (1981b).

A. Mouse α-Fetoprotein Gene

Law and Dugaiczyk (1981) have analyzed a full-size mouse AFP cDNA clone and found that mouse AFP mRNA contains 57 nucleotides that code for 19 amino acids of the signal sequence (Peters et al., 1979), 1758 nucleotides that code for 586 amino acids of mature AFP, and 153 nucleotides that make up the 3' noncoding region. Another mouse cDNA clone, isolated by Tilghman and her associates, gives a similar though not identical nucleotide sequence (Gorin et al., 1981). Subsequent analysis of a 5' genomic AFP clone has revealed that the cap site is located 44 nucleotides upstream from the AUG initiation codon, and a TATAA sequence, thought to be required for precise initiation by RNA polymerase II, is found 30 nucleotides upstream from the cap site (Eiferman et al., 1981). The 5' noncoding region of the mouse AFP mRNA is characterized by the presence of a purine-rich sequence having partial sequence homology with other mammalian mRNA (Baralle and Brownlee, 1978; Kozak, 1978), and a sequence of 8 nucleotides complementary to a 3' terminal region of eukaryotic 18 S RNA (Azad and Deacon, 1980).

The 3' noncoding region contains the characteristic hexanucleotide, AATAAA, the poly(A) addition signal, located 21 nucleotides upstream from the poly(A) at the 3' end (Law and Dugaiczyk, 1981; Eiferman et al., 1981). Another characteristic sequence, TTTTCAACTAT, is found immediately to the left of the polyadenylation site (Benoist et al., 1980).

Amino acid sequences of mouse AFP were deduced from nucleotide sequences of AFP cDNA clones. Mouse AFP has cysteine residues that are characteristically spaced and form 15 disulfide bridges. These crosslinks generate a pattern of repeated loops, defining three structural domains of mouse AFP.

Mouse AFP contains three potential sites of N-glycosylation (Asn-X-Thr; Asn-X-Ser), two in Domain II, and one in Domain III (Gorin et al., 1981; Law and Dugaiczyk, 1981). No information is available regarding the number of carbohydrate chains in mouse AFP, but there are at least two molecular variants of mouse AFP with respect to lectin binding (Savu et al., 1977; Ruoslahti and Adamson, 1978; Kerchaert et al., 1979). It is possible that two or all three of the glycosylation sites contain carbohydrate chains. These sites are not present in bovine and human albumin.

Tilghman and associates (see Kioussis et al., 1981) have cloned mouse AFP gene sequences from genomic libraries derived from BALB/c mouse DNA. The mouse AFP gene is comprised of 15 coding segments (exons), interrupted by 14 intervening sequences (introns), spanning a total length of 22-kilobase (kb) DNA which is 10 times larger than AFP mRNA. Approximate sizes of the exons and introns were estimated by electron microscopy. The 12 internal exons are composed of three similar sets of 4 exons of sizes approximately 140, 220, 140, and 100 base pairs (bp). It is proposed that this triplicated pattern is the basis for the three repeated domains of the mouse AFP polypeptide described in the preceding paragraph (Eiferman et al., 1981). These general features of the mouse AFP gene are strikingly similar to those of the mouse albumin gene (Kioussis et al., 1981). The albumin gene is also comprised of 15 exons; each has a close parallel, in terms of size, in the AFP gene. In contrast, the sizes of the 14 corresponding introns show much greater variation (see below for further comparisons between AFP and albumin).

A middle repetitive "foldback" B1 or Alu-like element is present in the first intron of the mouse AFP gene (Young et al., 1982). It is 220 nucleotides long and located within 400 bp of the initiation site for AFP transcription, on the opposite strand to that encoding the AFP gene. This sequence is transcribed in monkey kidney cells when a "minigene" consisting of the first two exons including the first intron, along with 0.9 kb of 5' flanking sequence and the last exon along with 0.4 kb of 3' flanking sequence, is constructed, inserted into an SV40 vector, and used to infect African green monkey kidney cells. Mouse cells that are actively transcribing the AFP gene, however, do not seem to contain the transcripts of the Alu repeat. The biological significance of the Alu repeat in the AFP gene is not clear at present.

Church et al. (1982) have recently shown that a cloned mouse AFP cDNA sequence injected into the pronucleus of fertilized mouse oocytes can be transferred to succeeding generations. In this study, about 50% of the oocytes that received the AFP cDNA went through culture to the late morula stage and were transferred to recipient mothers. Of the 22 live offspring born, 7 were found to contain the transferred DNA sequence as determined by dot hybridization. Four of the offspring of the next generation were also positive by dot hybridization analysis, suggesting that the injected AFP cDNA sequence had integrated into the host genome.

B. Rat α-Fetoprotein Gene

Jagodzinski *et al.* (1981) have isolated three overlapping rat AFP cDNA clones from which 85% of the total mRNA sequence was determined. In comparison with mouse AFP mRNA, an 88% homology was found in the coding sequence and a 72% homology in the 3′ noncoding sequence. The putative polyadenylation signal sequence, AATAAA, is located 21 nucleotides upstream from the polyadenylation site. The 11 bp sequence immediately to the left of the polyadenylation site, TTTTCAACTGT, is similar to the corresponding sequence in the mouse AFP mRNA with only 1 different nucleotide.

So far, six introns have been located in the rat AFP gene (Jagodzinski *et al.,* 1981). Five of these splice sites are identical with those in the rat albumin gene. The sixth is slightly off, due to its location in the noncoding region. No further details of the structure of the rat AFP gene have been published.

The amino acid sequence of rat AFP deduced from the cDNA nucleotide sequence shows an 85% homology to the mouse AFP. It has regularly spaced cysteine residues forming 15 disulfide bridges in the same manner as in mouse AFP. Three potential sites of N-glycosylation are present in rat AFP, two of which are located in Domain II at the same positions as in mouse AFP. The third one is present in Domain I. This is in contrast to the third glycosylation site in mouse AFP which is present in Domain III. By chemical analysis, rat AFP has been reported to contain two N-linked carbohydrate chains (Bayard and Kerckaert, 1981; Krusius and Ruoslahti, 1982).

C. Human α-Fetoprotein Gene

Morinaga *et al.* (1982a) cloned human AFP cDNA corresponding to 841 nucleotides at the 3′ end of the mRNA. More recently, they have obtained cDNA clones that cover the entire coding sequence (Morinaga *et al.,* 1983). The amino acid sequence was deduced from the nucleotide sequence, which revealed 19 amino acids in the signal sequence and 590 amino acids in mature AFP. As in mouse AFP, there are 15 regularly spaced disulfide bridges, which generate a folding structure having three repeating domains. Unlike mouse AFP, human AFP has only one potential site for N-glycosylation. This is in agreement with the observation of Yoshima *et al.* (1980) that human AFP has one asparagine-linked sugar chain as analyzed by chemical means. In comparison with mouse AFP, 66% of the amino acid sequence was conserved, with the highest homology (72%) in Domain III followed by Domain II (67%) and Domain I (59%).

Sakai *et al.* (1983) have also analyzed genomic human AFP clones and found that the human AFP gene consists of 15 exons, most of which are identical in size to the corresponding exons in the mouse AFP gene. The sizes of introns, on the other hand, are different from those of the corresponding introns in the mouse AFP gene. As in the mouse AFP gene, all exon/intron junctions in the human AFP gene conform to the "GT–AG" rule (Breathnach *et al.*, 1978).

D. Comparisons between α-Fetoprotein and Albumin Genes

The genetic basis for the structural and functional similarities between AFP and albumin and the reciprocal expression of their genes during development has been a subject of great interest. Some of the recent findings related to this problem are summarized as follows.

1. The mouse AFP amino acid sequence, deduced from the nucleotide sequence of cloned cDNA, shows a 34% homology to human or bovine albumin amino acid sequence, analyzed by conventional amino acid sequencing procedures (Gorin *et al.*, 1981; Law and Dugaiczyk, 1981). Somewhat higher homologies are observed in intraspecies comparisons. In the case of the rat, 521 of the 585 possible total number of amino acid residues of mature rat AFP, deduced from the nucleotide sequence, show a 35% homology to those of rat albumin, also deduced from the nucleotide sequence (Jagodzinski *et al.*, 1981). In the case of the human, the amino acid sequence of the mature AFP (Morinaga *et al.*, 1983) has a 39% homology to that of the mature albumin (Lawn *et al.*, 1981; Dugaiczyk *et al.*, 1982), with the highest homology in Domain III (48%) followed by Domain II (40%) and Domain I (30%).

2. There are 15 disulfide bridges in mouse, rat, and human AFP, and 17 in bovine and human albumin. All disulfide bridges in the AFP molecules are located at the same positions as those in albumin, forming three groups of loops or triplicate domains initially proposed for albumin (Brown, 1976; McLachlan and Walker, 1977).

3. Analyses of genomic AFP clones of the mouse (Kioussis *et al.*, 1981) and the human (Sakai *et al.*, 1983) show that there are 15 exons in these genes. These exons are similar in size to corresponding exons in the mouse albumin gene. The internal 12 exons of either gene are composed of three similar sets of 4 exons (Eiferman *et al.*, 1981). This provides a genetic basis for the three-domain structure of AFP and albumin.

4. It has been proposed that the three protein domains are generated by the triplication of a primordial domain consisting of 5 exons (Eiferman *et al.*, 1981). A series of at least three intragenic duplication events leads the primordial gene into the 15-exon/14-intron/3-domain ancestor of AFP and albumin genes. Subsequently, an intergenic duplication results in the appearance of AFP and albumin as distinct genes.

5. Physical linkage of AFP and albumin genes in the mouse genome has recently been established. D'Eustachio *et al.* (1981), using Southern blot hybridization, have analyzed AFP and albumin DNA sequences in several mouse × hamster somatic cell hybrids containing various combinations of mouse chromosomes together with a constant set of hamster chromosomes. Both AFP and albumin genes were detected in those hybrids retaining mouse chromosome 5. Subsequent analysis of genomic clones that contain sequences flanking the 5' end of the AFP gene and the 3' end of the albumin gene shows that the genes are in tandem with the albumin gene 13.5 kb upstream from the AFP gene (Ingram *et al.*, 1981).

III. METHYLATION OF THE α-FETOPROTEIN GENE

There has been considerable interest in the possibility that DNA methylation may play an important role in regulating gene expression and establishing stable patterns of genetic expression in higher organisms (for review, see Razin and Riggs, 1980). In mammalian DNA almost all methylation is at the 5 position of cytosine in the sequence 5'-CpG. Methylation of specific genes has been studied using restriction endonucleases *Msp*I and *Hpa*II. Both enzymes cleave the sequence CCGG, but the methylated sequence CmCGG can be cleaved only by *Msp*I (Waalwijk and Flavell, 1978). Thus, comparisons of the DNA cleavage patterns obtained with these enzymes can reveal the presence or absence of CmCGG. This approach has been used to study the methylation of globin genes (McGhee and Ginder, 1979; van der Ploeg and Flavell, 1980; Shen and Maniatis, 1980), ovalbumin gene (Mandel and Chambon, 1979), and metallothionein gene (Compere and Palmiter, 1981). In all cases, lower levels of methylation have been found in tissues expressing the gene than in those not expressing the gene.

Andrews *et al.* (1982a) studied the possibility that AFP gene activity is associated with reduced methylation of DNA. Expression of the mouse AFP gene in embryonic, adult, and neoplastic tissues was assayed by dot hybridization analysis of AFP mRNA. The extent of methylation of the AFP gene in these tissues was assessed by comparing DNA fragments released by *Hpa*II and *Msp*I by Southern blot hybridization using cloned AFP cDNA as a probe. The extent of methylation of the AFP gene was found to be higher in those tissues not expressing the AFP gene (yolk sac mesoderm, adult liver, brain, and non-AFP-producing hepatoma) than in those tissues expressing the gene (yolk sac endoderm, fetal liver, and AFP-producing hepatoma).

Characteristic patterns of methylation of the AFP gene were obtained in endoderm (AFP-producer) and mesoderm (non-AFP-producer) of the yolk sac of day 15 gestation. This is consistent with the hypothesis that DNA methylation may play a role in cellular differentiation (Jones and Taylor, 1980; van der Ploeg and Flavell, 1980; Razin and Riggs, 1980; Manes and Menzel, 1981). As it is known that the DNA in mammalian sperm and early embryos is highly methylated (Waalwijk and Flavell, 1978; Mandel and Chambon, 1979; Desrosiers *et al.*, 1979), one model proposes that differentiated states are accomplished and maintained by tissue-specific methylation which is generated by specific demethylation events (see Razin and Riggs, 1980).

The increased level of AFP gene methylation in adult liver correlated with the decrease in AFP synthesis, but the exact basis for this change is difficult to assess. It may reflect changes in the population of liver cells producing AFP rather than decreases in AFP gene activity in the cells. About 30–40% of cells in fetal liver contain AFP, whereas less than 5% of adult liver cells contain AFP after liver injury (Engelhardt *et al.*, 1971; Adinolfi *et al.*, 1975; Dziadek and Adamson, 1978; Kuhlmann, 1981; Marceau *et al.*, 1982). This interpretation is consistent with the observation that a large portion of the AFP gene sequences were resistant to *Hpa*II in both fetal and adult liver. However, we cannot rule out the possibility that specific methylation events occur in the AFP gene during liver development.

It should be noted that a few hypomethylated *Hpa*II sites are also present in the AFP gene of inactive tissues studied. It has been reported with other genes that certain CCGG sites are not methylated, regardless of the level of activity of the gene (van der Ploeg and Flavell, 1980; Shen and Maniatis, 1980; Mandel and Chambon, 1979). In addition, we have recently observed that at least one CCGG site in yolk sac and fetal liver is modified in such a manner as to become highly resistant to *Msp*I. AFP gene methylation is thus complex, and further work is necessary to elucidate its significance in relation to AFP gene expression.

IV. STABILITY OF α-FETOPROTEIN mRNA

Availability of cloned AFP cDNA has made possible the measurement of the rate of decay of AFP mRNA relative to that of other mRNA. The amount of radioactive AFP mRNA is assayed by hybridization to cloned AFP cDNA which is immobilized on nitrocellulose filters (Gillespie, 1968). Innis and Miller (1979) labeled Morris rat hepatoma 7777 cells with [³H]uridine for 3 hr and then chased for 60 hr with excess uridine, cytosine, and glucosamine (Scholtissek, 1971;

Levis and Penman, 1977). At various time intervals, portions of the cells were lysed, and RNA was extracted and assayed for AFP mRNA. The disappearance of radioactive AFP mRNA proceeded as a linear logarithmic function with a half-life of 40 hr. This is more than five times larger than the half-life of total poly(A)$^+$ RNA in Morris rat hepatoma 7777.

AFP mRNA in mouse yolk sac also shows high stability (Andrews *et al.*, 1982b). Mouse yolk sac was labeled *in vitro* with [^{3}H]uridine for 4 hr and chased for 30 hr in a manner similar to that described above for Morris rat Hepatoma 7777. Assays of radioactive AFP mRNA by hybridization to cloned mouse AFP cDNA bound to nitrocellulose filters showed only a small amount of decay of AFP mRNA during the 30-hr chase period. By extrapolating the decay curve, the half-life of AFP mRNA in mouse yolk sac was estimated to be at least 60 hr. By contrast, the majority of poly(A)$^+$ RNA in mouse yolk sac decayed with a half-life of 6 hr. No significant changes in the stability of AFP mRNA in mouse yolk sac were observed between day 11.5 and day 17.5 of gestation.

V. α-FETOPROTEIN mRNA DURING DEVELOPMENT

A. Liver

Developmental changes in AFP gene expression were first studied by assaying AFP mRNA activity in cell-free systems. Fetal mouse liver contained a higher level of AFP mRNA than adult liver (Koga *et al.*, 1974; Tamaoki *et al.*, 1974, 1976; Koga and Tamaoki, 1974; Iio and Tamaoki, 1976). In contrast, the level of albumin mRNA increased with development (Koga and Tamaoki, 1974; Tamaoki *et al.*, 1976). Similar results were obtained with rat liver and yolk sac (Liao *et al.*, 1980). Intracellular distribution of AFP mRNA and albumin mRNA in fetal mouse liver was 2% in the nucleus and 98% in the cytoplasm, of which more than 90% was in the polysome fraction. The major portion of albumin mRNA in adult mouse liver was also associated with polysomes (Iio and Tamaoki, 1976). These results suggested that AFP and albumin mRNAs, once formed, are quickly and efficiently utilized for protein synthesis. It was thus concluded that AFP and albumin syntheses are not controlled at the level of translation, but rather by the amount of corresponding functional mRNAs.

Assays of AFP and albumin mRNA levels by molecular hybridization with specific cDNA probes have confirmed the reciprocal changes in neonatal liver (Sala-Trepat *et al.*, 1979a; Nishi *et al.*, 1979; Liao *et al.*, 1980; Miura *et al.*, 1979; Tilghman and Belayew, 1982), but similar assays in fetal liver have so far yielded different results (Liao *et al.*, 1980; Tilghman and Belayew, 1982). Intracellular distribution of AFP and albumin mRNA sequences as determined by cDNA probes established that less than 2% of these sequences were present in the nuclear or nonpolysomal cytoplasmic components (Nahon *et al.*, 1982). It is clear that accumulation of nonfunctional mRNA molecules does not occur in adult liver. The major control for AFP and albumin gene expression must be found at the level of transcription or posttranscriptional processing.

B. Yolk Sac

AFP mRNA in mouse and rat yolk sac exhibits characteristic changes during development. In the mouse yolk sac, AFP mRNA increases with the gestational age reaching the highest level (20–30% of total mRNA) at day 15.5, and gradually decreases thereafter (Nishi *et al.*, 1979; Janzen *et al.*, 1982). The number of AFP mRNA molecules per cell at day 14.5 and 16.6 yolk sac was estimated to be about 40×10^3/cell (Janzen *et al.*, 1982). AFP mRNA in the rat yolk sac shows a similar developmental change, with the highest level detected at day 17 (Liao *et al.*, 1980). In the yolk sac, as in the case of liver, assays of polysome-associated AFP mRNA by cell-free translation and by cDNA hybridization yield basically the same pattern of developmental changes.

C. Effect of Hormones

Various hormones, when administered to newborn rats, alter the level of circulating AFP (Bélanger *et al.*, 1975). For instance, dexamethasone drastically decreases the serum AFP level. The cause of this reduction was investigated by analyzing the level of AFP mRNA by cDNA hybridization in mice (Commer *et al.*, 1979) and rats (Bélanger *et al.*, 1981). The level of AFP mRNA in the liver of dexamethasone-treated mice decreased almost in parallel with the decrease of the circulating AFP. This indicates that dexamethasone exerts the inhibitory effect at the transcriptional level.

VI. α-FETOPROTEIN mRNA IN LIVER REGENERATION AND NEOPLASIA

As pointed out by Ruoslahti and Seppälä (1979), the correlation between serum AFP levels and liver cell proliferation appears to be straightforward, but there are many observations that complicate this correlation or place it, in its simplest form, into question. Some of the most pertinent of these observations include the following: (1) Elevations of serum AFP are relatively small in liver regeneration after partial hepatectomy in young adult rats although more than 90% of hepatocytes undergo DNA synthesis; (2) in older rats, AFP changes after partial hepatectomy are negligible although more than 75% of hepatocytes go into DNA synthesis; (3) serum AFP levels are generally higher in liver regeneration after

CCl_4 injury than in partially hepatectomized rats, despite the similar rates of DNA synthesis in the two regenerative processes; (4) chemicals such as ethionine may induce an increase in serum AFP in the absence of hepatocyte proliferation; (5) during the preneoplastic stage of liver carcinogenesis there may be very large increases in serum AFP without concomitant hepatocyte replication; and (6) AFP synthesis in human and rat hepatomas is variable and not related to the degree of cell replication in the tumor (for individual references, see Ruoslahti and Seppälä, 1979; Hirai, 1979; Stillman and Sell, 1979).

Measurements of AFP mRNA have made possible a more direct examination of AFP gene expression in regenerative and neoplastic growth of the liver. However, despite the advances, many of the questions regarding the relationship between AFP gene expression and hepatocyte replication and transformation remain unanswered. Nevertheless, it is important to examine the relationship between the expression of the AFP gene in adult liver and possible changes in the differentiated state of the hepatocyte.

A. Normal Adult Liver

AFP mRNA can be detected in the liver of adult rats and mice by hybridization of cellular RNAs with AFP cDNA probes (Innis and Miller, 1979; Nishi et al., 1979; Bélanger et al., 1979; Sala-Trepat et al., 1979a; Miura et al., 1979; Liao et al., 1980; Atryzek et al., 1980; Tilghman and Belayew, 1982). This suggests that the liver is the source of serum AFP which is present in small amounts in normal adult animals. The probes used in this study were synthesized with reverse transcriptase from AFP mRNA purified from rat or mouse tissues (hepatomas or yolk sac), or they were cDNA inserts of plasmids containing mouse AFP cDNA sequences. Hybridization of these probes with polysomal polyadenylated RNA (RNA excess) indicates that AFP mRNA is present in liver polysomal RNA of adult rats and mice, but an accurate quantitation of the amount of AFP mRNA present in the RNA preparations is difficult to achieve with this method because of the very small concentration of AFP mRNA in adult liver. Petropoulos et al. (1982) recently used a hybridization–titration procedure (DNA excess) to determine the percentage of AFP mRNA present in polysomal RNA of young adult rats (45 days old) and found that in these rats AFP mRNA constitutes approximately 0.0058% of liver poly(A)$^+$ RNA. Liver polysomes also contain AFP mRNA molecules which are not adenylated. In normal liver, the number of nonadenylated AFP mRNA molecules is at least equal to, and probably exceeds, the number of adenylated AFP mRNA sequences. The total amount of AFP mRNA (A$^+$ and A$^-$) corresponds to approximately 0.00014% of the mass of polysomal RNA. The biological significance for the existence of two forms of AFP mRNA is not completely understood, but similar findings have been reported for other messengers. The nonadenylated molecules are present in polysomes and do not seem to constitute a precursor pool for the adenylated messenger. More likely, the nonadenylated molecules represent stages in the degradation of poly(A)$^+$ AFP mRNA (Bergmann and Brawerman, 1977; Darnell, 1982) and may contain short poly(A) tails, probably with less than 10 adenylic acid residues, which are not long enough to bind to poly(U)–Sepharose. Poly(A)$^-$ AFP mRNA directs the *in vitro* synthesis of peptides that are precipitable by antibodies to AFP (Tamaoki *et al.*, 1976; Sala-Trepat *et al.*, 1979a), but the product synthesized appears to be incomplete. This may be a consequence of lower stability of poly(A)$^-$ AFP mRNA in the translation system (Tamaoki *et al.*, 1976; Miura *et al.*, 1979).

Although one can detect AFP mRNA in normal adult liver with hybridization techniques, the cellular localization of the messenger in the liver is unknown. AFP mRNA might be a very rare mRNA present in most liver cells, or it might be a moderately abundant messenger in a small number of hepatocytes or other cell types. This question can probably be resolved by *in situ* nucleic acid hybridization techniques and by measurements of AFP mRNA content in isolated cell fractions that correspond to the various cell populations of mammalian liver. Unfortunately, no reports have appeared on the use of these methods for AFP mRNA analysis.

Kuhlmann (1979a), who has used immunocytochemical methods for AFP detection with light and electron microscopy, did not find AFP-staining cells in liver of rats after the fourth week of postnatal growth, an observation also reported by Guillouzo *et al.* (1978) and Abelev (1978). AFP-staining cells were also not detected in the livers of C3H and BALB/c/J mice, although in one study some hepatocytes of BALB/c/J were found to be AFP positive (Kuhlmann, 1979b, 1981). Using the hemolytic plaque assay, Marceau *et al.* (1982) found no AFP-producing cells in hepatocytes isolated by collagenase perfusion from 23-day- and 12-week-old rats.

The apparent contradiction between the results of nucleic acid hybridization studies and the work with immunochemical methods might be explained if the amounts of AFP synthesized in normal liver are too small to permit detection by immunochemical methods. One should not assume, however, that the discrepancies between results obtained with hybridization or immunochemical methods indicate that AFP is produced in small amounts by most hepatocytes. Since the limits of the sensitivity of the immunochemical methods are not clearly defined one cannot exclude the possibility that AFP mRNA is present in only a small number of cells. Another possible but less likely explanation for the lack of agreement between AFP mRNA measurements and AFP detection in the liver is that AFP might be rapidly secreted from the liver after its synthesis. Thus, whereas the mRNA is detectable, the hepatic levels of the protein would be very low.

B. Liver Regeneration

Because AFP is a typical "oncodevelopmental" protein, studies of AFP mRNA in liver regeneration and carcinogenesis have a direct bearing on our understanding of the biological nature of these growth processes. Many investigators have assumed that a process variably called "retrodifferentiation" or "dedifferentiation" takes place during regenerative liver growth after partial hepatectomy (Uriel, 1979).

The concept of retrodifferentiation has a very different meaning depending on the time during liver regeneration in which genes for oncodevelopmental proteins might be expressed. If the mRNAs for these proteins are mostly synthesized before DNA synthesis, one may consider the activation of genes coding for oncodevelopmental proteins as part of a reprogramming of gene expression which might be required for DNA replication. On the other hand, if mRNAs coding for oncodevelopmental proteins are synthesized primarily during or after the peak of DNA synthesis in liver regeneration, the expression of these genes might be a characteristic of newly replicated hepatocytes. Depending on the magnitude of the increase of a specific mRNA in regenerating liver in comparison with embryonic growth, one might view the expression of the corresponding gene during regeneration as a modulation phenomenon rather than an indication of true regression from the differentiated state.

Important questions regarding the expression of AFP gene during liver regeneration include the following: (1) Is there a preferential increase in AFP mRNA over other messengers during liver regeneration, that is, is the proportion of AFP mRNA in hepatic polysomal RNA changed during regenerative growth? (2) What is the relationship between changes in the proportion of AFP mRNA in hepatic polysomal RNA and the timing of DNA synthesis during the regenerative process? (3) Are the changes in AFP mRNA in regenerating liver large enough in comparison with the levels of AFP mRNA detected during embryonic growth? (4) Are the levels of AFP mRNA in regenerating liver after partial hepatectomy similar to those present during regenerative growth following chemical injury?

1. Rat Liver

It is well known that the rates of DNA synthesis in regenerating liver after partial hepatectomy or CCl_4 injury are similar (depending on the CCl_4 dosage) but that the timing of DNA synthesis in the two processes differ. In partially hepatectomized young adult rats (120–200 g), DNA synthesis does not start until 12–14 hr after the operation and reaches a maximum at 22–24 hr. In contrast, after CCl_4 administration to rats, little DNA synthesis takes place in the first 24 hr, and highest levels of DNA synthesis are reached at 48 hr. In the prereplicative stage of liver regeneration after partial hepatectomy (first 12–14 hr) there are large increases in the synthesis of hepatic mRNA and rRNA. In CCl_4 injury, a relatively small increase of hepatic mRNA occurs between 24 and 28 hr after administration of the chemical. In chemical injury, the first 24–36 hr are characterized by cell death and fat accumulation.

Petropoulos *et al.* (1982) recently completed a study of AFP mRNA levels during liver regeneration induced by partial hepatectomy or CCl_4 administration in rats. Whereas the total amount of polysomal poly(A)$^+$ RNA in the liver doubles at 12 hr after partial hepatectomy (Atryzek and Fausto, 1979), the proportion of AFP mRNA in poly(A)$^+$ polysomal RNA increases by only 16% over the corresponding value in livers of normal or sham-operated rats. The highest value for the proportion of AFP mRNA in liver polysomal poly(A)$^+$ RNA was detected 48 hr after partial hepatectomy, that is, 24 hr after the peak of DNA synthesis had taken place. The magnitude of the increase of AFP mRNA in regenerating liver after partial hepatectomy in young adult rats is small. The messenger levels in regeneration are 200- to 1000-fold lower than those detected in the liver of fetal and newborn rats.

In CCl_4 injury (as is the case for partial hepatectomy) the highest values for the proportion of AFP mRNA in liver polysomal poly(A)$^+$ RNA are reached 24 hr after the peak of DNA synthesis. The AFP mRNA levels in CCl_4 injury are about twofold higher than those detected after partial hepatectomy. Chiu *et al.* (1981) have also shown that after CCl_4 injury, AFP mRNA is highest after the peak of DNA synthesis has occurred and that dexamethasone injections inhibit the increase of liver AFP mRNA induced by CCl_4 injury.

Nonadenylated AFP mRNA also increases during liver regeneration after partial hepatectomy or CCl_4 administration. The ratios between adenylated/ nonadenylated AFP mRNA molecules in polysomal RNA are approximately 1 at 48 hr after partial hepatectomy and 1.7 at 72 hr after CCl_4 administration.

The major conclusions obtained from these studies are the following: (1) Little preferential synthesis of AFP mRNA over that of other hepatic mRNAs takes place before the start of DNA synthesis during liver regeneration; (2) the highest levels in the proportion of AFP mRNA in polysomal RNA are achieved 24 hr after the peak of DNA synthesis has occurred and are thus probably associated with dividing or newly divided cells; (3) even the highest levels of AFP mRNA detected in regenerating liver (partial hepatectomy or CCl_4 injury) are much below those present during fetal development and early neonatal growth.

These conclusions must be qualified because one does not know the cellular distribution of the AFP mRNA in the regenerating liver. There are uncertainties regarding the results obtained with immunochemical methods for AFP detection in regenerating livers. Using the hemolytic plaque assay, Marceau *et al.* (1982) did not detect AFP-positive cells in the regenerating liver of adult rats after partial hepatectomy; in 21-day-old rats, approximately 5% of hepatocytes were positive. Similar values have been reported for mouse liver by Engelhardt *et al.*

(1976) who detected AFP by immunofluorescence. With the peroxidase method, Kuhlmann (1981) detected AFP-positive cells in regenerating mouse liver after partial hepatectomy and CCl_4 injury. However, AFP-positive cells were found only in regenerating rat liver after chemical injury but not following partial hepatectomy. Abelev (1978) concluded that AFP-positive cells are scattered, single cells located around the central veins in the liver of partially hepatectomized mice and adjacent to necrotic areas in animals that have received CCl_4. More recently, Baranov et al. (1982) reported that after CCl_4 administration to mice, AFP was detected only in a small percentage of cells, located in a narrow zone bordering the necrotic foci. These cells were identified as typical differentiated hepatocytes and small hepatocytes similar to "oval" cells.

Several investigators have tried to determine the phase of the cell cycle in which AFP is synthesized. The methods used generally include a combination of immunocytochemical detection for AFP and DNA labeling with thymidine. Abelev (1978) concluded that AFP synthesis is induced before cell replication in hepatocytes of regenerating liver. On the other hand, Guillouzo et al. (1979) reported that cells that have completed mitosis constitute the bulk of AFP-synthesizing cells in the liver of postneonatal rats. Tsukada and Hirai (1976) observed that the synthesis of AFP in a rat hepatoma cell line, AH66 C-4, took place from late G_1 phase to the end of the S phase.

Taking together the results of hybridization studies to measure AFP mRNA levels and of the immunocytological work to localize AFP in liver cells during regeneration, the following alternative conclusions are possible: (1) Only a small proportion of hepatocytes (around 5% in young rats and mice) synthesize AFP during liver regeneration after partial hepatectomy; (2) AFP might be synthesized by most hepatocytes, but in very small amounts; (3) most hepatocytes synthesize AFP during liver regeneration but only for a very short time period so that at any given time during the regenerative process only a small fraction of cells is actively synthesizing AFP. The last alternative is the least likely one because it would require that AFP mRNA be rapidly degraded once synthesized by liver cells during liver regeneration. This is, however, not consistent with the kinetics of AFP mRNA accumulation in regenerating liver. If the first alternative listed above is correct, that is, that the fraction of cells that synthesize AFP in regeneration is very small, the obvious implication is that a period of "retrodifferentiation," characterized by increased transcription of the AFP gene, does not take place in the greater majority of hepatocytes during liver regeneration. Finally, if one assumes that AFP is synthesized in small quantities (below the detection level of immunocytochemical methods) by most hepatocytes (Alternative 2) the results of the hybridization studies indicate that most of the increase in the cellular levels of AFP mRNA might occur in hepatocytes that have already undergone at least one round of replication.

2. Inbred Strains of Mice: rif and raf Control

Olsson et al. (1977) and Jalanko (1979) have shown that BALB/c/J strain mice homozygous for the recessive trait raf (regulation of α-fetoprotein) have high serum AFP levels. Belayew and Tilghman (1982) recently investigated the mechanism that may account for the differences in AFP serum levels between BALB/c/J (high producers), C3H/He (intermediate producers), and C57BL/6 (low producers) mice in normal animals and during liver regeneration induced by CCl_4 injury. These authors found that the concentration of AFP mRNA in the liver of mice of these three strains is similar up to the end of the first week of life. During the next 2 weeks, the rate of AFP mRNA decrease is approximately 5 times slower in BALB/c/J mice than in the other two strains. In adult BALB/c/J mice the basal level of AFP mRNA in the liver is approximately 1 ng/mg of total RNA, and it is about 10 times lower in C3H and C57BL/6 mice. Thus, the raf gene, which controls the basal level of serum AFP in BALB/c/J mice, exerts its effects through the regulation of AFP mRNA concentration in the liver. Three days after CCl_4 administration, AFP mRNA was 15-fold higher in BALB/c/J than in C3H mice. Moreover, C3H mice had AFP mRNA levels that were about 10-fold higher than C57BL/6 animals. These differences are very close to those described by Jalanko (1979) for serum AFP concentrations in these three strains. By using appropriate genetic crosses, Belayew and Tilghman (1982) demonstrated that AFP inducibility is controlled by a different gene which was named rif (regulation of inducibility of AFP). Rif and raf are separate, unlinked loci and are not in the genomic region of the AFP or the albumin gene. Belayew and Tilghman (1982) concluded that neither of these loci act in early embryonic development to control AFP mRNA levels. Raf gene effects are first noted 2 weeks postnatally whereas the rif gene acts in the induction of AFP mRNA in adult liver. These authors also point out that there is no reciprocal switch between the AFP and albumin genes in adult mice. This conclusion is of major importance for the evaluation of the concept of retrodifferentiation (see above).

C. Transplantable Hepatomas

Sell et al. (1979) have shown that there is wide variation in the level of AFP mRNA in various hepatomas. It is well known that the Morris rat hepatoma 7777 has a high AFP mRNA content. Sell et al. (1979) showed that this is also the case for another Morris hepatoma (8994) although AFP mRNA levels in this tumor are about fivefold lower than those in 7777. Two other hepatomas (hepatomas 311 and 252) did not contain detectable levels of the messenger. Similarly, Zajdela rat ascites hepatoma cells have negligible levels of AFP mRNA (Morinaga et al., 1982b), although it had been previously reported that these cells might synthesize but not secrete AFP (Sarcione and Smalley, 1976). Prior to these studies Sell and

Morris (1974) had shown that of 38 strains of rat hepatomas investigated, only 7 were high AFP producers. Of 78 strains of AH rat hepatomas studied by Hirai (1979) approximately one-third did not synthesize AFP above normal levels.

The variation in AFP synthesis and AFP mRNA content of various hepatomas indicates that the expression of the AFP gene is not essential for hepatoma growth and constitutes an independent trait not directly related to malignancy, to the tumor growth rate, or to the degree of differentiation of hepatoma cells. Obviously, these observations do not exclude the possibility that the expression of AFP gene is important during the initial process of cell transformation when these hepatomas were first established.

The organization of the AFP gene and probably its regulation in those hepatomas that produce AFP appear to be similar to that of normal tissues. Sala-Trepat *et al.* (1979b) found no evidence for amplification, deletion, or rearrangement of the AFP gene in Morris hepatoma 7777. In this hepatoma, transcriptional control appears to be the major mechanism regulating AFP synthesis, and AFP mRNA represents 3–5% of the polysomal poly(A)$^+$ RNA. The rate of AFP synthesis in Morris hepatoma 7777 cultured *in vitro* is directly related to the amount of AFP mRNA in the cells (Innis and Miller, 1979). Thus, when the *in vivo* fractional rate of AFP synthesis declines from 4.8 to 1.8%, a decrease to 38% of the maximal rate, the hybridizable AFP mRNA per cell declines to 44% of its maximal value. Translation of hepatoma mRNA *in vitro* produces the same proportion of AFP as observed *in vivo* at all cell densities which indicates that there is no process that specifically activates or inhibits translation of AFP mRNA. The close agreement between estimates of AFP mRNA obtained by hybridization and by *in vivo* translation assays shows that a significant pool of inactive AFP mRNA does not exist in the cells.

Andrews *et al.* (1982a) have shown that the sequence organization and reiteration frequency of the AFP gene was the same in mouse DNA and in hepatomas, regardless of the tumor capacity to synthesize AFP. However, the extent of methylation of the AFP gene was inversely related to AFP mRNA content and AFP production. Thus, methylation of specific sites may be an important regulatory control in AFP gene expression. It is not clear at this time why hypomethylation is present in some but not all hepatomas, or whether the pattern of methylation might be regulated in some manner by the *rif* gene. Hypomethylation of the AFP gene might be a random event which determines the capacity of a hepatoma to produce AFP, or it may reflect the pattern of gene expression that existed in the cells that gave origin to the tumor.

D. Hepatocarcinogenesis

Measurements of AFP mRNA in primary hepatocarcinogenesis have been done in rats fed a choline-deficient diet containing 0.05 or 0.1% ethionine

(Atryzek *et al.*, 1980; Petropoulos *et al.*, 1982). In this system, well-differentiated hepatocellular carcinomas develop 20–25 weeks after the start of the feeding regime. The preneoplastic stage is characterized by a massive proliferation of oval cells (Shinozuka *et al.*, 1978; Schultz-Ellison *et al.*, 1981). These cells, which have not yet been properly characterized but appear to derive from biliary duct cells, proliferate 4–6 weeks after the start of the feeding and subsequently form normal and atypic bile ducts which are surrounded by conective tissue. Shinozuka *et al.* (1978) have shown by immunofluorescence that AFP is localized in the oval cells but not in typical hepatocytes in the preneoplastic stage of carcinogenesis in these animals.

Atryzek *et al.* (1980) and Petropoulos *et al.* (1982) have shown that a very large increase in hepatic AFP mRNA content takes place in animals receiving the choline-deficient, ethionine diet. AFP mRNA increases by 30- to 50-fold after 4 weeks of feeding and it is still high at 16 weeks, when it constitutes approximately 0.25% of the liver polysomal poly(A)$^+$ RNA. The AFP mRNA content per cell is highest at 4 weeks when it reaches values of 80–100 molecules per cell. Nonadenylated sequences in liver polysomal AFP mRNA also increase in this carcinogenesis system, but the increase is not proportional to that of the adenylated AFP mRNA molecules. The ratio between A$^+$ and A$^-$ AFP molecules is approximately 5 in the livers of rats that were fed the carcinogenic diet for 9 weeks (Petropoulos *et al.*, 1982).

The increase in hepatic AFP mRNA abundance in preneoplastic livers occurs without a corresponding large change in the content of polysomal poly(A)$^+$ RNA in the liver. Nucleic acid hybridization studies have shown that globin mRNA abundance is not changed during choline-deficiency–ethionine carcinogenesis (Atryzek *et al.*, 1980). Moreover, most if not all of the sequences of polysomal poly(A)$^+$ mRNA from preneoplastic livers are present in normal liver. During carcinogenesis there are shifts in the abundance of some messengers, resulting in a decreased frequency of some poly(A)$^+$ polysomal RNA species in preneoplastic livers (Atryzek *et al.*, 1980). Thus, the changes in AFP mRNA that occur in these animals do not follow the overall pattern detected in mRNA populations at this stage of carcinogenesis.

The levels of hepatic AFP mRNA during carcinogenesis induced by the choline-deficient, ethionine diet are 20–50 times higher than those found during liver regeneration after partial hepatectomy. Yet, in preneoplastic livers, at the time in which the highest levels of AFP mRNA are detected, there is a massive proliferation of oval cells but very little hepatocyte replication (Shinozuka *et al.*, 1978; Schultz-Ellison *et al.*, 1981). From these findings and the immunofluorescence data, it is reasonable to conclude that the increase in AFP mRNA in animals receiving the choline-deficient, ethionine diet is related to the proliferation and/or differentiation of oval cells and is not likely to be associated with the replication or dedifferentiation of mature hepatocytes. To properly understand the regulation of the expression of the AFP gene in this system it is essential to elucidate the origin and differentiation potential of oval cells. Although these cells are related, at least morphologically, to biliary cells, ductular cell proliferation in normal liver is not accompanied by significant changes in hepatic AFP mRNA. It is possible that biliary cells and oval cells represent different paths in the differentiation of stem cells which might exist in the normal liver. In their transition to hepatocytes, oval cells might be capable of producing AFP. It would be important to know if oval cells in preneoplastic liver (and their precursor cells in normal liver, if they exist) have a hypomethylated AFP gene. It must be pointed out, however, that oval cell proliferation is not an obligatory stage in liver carcinogenesis. Moreover, although the increase in serum AFP levels generally correlates well with the presence of oval cells during carcinogenesis, serum AFP increases in rats which have received a small dose of the carcinogen 2-acetylaminofluorene are not sufficient to cause morphological alterations in the liver (Becker and Sell, 1974).

ACKNOWLEDGMENTS

We would like to thank Dr. Shirley Tilghman for making available to us unpublished material from her laboratory, and Chris Petropoulos for lively discussion. Supported by grants from the National Cancer Institute of Canada, the Medical Research Council of Canada, and the Alberta Heritage Savings Trust Fund (T.T.) and by Grant CA23226 from the National Cancer Institute, United States (N.F.).

REFERENCES

Abelev, G. I. (1978). Experimental study of alpha-fetoprotein re-expression in liver regeneration and hepatocellular carcinomas. *In* "Cell Differentiation and Neoplasia" (G. F. Saunders, ed.), pp. 257–269. Raven Press, New York.

Adinolfi, A., Adinolfi, M., and Lesso, M. H. (1975). Alpha-fetoprotein during development and in disease. *J. Med. Genet.* **12**, 138–151.

Andrews, G. K., Dziadek, M., and Tamaoki, T. (1982a). Expression and methylation of the mouse α-fetoprotein gene in embryonic, adult, and neoplastic tissues. *J. Biol. Chem.* **257**, 5148–5153.

Andrews, G. K., Janzen, R. G., and Tamaoki, T. (1982b) Stability of α-fetoprotein messenger RNA in mouse yolk sac. *Dev. Biol.* **89**, 111–116.

Atryzek, V., and Fausto, N. (1979) Accumulation of polyadenylated mRNA during liver regeneration. *Biochemistry* **18**, 1281–1287.

Atryzek, V., Tamaoki, T., and Fausto, N. (1980). Changes in polysomal polyadenylated RNA and alpha-fetoprotein messenger RNA during hepatocarcinogenesis. *Cancer Res.* **40**, 3713–3718.

Azad, A. A., and Deacon, N. J. (1980). The 3'-terminal primary structure of five eukaryotic 18S RNAs determined by the direct chemical method of sequencing. The highly conserved sequences includes an invariant region complementary to eukaryotic 5S RNA. *Nucleic Acids Res.* **8**, 4365–4376.

Baralle, F. E., and Brownlee, G. G. (1978). AUG is the only recognizable signal sequence in the 5' non-coding regions of eukaryotic mRNA. *Nature (London)* **274,** 84–87.

Baranov, V. N., Engelhardt, N. V., Lazareva, M. N., and Gusen, A. I. (1982). Ultrastructural localization of alpha-fetoprotein synthesis in the regenerating mouse liver. *Bull. Exp. Biol. Med. (Engl. Transl.)* **93,** 96–100.

Bayard, B., and Kerckaert, J.-P. (1981). Uniformity of carbohydrate chains within molecular variants of rat α-fetoprotein with distinct affinity for concanavalin A. *Eur. J. Biochem.* **113,** 405–414.

Becker, F. F., and Sell, S. (1974). Early elevation of α-fetoprotein in N-2-fluorenylacetamide hepatocarcinogenesis. *Cancer Res.* **34,** 2489–2494.

Bélanger, L., Hamel, D., Lachance, L., Dufour, D., Tremblay, M., and Gagnon, P. M. (1975). Hormonal regulation of α-fetoprotein. *Nature (London)* **256,** 657–659.

Bélanger, L., Commer, P., and Chiu, J. F. (1979). Isolation of rat α-fetoprotein messenger RNA from Morris hepatoma 7777. *Cancer Res.* **39,** 2141–2148.

Bélanger, L., Frain, M., Baril, P., Gingras, M.-C., Bartkowiak, J., and Sala-Trepat, J. M. (1981). Glucocorticosteroid suppression of α-fetoprotein synthesis in developing rat liver. Evidence for selective gene repression at the transcriptional level. *Biochemistry* **20,** 6665–6676.

Belayew, A., and Tilghman, S. M. (1982). A genetic analysis of α-fetoprotein synthesis in the mouse. *Mol. Cell. Biol.* **2,** 1427–1435.

Benoist, C., O'Hare, K., Breathnach, R., and Chambon, P. (1980). The ovalbumin gene-sequence of putative control regions. *Nucleic Acids Res.* **8,** 127–143.

Bergmann, I. E., and Brawerman, G. (1977). Control of breakdown of the polyadenylate sequence in mammalian polyribosomes: Role of poly(adenylic acid)-protein interactions. *Biochemistry* **16,** 259–264.

Breathnach, R., Benoist, C., O'Hare, K., Gagnon, F., and Chambon, P. (1978). Ovalbumin gene: Evidence for a leader sequence in mRNA and DNA sequences at the exon-intron boundaries. *Proc. Natl. Acad. Sci. U.S.A.* **75,** 4853–4857.

Brown, J. R. (1976). Structural origins of mammalian albumin. *Fed. Proc., Fed. Am. Soc. Exp. Biol.* **35,** 2141–2144.

Chiu, J. F., Gabryelak, T., Commers, P., and Massari, R. (1981). The elevation of α-fetoprotein messenger RNA in regenerating rat liver. *Biochem. Biophys. Res. Commun.* **98,** 250–254.

Church, R. B., Tamaoki, T., Janzen, R. G., Andrews, G. K., and Morinaga, T. (1982). The fate of alpha-fetoprotein DNA following injection into mouse embryos. *Abstr. Int. Cong. Biochem., 12th, 1982* p. 92.

Commer, P., Schwartz, C., Tracy, S., Tamaoki, T., and Chiu, J. F. (1979). Dexamethasone inhibits α-fetoprotein gene expression in developing mouse liver. *Biochem. Biophys. Res. Commun.* **89,** 1294–1299.

Compere, S. J., and Palmiter, R. D. (1981). DNA methylation controls the inducibility of the mouse metallothionein-I gene in lymphoid cells. *Cell* **25,** 233–240.

Darnell, J. E., Jr. (1982). Variety in the level of gene control in eukaryotic cells. *Nature (London)* **297,** 265–371.

Desrosiers, R. C., Mulder, C., and Fleckenstein, B. (1979). Methylation of *Herpesvirus saimiri* DNA in lymphoid tumor cell lines. *Proc. Natl. Acad. Sci. U.S.A.* **76,** 3839–3843.

D'Eustachio, P., Ingram R. S., Tilghman, S. M., and Ruddle, F. H. (1981). Murine α-fetoprotein and albumin: Two evolutionarily linked proteins encoded on the same mouse chromosome. *Somatic Cell Genet.* **7,** 289–294.

Dugaiczyk, A., Law, S. W., and Dennison, O. E. (1982). Nucleotide sequence and the encoded amino acids of human serum albumin mRNA. *Proc. Natl. Acad. Sci. U.S.A.* **79,** 71–75.

Dziadek, M., and Adamson, E. (1978). Localization and synthesis of alpha-fetoprotein in post-implantation mouse embryos. *J. Embryol. Exp. Morphol.* **43,** 289–313.

Eiferman, F. A., Young, P. R., Scott, R. W., and Tilghman, S. M. (1981). Intragenic amplification and divergence in the mouse α-fetoprotein gene. *Nature (London)* **294**, 713–718.

Engelhardt, N. V., Goussev, A. L., Shipova, T., and Abelev, G. I. (1971). Immunofluorescent study of alpha-fetoprotein (AFP) in liver and liver tumours. I. Technique of AFP localization in tissue sections. *Int. J. Cancer* **7**, 198–206.

Engelhardt, N. V., Lazareva, M. N., Uryvaeva, I. V., Factor, V. M., Poltoranina, V. S., Gleiberman, A. S., Brodsky, V. Ya., and Abelev, G. I. (1976). Alpha-fetoprotein in adult differentiated hepatocytes of the regenerating mouse liver. *In* "Onco-Developmental Gene Expression" (W. H. Fishman and S. Sell, eds.), pp. 533–539. Academic Press, New York.

Gillespie, D. (1968). The formation and detection of DNA-RNA hybrids. *In* "Methods in Enzymology" (K. Moldave and L. Grossman, eds.), Vol. 12B, pp. 641–668. Academic Press, New York.

Gorin, M. B., Cooper, D. L., Eiferman, F., van de Rijn, P., and Tilghman, S. M. (1981). The evolution of α-fetoprotein and albumin. I. A comparison of the primary amino acid sequences of mammalian α-fetoprotein and albumin. *J. Biol. Chem.* **256**, 1954–1959.

Guillouzo, A., Bélanger, L., Beaumont, C., Valet, J. P., Briggs, R., and Chiu, J. F. (1978). Cellular and subcellular immunolocalization of α-fetoprotein and albumin in rat liver. Reevaluation of various experimental conditions. *J. Histochem. Cytochem.* **26**, 948–959.

Guillouzo, A., Boisnard-Rissel, M., Bélanger, L., and Bourel, M. (1979). α-Fetoprotein production during the hepatocyte growth cycle of developing rat liver. *Biochem. Biophys. Res. Commun.* **91**, 327–331.

Hirai, H. (1979). Model system of AFP and CEA expression. *Methods Cancer Res.* **25**, 39–97.

Iio, T., and Tamaoki, T. (1976). Intracellular distribution of α-fetoprotein and albumin messenger RNAs in developing mouse liver. *Can. J. Biochem.* **54**, 408–412.

Ingram, R., Scott, R. W., and Tilghman, S. M. (1981). α-Fetoprotein and albumin genes are in tandem in the mouse genome. *Proc. Natl. Acad. Sci. U.S.A.* **78**, 4694–4598.

Innis, M. A., and Miller, D. L. (1979). Alpha-fetoprotein gene expression. Control of alpha-fetoprotein mRNA levels in cultured rat hepatoma cells. *J. Biol. Chem.* **254**, 9148–9154.

Jagodzinski, L. L., Sargent, T. D., Yang, M., Glackin, C., and Bonner, J. (1981). Sequence homology between RNAs encoding rat α-fetoprotein and rat serum albumin. *Proc. Natl. Acad. Sci. U.S.A.* **78**, 3521–3525.

Jalanko, H. (1979). Alpha-fetoprotein and gamma-glutamyltranspeptidase in mice. Effect of *raf* gene. *Int. J. Cancer* **24**, 394–397.

Janzen, R. G., Andrews, G. K., and Tamaoki, T. (1982). Synthesis of secretory proteins in developing mouse yolk sac. *Dev. Biol.* **90**, 18–23.

Jones, P. A., and Taylor, S. M. (1980). Cellular differentiation, cytidine analogs and DNA methylation. *Cell* **20**, 85–93.

Kerchkaert, J.-P., Bayard, B., and Biserte, G. (1979). Microheterogeneity of rat, mouse and human α-fetoprotein as revealed by polyacrylamide gel electrophoresis and by crossed immunoaffino-electrophoresis with different lectins. *Biochim. Biophys. Acta* **576**, 99–108.

Kioussis, D., Eiferman, F., van de Rijn, P., Gorin, M. B., Ingram, R. S., and Tilghman, S. M. (1981). The evolution of α-fetoprotein and albumin. II. The structures of the α-fetoprotein and albumin genes in the mouse. *J. Biol. Chem.* **256**, 1960–1967.

Koga, K., and Tamaoki, T. (1974). Developmental changes in the synthesis of α-fetoprotein and albumin in the mouse liver. Cell-free synthesis by membrane-bound polyribosomes. *Biochemistry* **13**, 3024–3028.

Koga, K., O'Keefe, D. W., Iio, T., and Tamaoki, T. (1974). Transcriptional control of α-fetoprotein synthesis in developing mouse liver. *Nature (London)* **252**, 495–497.

Kozak, M. (1978). How do eukaryotic ribosomes select initiation regions in messenger RNA? *Cell* **15**, 1109–1123.

Krusius, T., and Ruoslahti E. (1982). Carbohydrate structure of the concanavalin A molecular variants of α-fetoprotein. *J. Biol. Chem.* **257**, 3453–3457.

Kuhlmann, W. D. (1979a). Immuno-electron microscopy of α-fetoprotein during normal development of rat hepatocytes. *Ultrastruct. Res.* **68**, 109–117.

Kuhlmann, W. D. (1979b). Immunoperoxidase labelling of α-fetoprotein (AFP) in normal and regenerating livers of a low and ʔ high AFP producing mouse strain. *Histochemistry* **64**, 67–75.

Kuhlmann, W. D. (1981). Alpha-fetoprotein: Cellular origin of a biological marker in rat liver under various experimental conditions. *Virchows Arch. A: Pathol. Anat. Histol.* **393**, 9–26.

Law, S., and Dugaiczyk, A. (1981). Homology between the primary structure of α-fetoprotein, deduced from a complete cDNA sequence, and serum albumin. *Nature (London)* **291**, 201–205.

Lawn, R. M., Adelman, J., Bock, S. C., Franke, A. E., Houck, C. M., Najarian, R. C., Seeburg, P. N., and Wion, K. (1981). The sequence of human serum albumin cDNA and its expression in *E. coli. Nucleic Acids Res.* **9**, 6103–6114.

Levis, R., and Penman, S. (1977). The metabolism of poly(A)$^+$ and poly(A)$^-$ hnRNA in cultured *Drosophila* cells studied with a rapid uridine pulse-chase. *Cell* **1**, 105–113.

Liao, W. S. L., Conn, A. R., and Taylor, J. M. (1980). Changes in rat α-fetoprotein and albumin mRNA levels during fetal and neonatal development. *J. Biol. Chem.* **255**, 10036–10039.

McGhee, J. D., and Ginder, G. D. (1979). Specific DNA methylation sites in the vicinity of the chicken β-globin genes. *Nature (London)* **280**, 419–420.

McLachlan, A. D., and Walker, J. E. (1977). Evolution of serum albumin. *J. Mol. Biol.* **112**, 543–558.

Mandel, J. L., and Chambon, P. (1979) DNA methylation: Organ specific variations in the methylation pattern within and around ovalbumin and other chicken genes. *Nucleic Acids Res.* **7**, 2081–2103.

Manes, C., and Menzel, P. (1981). Demethylation of CpG sites in DNA of early rabbit trophoblast. *Nature (London)* **293**, 589–590.

Marceau, N., Deschenes, J., and Valet, J. P. (1982). Effect of hepatocyte proliferation and neoplastic transformation on alpha-fetoprotein (AFP) and albumin (ALB) production per cell. Influence of cell specialization and cell size. *Oncodev. Biol. Med.* **3**, 49–63.

Miura, K., Law, S. W. T., Nishi, S., and Tamaoki, T. (1979). Isolation of α-fetoprotein messenger RNA from mouse yolk sac. *J. Biol. Chem.* **254**, 5515–5521.

Morinaga, T., Sakai, M., Wegmann, T. G., and Tamaoki, T. (1982a). α-Fetoprotein messenger RNA in human embryonal carcinoma grown in nude mice and cloning of its complementary DNA. *Oncodev. Biol. Med.* **3**, 301–313.

Morinaga, T., Andrews, G. K., Panrucker, D. E., Lorscheider, F. L., and Tamaoki, T. (1982b). Synthesis and secretion of α-fetoprotein and acute-phase α$_2$ macroglobulin by Zajdela rat ascite hepatoma. *Oncodev. Biol. Med.* **3**, 23–29.

Morinaga, T., Sakai, M., Wegmann, T. G., and Tamaoki, T. (1983). The primary structures of human α-fetoprotein and its messenger RNA. *Proc. Natl. Acad. Sci. U.S.A.* **80**, 4604–4608.

Nahon, J.-L., Gal, A., Frain, M., Sell, S., and Sala-Trepat, J. M. (1982). No evidence for post-transcriptional control of albumin and α-fetoprotein gene expression in developing rat liver and neoplasia. *Nucleic Acids Res.* **10**, 1895–1911.

Nishi, S., Miura, K., and Tamaoki, T. (1979). Mouse α-fetoprotein messenger RNA: Characterization and quantitation. *In* "Carcino-Embryonic Proteins" (F.-G. Lehmann, ed.), Vol. 1, pp. 137–143. Elsevier/North-Holland, New York.

Olsson, M., Lindahl, G., and Ruoslahti, E. (1977). Genetic control of α-fetoprotein synthesis in the mouse. *J. Exp. Med.* **145**, 819–827.

Peters, E. H., Nishi, S., Miura, K., Lorscheider, F. L., Dixon, G. H., and Tamaoki, T. (1979). *In vitro* synthesis of murine pre-alpha-fetoprotein. *Cancer Res.* **39**, 3702–3706.

Petropoulos, C., Andrews, G. K., Tamaoki, T., and Fausto, N. (1983). Alpha-fetoprotein and albumin mRNA levels in liver regeneration and carcinogenesis. *J. Biol. Chem.* **258**, 4901–4906.

Razin, A., and Riggs, A. D. (1980). DNA methylation and gene function. *Science* **210**, 604–610.

Ruoslahti, E., and Adamson, E. (1978). Alpha-fetoprotein produced by the yolk sac and the liver are glycosylated differently. *Biochem. Biophys. Res. Commun.* **85**, 1622–1630.

Ruoslahti, E., and Seppälä, M. (1979). Alpha-fetoprotein in cancer and fetal development. *Adv. Cancer Res.* **29**, 275–346.

Sakai, M., Morinaga, T., Urano, Y., Stockton, J., and Tamaoki, T. (1983). Human α-fetoprotein: Characterization of genomic DNA organization. *Fed. Proc., Fed. Am. Soc. Exp. Biol.* **42**, 1760.

Sala-Trepat, J. M., Dever, J., Sargent, T. D., Thomas, K., Sell, S., and Bonner, J. (1979a). Changes in expression of albumin and α-fetoprotein genes during rat liver development and neoplasia. *Biochemistry* **18**, 2167–2178.

Sala-Trepat, J. M., Sargent, T. D., Sell, S., and Bonner, J. (1979b). α-Fetoprotein and albumin genes of rats: No evidence for amplification, deletion or rearrangement in rat liver carcinogenesis. *Proc. Natl. Acad. Sci. U.S.A.* **76**, 695–699.

Sarcione, E. J., and Smalley, J. R. (1976). Intracellular synthesis of α-fetoprotein and fibrinogen without secretion by Zajdela rat ascites hepatoma cells. *Cancer Res.* **36**, 3203–3206.

Sargent, T. D., Yang, M., and Bonner, J. (1981a). Nucleotide sequence of cloned rat serum albumin messenger RNA. *Proc. Natl. Acad. Sci. U.S.A.* **78**, 243–246.

Sargent, T. D., Jagodzinski, L. L., Yang, M., and Bonner, J. (1981b). Fine structure and evolution of the rat serum albumin gene. *Mol. Cell. Biol.* **1**, 871–883.

Savu, L., Benassayag, C., Vallette, G., Nunez, E., and Jaylee, M.-F. (1977). Purification and estrogen binding properties of mouse α-fetoprotein and of two forms of the protein with different affinities for concanavalin-A. *Biochimie* **59**, 323–328.

Scholtissek, C. (1971). Detection of an unstable RNA in chick fibroblasts after reduction of the DTP pool by glucosamine. *Eur. J. Biochem.* **24**, 358–365.

Schultz-Ellison, G., Atryzek, V., and Fausto, N. (1981). On the existence of variant forms of gamma-glutamyl transferase in regenerating and neoplastic liver. *Oncodev. Biol. Med.* **2**, 109–116.

Sell, S., and Morris, H. P. (1974). Rat α-fetoprotein relationship to growth rate and chromosome composition of Morris hepatoma. *Cancer Res.* **34**, 1413–1417.

Sell, S., Thomas, K., Michalson, M., Sala-Trepat, J., and Bonner, J. (1979). Control of albumin and α-fetoprotein expression in rat liver and in some transplantable hepatocellular carcinomas. *Biochim. Biophys. Acta* **564**, 173–178.

Shen, C.-K. J., and Maniatis, T. (1980). Tissue-specific DNA methylation in a cluster of rabbit β-like globin genes. *Proc. Natl. Acad. Sci. U.S.A.* **77**, 6634–6638.

Shinozuka, H., Lombardi, B., Sell, S., and Iammarino, R. M. (1978). Early histological and functional alterations of ethionine liver carcinogenesis in rats fed a choline-deficient diet. *Cancer Res.* **38**, 1092–1098.

Stillman, D., and Sell, S. (1979). Models of chemical hepatocarcinogenesis and oncodevelopmental gene expression. *Methods Cancer Res.* **25**, 135–168.

Tamaoki, T., Thomas, K., and Schindler, I. (1974). Cell-free studies of developmental changes in synthesis of α-fetoprotein and albumin in the mouse liver. *Nature (London)* **249**, 269–271.

Tamaoki, T., Miura, K., Lin, T., and Banks, P. (1976). Developmental changes in α-fetoprotein and albumin messenger RNAs. *In* "Onco-Developmental Gene Expression" (W. H. Fishman and S. Sell, eds.), pp. 115–122. Academic Press, New York.

Tilghman, S., and Belayew, A. (1982). Transcriptional control of the murine albumin/α-fetoprotein locus during development. *Proc. Natl. Acad. Sci. U.S.A.* **79,** 5254–5257.

Tsukada, Y., and Hirai, H. (1976). Studies on alpha-fetoprotein in culture of rat hepatoma cells. *In* "Onco-Developmental Gene Expression" (W. H. Fishman and S. Sell, eds.), pp. 639–646. Academic Press, New York.

Uriel, J. (1979). Retrodifferentiation and the fetal patterns of gene expression in cancer. *Adv. Cancer Res.* **29,** 127–174.

van der Ploeg, L. H. T., and Flavell, R. A. (1980). DNA methylation in the human γδβ-globin locus in erythroid and nonerythroid tissues. *Cell* **19,** 947–958.

Waalwijk, C., and Flavell, R. A. (1978). *Msp*I, an isoschizomer of *Hpa*II which cleaves both unmethylated and methylated *Hpa*II sites. *Nucleic Acids Res.* **5,** 3231–3236.

Yoshima, H., Mizuochi, T., Ishii, M., and Kobata, A. (1980). Structure of the asparagine-linked sugar chain of α-fetoprotein purified from human ascites fluid. *Cancer Res.* **40,** 4276–4281.

Young, P., Scott, R. W., Hamer, D. H., and Tilghman, S. M. (1982). Construction and expression *in vivo* of an internally deleted mouse α-fetoprotein gene: Presence of a transcribed Alu-like repeat within the first intervening sequence. *Nucleic Acids Res.* **10,** 3079–3116.

7

Organization and Expression of Eukaryotic Ribosomal Protein Genes

HOWARD M. FRIED

Department of Biochemistry and Nutrition
University of North Carolina School of Medicine
Chapel Hill, North Carolina

JONATHAN R. WARNER

Departments of Cell Biology and Biochemistry
Albert Einstein College of Medicine
Bronx, New York

RECOMBINANT DNA AND
CELL PROLIFERATION

I. INTRODUCTION

Some of the earliest molecular studies of the control of cell proliferation concerned the regulation of ribosome synthesis in *Escherichia coli* (Maaloe and Kjeldgaard, 1966). Over the years the structure of the ribosome has been clarified (Nomura *et al.,* 1974; Chambliss *et al.,* 1980; Wittman, 1982), and the role that it plays in protein synthesis and cell growth has been established (Chambliss *et al.,* 1980; Nomura *et al.,* 1974).

Whereas the eukaryotic ribosome has not been as thoroughly characterized as that of *E. coli,* its overall structure is similar. Briefly, the eukaryotic ribosome is an assembly of four RNA molecules and 70–80 proteins, each present in a single copy, although there may be more than one copy of one or a few proteins (Wool, 1979). The concentration of ribosomes within a cell is closely tied to cell growth, and the biosynthesis of ribosomal components is one of the most sensitive indicators of an alteration of the environmental conditions of a cell (reviewed by Warner *et al.,* 1980).

From studies of ribosome synthesis in eukaryotic cells, as well as by analogy to the findings in prokaryotic cells (Lindahl and Zengel, 1982), three levels of coordination in the synthesis of ribosomal components have been identified:

1. *Coordination among the ribosomal proteins:* Most ribosomal proteins are present in a single copy per ribosome (Hardy, 1975). In *E. coli* (Dennis, 1974), yeast (Gorenstein and Warner, 1976), and HeLa cells (Warner, 1977), the synthesis of ribosomal proteins has indeed been shown to be equimolar.

2. *Coordination between ribosomal RNA and ribosomal proteins:* Similarly, ribosomal proteins and rRNA should be synthesized in equimolar amounts. Since three of the four rRNAs (excluding 5 S RNA) derive from a single transcript (Miller, 1981), this condition can be met if the synthesis of ribosomal proteins is coupled to the transcription of rRNA genes and if the processing steps leading to each of the three RNAs are equally efficient. The synthesis of 5 S RNA, in higher eukaryotes at least, is not coordinated with that of the rest of the components of the ribosome (Zylber and Penman, 1971) but is regulated by the availability of a 5 S binding protein (Pelham and Brown, 1980).

3. *Coordination between ribosome synthesis and cell growth:* For any given cell type each nutritional condition, be it the carbon source for yeast or the presence of growth factors for cultured cells, leads to a characteristic rate of growth and to a corresponding rate of ribosome synthesis sufficient to balance turnover and provide for new cells. Therefore, one of the first consequences of an alteration in growth conditions is a rapid change in the synthesis of ribosomal components.

Little is known about the molecular mechanisms involved in maintaining these three levels of coordination. In an attempt to provide some experimental tools a number of laboratories have set out to isolate cloned DNA segments containing the genes for ribosomal proteins. This chapter summarizes the methods employed for isolating the clones, the information obtained about the structure and the distribution of the genes, and the attempts to use these genes to understand the three levels of coordination. We confine ourselves mostly to work on lower eukaryotes. Chapter 10 by Meyuhas discusses higher eukaryotes.

II. ISOLATION OF EUKARYOTIC RIBOSOMAL PROTEIN GENES

A. General Considerations

Several aspects of ribosomal proteins have made the task of cloning their respective genes somewhat less than straightforward. Unlike many developmentally regulated proteins whose mRNAs can be purified from differentiated cells, ribosomal proteins constitute a relatively invariant population whose mRNAs are of only moderate abundance (Hereford and Rosbash, 1977). Since individual ribosomal protein mRNAs make up only 0.05–0.2% of total cellular mRNA, none has been purified sufficiently for use as a specific hybridization probe to identify a cloned DNA segment containing a ribosomal protein gene. Also, a paucity of amino acid sequence information for eukaryotic ribosomal proteins precludes the synthesis of DNA oligonucleotides which could be used as specific hybridization probes. Finally, only a few genetically identifiable eukaryotic ribosomal proteins are known. This deficiency, coupled with the inability to introduce recombinant DNA into cells of most species, has limited the number of genes that one could isolate by selection of transformed cells having a phenotype attributable to a cloned ribosomal protein gene.

Nevertheless, these drawbacks have not been insurmountable. Ribosomal proteins are among the smallest of the cell's proteins. Consequently, their mRNAs are fairly small in size and can be translated with good fidelity *in vitro*. Using *in vitro* translation as an assay, many investigators have demonstrated significant enrichment of ribosomal protein mRNA in RNA preparations below 12 S in size (Woolford *et al.*, 1979; Bollen *et al.*, 1981; Pierandrei-Amaldi and Beccari, 1980). Furthermore, since ribosomal proteins are generally basic in charge, most can be separated easily from other proteins in acrylamide gels run at the appropriate pH (Kaltschmidt and Wittman, 1970; Mets and Bogorad, 1974; Gorenstein and Warner, 1976). These features have facilitated identification by *in vitro* translation of cloned DNA segments complementary to mRNA for ribosomal proteins. Finally, eukaryotes contain more than 70 different ribosomal proteins. Thus, a cloning strategy with less than optimal specificity may still have a reasonable probability of identifying at least one ribosomal protein gene. As will be described, these considerations have led thus far to the isolation of genes for ribosomal proteins of *Saccharomyces, Drosophila, Xenopus,* and *Mus.*

B. Cloning Strategies for Eukaryotic Ribosomal Protein Genes

1. Hybridization–Translation Techniques

The basic strategy to obtain ribosomal protein genes was utilized first by Woolford *et al.* (1979) with the yeast *S. cerevisiae.* These investigators prepared small-sized mRNA, enriched 5- to 10-fold for ribosomal protein mRNA. The RNA, labeled *in vitro* with ^{32}P, was hybridized to filter-bound plasmids carrying yeast DNA segments. The same collection of plasmids was also hybridized to ^{32}P-labeled RNA prepared from unfractionated mRNA. Plasmids whose hybridization signals from the enriched [^{32}P]mRNA were stronger than those from total [^{32}P]mRNA were selected for further analysis. To analyze the sequences contained within such plasmids, Woolford and Rosbash (1979) developed an assay based on the observation that in high concentrations of formamide an RNA–DNA duplex is more stable than a DNA–DNA duplex of the same sequence (Casey and Davidson, 1977). On incubation of a mixture of RNA with duplex DNA in high concentrations of formamide at the appropriate temperature, a region of the DNA can denature, allowing a complementary RNA molecule to form a hybrid with one of the DNA strands. If DNA sequences on either side of the denatured region are of greater G + C content, (a condition generally met by bacterial cloning vectors now in use) these segments can remain paired. The resulting molecule contains an ''R-loop,'' double-stranded DNA interrupted by a loop consisting of a segment of single-stranded DNA and a segment of a DNA–RNA hybrid (Thomas *et al.,* 1976). Woolford *et al.* (1979) showed that R-loop hybridization conditions could be used to select sufficient mRNA complementary to plasmid DNA such that, after separating the R-looped DNA from unhybridized RNA by gel filtration and releasing the hybridized RNA by heating, this RNA can be translated *in vitro* to identify the protein encoded by the cloned sequence. For ribosomal proteins, the *in vitro* translation products were analyzed by two-dimensional get electrophoresis. Of the 25 plasmids examined by this procedure, 5 hybridized to a ribosomal protein mRNA. Two of the 5 plasmids contained different yeast DNA segments which hybridized to mRNA encoding the same protein, suggesting that the gene for this protein is duplicated (see Section III,C). In a similar fashion Bollen *et al.* (1981) have isolated recombinant DNA plasmids containing sequences complementary to ribosomal proteins mRNAs of *S. carlsbergensis.* Out of 140 plasmids which hybridized to a size-fractionated mRNA probe, 15 carried ribosomal protein genes.

The R-loop translation method was used in a slightly different way by Fried *et al.* (1981) to isolate additional yeast ribosomal protein genes. These investigators did not limit the RNA to small mRNA however, since this restriction would have precluded the isolation of genes for the larger ribosomal proteins. Thus, total yeast poly(A)$^+$ mRNA was used as a template to prepare ^{32}P-labeled cDNA which was hybridized to plaques from a clone bank of yeast DNA in a bacteriophage λ vector (Benton and Davis, 1977). From 88 phage which produced hybridization signals of varying strength and were analyzed by R-loop *in vitro* translation, 14 ribosomal protein genes were identified, a number of which were indeed found to encode large ribosomal proteins. The success of screening with total RNA was probably due to the small size of the yeast genome, 14,000 kilobases (kb) (Fangman and Zakian, 1982), as well as to the relative abundance of ribosomal protein mRNAs.

Similar screening protocols have also been employed to isolate ribosomal protein genes from *Drosophila.* Vaslet *et al.* (1980) prepared cDNA to a preparation of 6–12 S poly(A)$^+$ mRNA and, after selecting 10 λ phage-carrying *Drosophila* DNA complementary to the probe, R-loop translation analysis revealed one phage that contained a ribosomal protein gene. Fabijanski and Pellegrini (1982b) have isolated a second *Drosophila* ribosomal protein gene by the same procedure, although they selected mRNA for *in vitro* translation by binding cloned DNA to filters. Since the *Drosophila* genome is more than 10 times the size of the yeast genome, it is probably essential to use a partially enriched probe to screen for ribosomal protein genes by this procedure.

Since the yeast and *Drosophila* genomes are relatively small, the screening of clone banks encompassing an entire genome is not such an unreasonable task. However, to clone ribosomal protein genes from organisms with far greater genetic complexity, some modifications to the aforementioned protocol are necessary. Thus, Bozzoni *et al.* (1981) working with *Xenopus,* and Meyuhas and Perry (1980) working with mouse cells, each isolated mRNA enriched for ribosomal protein messages and constructed cDNA clones in a bacterial plasmid vector (Williams, 1981). As the template, Bozzoni *et al.* (1981) isolated size-fractionated poly(A)$^+$ mRNA from immature *Xenopus* oocytes, a cell type known for its massive accumulation of ribosomes; by quantitating *in vitro* translation products, approximately 10–20% of the mRNA was estimated to be ribosomal protein mRNA (Pierandrei-Amaldi and Beccari, 1980). Similar analysis by Meyuhas and Perry (1980) of a low-molecular-weight mouse RNA fraction suggested that 20% of the mRNA was ribosomal protein mRNA. The cDNA plasmids were rehybridized to the enriched mRNA fractions, and those giving strong or moderate signals were selected. Both groups employed denatured plasmid DNA immobilized on nitrocellulose or diazobenzoxymethyl cellulose discs to select mRNA for *in vitro* translation. This work resulted in the isolation of cDNA clones for six *Xenopus* ribosomal proteins and nine mouse ribosomal proteins.

The strategies described so far have in common the utilization of a ''positive'' translation assay, i.e., one identifies the product of translation of a hybrid-selected mRNA. Solution hybridization (R-loop version) is more sensitive, since reaction kinetics, which can be calculated, are rapid and allow one to assay mRNAs even with abundancies well below that of ribosomal protein mRNAs (Woolford and Rosbash, 1979). Hybridization efficiency is high, permitting the use of less input mRNA compared to filter hybridization to achieve the same degree of recovery of a particular RNA species, which can be nearly 100% for R-loop hybridization. One drawback to R-loop hybridization is that the melting temperature of the cloned DNA must be below that of the plasmid or phage vector, a condition that may not be met for transcribed sequences high in G + C. One usually performs the R-loop incubation over a range of temperatures to ensure encountering the optimal hybridization conditions.

2. Cloning Ribosomal Protein Genes by Selection in Vitro

A second but less generally applicable strategy for cloning ribosomal protein genes is to transform cells with DNA derived from a strain that is resistant to an inhibitor of protein synthesis. Subsequent selection for resistant transformants leads to recovery of the resistance gene, which may encode a ribosomal protein. In *S. cerevisiae* three such genetic loci were known, conferring resistance to trichodermin (*tcm*1) (Grant *et al.*, 1976), cycloheximide (*cyh*2) (Mortimer and Hawthorne, 1966), and cryptopleurine (*cry*1) (Skogerson *et al.*, 1973; Grant *et al.*, 1974). Whereas in *E. coli* resistance to similar inhibitors can result from mutations in specific ribosomal proteins, the products of these yeast genes were not known, since no resistance mutation had led to an electrophoretic alteration in a particular protein. Nevertheless, *in vitro* translation assays showed that ribosomes as well as subunits (60 S for *tcm*1 and *cyh*2 and 40 S for *cry*1) isolated from resistant strains are resistant to the drugs (Schindler *et al.*, 1974; Rao and Grollman, 1967; Grant *et al.*, 1974). Furthermore, diploid strains heterozygous for any one of these resistance genes exhibit resistance at drug concentrations intermediate between wild-type and completely resistant strains, suggesting that such diploids produce a mixed population of wild-type and resistant ribosomes (Cooper *et al.*, 1967; Jimenez *et al.*, 1972). These observations implied that *tcm*1, *cyh*2, and *cry*1 encode ribosomal proteins rather than enzymes involved in modifying either the ribosome or the antibiotic.

To isolate the gene for resistance to trichodermin, Fried and Warner (1981) prepared a plasmid clone bank of DNA derived from a *tcm*1 strain and introduced it into a drug-sensitive yeast strain. Because recipient cells contained a preexisting pool of sensitive ribosomes, it was anticipated that any cell acquiring *tcm*1 from the clone bank would still be sensitive unless time were allowed for dilution of the preformed ribosome population. Thus, the recipient strain was provided with a defective gene, *leu*2, for an amino acid biosynthetic enzyme; and a collection of plasmid-containing cells was selected via complementation by the *LEU*2 gene carried as part of the cloning vector. After a period of growth in the absence of leucine, transformants were plated in the presence of a low concentration of trichodermin to select those that had become partially resistant. The plasmid obtained from resistant colonies, pTCM, when reintroduced in yeast, caused all cells to become partially resistant. That the resistance due to the plasmid was in fact the result of it carrying *tcm*1 was verified after selecting fully resistant cells from cells carrying pTCM. Genetic mapping of the *leu*2 allele in cells that had become fully resistant showed that plasmid pTCM had recombined with genomic DNA at the *tcm*1 locus. Since DNA sequence homology is required for this recombination (Hinnen *et al.*, 1978), pTCM must contain *tcm*1.

Positive translation analysis of pTCM as well as of a small subcloned segment of the yeast DNA in the plasmid (see Section III, C) revealed that *tcm*1 encodes the large subunit protein L3 (rp1).* This finding was consistent with the previous localization of the effect of *tcm*1 to the large subunit.

3. Cloning Ribosomal Protein Genes by "Chromosome Walking"

A similar attempt to isolate *cyh*2, the gene specifying resistance to cycloheximide, was unsuccessful. As an alternative, Fried and Warner (1982) used *in vivo* selection to isolate a gene which by genetic criteria had been found to lie very close to *cyh*2. A particular allele of this nearby gene, *tsm*437, confers a temperature-sensitive phenotype upon cells; thus, the *tsm*437 locus was isolated by introducing a clone bank into a strain carrying this mutation and selecting for temperature-resistant transformants. Isolation of *tsm*437 provided a DNA fragment that could be hybridized to a different clone bank to identify other genomic segments overlapping *tsm*437. Eventually, a distal DNA fragment was obtained which, when assayed by the R-loop translation method, was found to encode the large subunit ribosomal protein L29 (rp44). This same DNA fragment was then isolated from a *cyh*2 strain. Introduction of this segment into a sensitive yeast strain resulted in transformants partially resistant to cycloheximide; as with pTCM, the plasmid carrying this fragment was shown by genetic tests to recombine at the *cyh*2 locus. Thus, the cloned fragment encoding protein L29 is in fact the *cyh*2 gene. At about the same time, Stocklein and Piepersberg (1980) described a yeast strain with a mutation in *cyh*2 which led to an electrophoretically

---

*The numbering of yeast ribosomal proteins is chaotic [see Warner (1982) and Otaka and Osawa (1981)]. Where possible we will use the Kruiswijk and Planta (1974) system, together with the system from this laboratory (Warner, 1982), e.g., rpl.

altered L29 protein. Thus, two independent approaches confirm the identification of L29 as the product of the *cyh2* gene.

Recently, Larkin and Woolford (1983) have used a similar "chromosome walking" strategy to isolate the cryptopleurine-resistance gene *cry*1 which lies near the mating-type locus. This gene encodes the 40 S subunit protein rp59.

In summary, three distinct strategies have been used for the isolation of ribosomal protein genes from eukaryotes. The strategies depend to varying degrees upon each other. Unless the product of a particular gene is already known to be a ribosomal protein, isolation of genes by *in vivo* selection still requires identification of the gene product, which is most easily ascertained by a positive-translation assay. On the other hand, since most ribosomal protein gene sequences appear to be duplicated or reiterated in the genome as will be seen, one cannot be certain that a genomic DNA fragment assayed by positive translation is actually a genuine, functional gene. Thus, a preexisting phenotype attributable to a ribosomal protein and the means to express this phenotype upon reintroduction of the gene into cells provide proof of function of the cloned gene.

Obviously, transformation as a means to isolate a ribosomal protein gene (or confirm its isolation) is limited to those genes for which a phenotype is known and organisms into which DNA can be introduced. However, one could hope to isolate a resistance gene of one organism by introducing a clone bank of its DNA into another. Success would depend on the ability of a heterologous ribosomal protein to assemble with the host ribosomes in such a manner so as to still confer resistance.

C. Criteria to Establish an Active Ribosomal Protein Gene within a Cloned Segment of DNA

Since a cloned genomic DNA fragment frequently includes all or parts of several transcription units, subcloning of the fragment is necessary to isolate the transcription unit in question. It is then possible to show by RNA blot analysis that the subcloned fragment hybridized to only a single mRNA species (or a family of related RNAs if nuclear RNA is included). Translation of mRNA hybridized to such a fragment generally yields a single protein, identifiable on a two-dimensional gel. However, in mouse (Meyuhas and Perry, 1980), *Xenopus* (Bozzoni *et al.*, 1982), and *Drosophila* (Fabijanski and Pellegrini, 1982a) there is evidence for hybridization of mRNA for several ribosomal proteins to a single DNA fragment.

When cloning a gene by selection for a specific phenotype, e.g., *tcm*1, it is essential to determine that the cloned DNA carried the desired genetic locus rather than a cryptic gene which produces fortuitously a similar phenotype. This confirmation can be accomplished by ligating the cloned gene to another identifiable marker gene, integrating the product into the genome by homologous recombination, and demonstrating by genetic means that the marker gene is now linked to the desired locus.

Beyond demonstrating that a cloned DNA fragment contains a particular genetic locus, it is also desirable to demonstrate a correspondence between that locus and the gene product assayed by positive translation, especially since only a portion of a gene need be present in a fragment to give a result in a positive translation assay. For example, in the case of *tcm*1, a 3.2-kb segment of the original 13.5-kb DNA fragment was subcloned and found both to hybridize to L3 mRNA and to confer resistance to trichodermin. This fragment could be bisected by the restriction enzyme *Sal*I, and, because both subfragments hybridized to L3 mRNA, the *Sal*I site must be in the L3 gene (Fried and Warner, 1981). Insertion of four bp by DNA polymerase at the *Sal*I site eliminated resistance to trichodermin (H. M. Fried, unpublished). Therefore, the restriction site is also in *tcm*1, i.e., *tcm*1 and the gene for protein L3 are the same.

In those cases where a ribosomal protein gene has been identified only by a positive translation assay, one needs to know if the DNA fragment includes the entire gene. Sometimes the presence of only part of a gene is apparent from the restriction map and subsequent subcloning, since one may find that the fragment that yields a signal in a positive translation assay is too small to encode the protein in question. In other cases it may be necessary to use S1 mapping (Berk and Sharp, 1977) to determine whether the ends of the transcript are within the cloned genomic fragment. Alternatively, in organisms susceptible to transformation, e.g., *Saccharomyces*, it is possible to introduce the cloned fragment into the cell in a suitable vector and to ask if it gives rise to mRNA. A similar analysis could be performed for DNA of higher eukaryotes by injecting it into *Xenopus* oocytes (Bakken *et al.*, 1982).

Eukaryotic organisms, unlike prokaryotes, often have multiple copies of a gene. For example, in the mouse genome there are many sequences homologous to each ribosomal protein gene (Chapter 10, this volume). In *Xenopus* there are two to five copies (Bozzoni *et al.*, 1982), and in *Saccharomyces* there are two copies of several genes (Woolford *et al.*, 1979; Fried *et al.*, 1981; Bollen *et al.*, 1981). It is important to establish whether the copy of a gene one has isolated is active within the cell. A gene that has been isolated by selection for a phenotype, e.g., *tcm*1 of yeast, is *ipso facto* active. Such genes, however, are likely to be present in only a single copy, since a mutation arising in one of two copies of a gene would not provide an easily detectable phenotype. Alternatively, one can introduce the gene back into the organism and ask if it is transcribed, as discussed in the previous paragraph. However, even though a gene may be functional when reintroduced into a cell, it does not follow that its expression is the same as when it is in its normal genomic location. For example, a cloned copy of one of the usually silent mating-type genes was expressed when carried into a yeast cell on a plasmid (Hicks *et al.*, 1979). Finally, for *S. cerevisiae* a number of investigators have described recombinant DNA procedures for replacing a wild-type chromosomal gene with a defective counterpart constructed *in vitro* (Struhl *et al.*, 1979; Rykowski *et al.*, 1981; McAlister and Holland, 1982; Shortle *et al.*, 1982). Such selective inactivation allows one to establish whether either one of two duplicated genes is sufficient for growth or to determine if a cloned gene can substitute in a strain in which both chromosomal genes in a set of duplicate genes have been inactivated (Grunstein *et al.*, 1982). The problem of determining which genes are active in the complex eukaryotic genome is more difficult, as is discussed by Meyuhas in Chapter 10.

III. CHARACTERISTICS OF EUKARYOTIC RIBOSOMAL PROTEIN GENES

A. Ribosomal Protein Genes Are Not Clustered

The origin of the three major ribosomal RNAs from a single transcription unit in almost all organisms attests to the importance of gene arrangement in ensuring equimolar production of these RNAs. In *E. coli* a major element in the coordinate expression of ribosomal protein genes is the clustering of most of the genes into a few large operons, each of which can be regulated at a single site. In yeast, the chromosomal arrangement of ribosomal protein genes was not known, but the mapping of the three drug-resistance loci to different chromosomes and the small size of ribosomal protein mRNAs as well as their independent translation (Gorenstein and Warner, 1979) suggested that their genes would be monocistronic. The isolation of a large number of ribosomal protein genes from yeast has provided an opportunity to establish that they are widely scattered.

Extensive regions of the genome surrounding ribosomal protein genes have been analyzed by R-loop translation and RNA blot hybridization. The five clones obtained by Woolford *et al.* (1979) ranged in size from 2 to 7 kb; each contained only one ribosomal protein gene. Four of these clones were used to isolate an additional 71 kb of the genome overlapping the original isolates in both directions, without finding any additional ribosomal protein genes (Woolford and Rosbash, 1981). Fried *et al.* (1981) and Fried and Warner (1981, 1982) have isolated 17 DNA segments averaging 8.5 kb in size (about 150 kb total). In every case but one, only a single ribosomal protein gene was identified within a given segment. A similar absence of multiple ribosomal protein genes within cloned segments has been reported for *S. carlsbergensis* (Bollen *et al.*, 1981). In both *S. cerevisiae* (Fried *et al.*, 1981) and *S. carlsbergensis* (Bollen *et al.*, 1981) one exceptional fragment contains two ribosomal protein genes separated by about 0.5 kb, one for the small subunit protein S16A (rp55), and one for rp28 of the large subunit. Whether such linkage is fortuitous or reflects a functional relationship is not known. Nevertheless, preliminary evidence suggests that the two genes are transcribed independently because the downstream gene has been subcloned and, upon introduction into yeast, causes an increase in the mRNA for the protein in question (N. Pearson, personal communication). Thus, it is clear that yeast ribosomal protein genes are indeed monocistronic and further, that they are widely dispersed throughout the genome.

Although methods are now available for mapping cloned genes by using integration vectors (Broach and Hicks, 1980), none of the ribosomal protein genes has yet been assigned a chromosomal location, except for the three drug-resistance loci previously assigned by classical methods. As discussed by Meyuhas in Chapter 10, Southern blot analysis of mouse–hamster hybrid cells containing various sets of mouse chromosomes indicates that mouse ribosomal protein genes are dispersed on many chromosomes (D'Eustachio *et al.*, 1981).

B. Nature of Sequences Surrounding Ribosomal Protein Genes

Not much is known about the DNA sequences that surround ribosomal protein genes or whether genes for other components of protein synthesis adjoin riboso-mal protein genes. The yeast genome is very extensively transcribed, with an average of one transcribed sequence every 3 kb (Kaback *et al.*, 1979). R-loop translation, electron microscopy, and Northern blot analyses of yeast segments containing ribosomal protein genes do not show any obvious deviation from this situation. Northern blot analysis demonstrated that under at least one condition affecting the synthesis of ribosomal protein mRNAs (see below), the synthesis of transcripts from adjacent genes is not affected, suggesting that ribosomal protein genes act independently with respect to neighboring genes (Woolford and Rosbash, 1981). Interestingly, the physical distance between ribosomal protein genes and nearby genes appears to be considerably greater than that determined by recombination frequency.For the yeast genome, various analyses have re-vealed a ratio of approximately 2.7–3 kb of DNA for 1 centimorgan (cM) (Strathern *et al.*, 1979; Shalit *et al.*, 1981). The genetic distance between *cyh2* and the gene *tsm437* is about 1.5 cM but the physical distance was found to be 8 kb, giving a ratio of 5.3 kb/cM. A similarly high ratio occurs near the *cry1* gene (Larkin and Woolford, 1983). It remains to be determined if the reduced recom-bination frequency of regions containing ribosomal protein genes has some phys-iological importance or is simply a statistical fluctuation. Frequent recombina-tion could open the way to a more rapid evolution of ribosomal proteins, a condition that may not be tolerated because of the interaction of a given protein with so many others in a precise assembly. Recombination could also alter control elements such that coordinate regulation would be compromised. In support of this notion, the analysis of the physical and genetic distance between the yeast cytochrome *c* gene and a transfer RNA gene (*SUP4*) yielded 5.6 kb/cM

(Shalit *et al.*, 1981), again demonstrating reduced recombination in an interval that includes a gene for a component of the protein synthesis apparatus.

C. Individual Ribosomal Protein Genes Are Present in Different Numbers

As discussed in Section II,C, some ribosomal protein genes are present in multiple copies. Hybridization of ribosomal protein gene fragments to Southern blots of yeast DNA revealed one or two additional homologous segments for 12 ribosomal protein genes; 6 genes, however, do not appear to hybridize to any other segment (Woolford *et al.*, 1979; Fried *et al.*, 1981; Fried and Warner, 1981, 1982; Bollen *et al.*, 1981). At this point it is not known whether all of the homologous segments are functional, but several factors would suggest that they are. No pseudogenes have yet been found in yeast; on the contrary, the genes for many proteins exist in more than one functional copy, e.g., enolase (Holland *et al.*, 1981), the histones H2A and H2B (Hereford *et al.*, 1979). The introns of the 2 genes for rp51 and their 5′ noncoding regions have diverged sufficiently that under certain conditions they do not cross-hybridize. Using this observation, N. Abovich and M. Rosbash (personal communication) have found that both genes are transcribed. In the case of yeast ribosomal protein S10, for which 2 genes appear to exist, two distinct precursor RNAs have been observed (Kim, 1982), suggesting that there are 2 transcriptionally active genes for this protein. This conclusion is supported by the finding that each copy of the gene, when introduced into the cell on a replicating plasmid, gives rise to increased levels of mRNA (W. Schwindinger, personal communication).

As described in the previous section it is now possible to create mutant alleles of yeast genes, even essential genes, for which no mutations were known. Selective inactivation of genes will allow one to determine if two or more genes for a ribosomal protein are functional *in vivo*. DNA sequence analysis will determine if microheterogeneity exists in the amino acid sequence of ribosomal proteins encoded by duplicate genes, a situation not uncommon for multigene families in higher eukaryotes.

The two *Drosophila* ribosomal protein genes isolated so far appear to be unique in the genome (Vaslet *et al.*, 1980; Fabijanski and Pellegrini, 1982b). In *Xenopus*, among the 6 genes that have been cloned, two to five copies of a given ribosomal protein gene are present (Bozzoni *et al.*, 1981). In the mouse, individual ribosomal protein genes exist in numbers from 7 to 20 (Monk *et al.*, 1981). Although it is not unusual to find proteins encoded by multigene families in higher eukaryotes, it is also not unusual to find some members of a family to be pseudogenes. Nevertheless, if some or all of these homologous sequences represent functional genes, then all eukaryotes must compensate for unequal gene dosage in order to maintain an equimolar synthesis of ribosomal proteins.

D. Many Ribosomal Protein mRNAs Are Transcribed as Precursors

To date, well over 65 nuclear genes have been cloned from yeast. Until recently, of those that encode proteins, only the gene for actin had been found to be transcribed into a precursor RNA containing an intervening sequence (Ng and Abelson, 1980; Gallwitz and Sures, 1980).

In the course of Northern blot analysis of ribosomal protein mRNAs in a temperature-sensitive yeast mutant, Rosbash *et al.* (1981) observed the appearance of a higher molecular weight transcript for two ribosomal protein genes. The appearance of these larger transcripts coincided with a disappearance of the mature ribosomal protein mRNAs from cells transferred to the restrictive temperature. The mutant that these investigators examined is one of a group of 10 nonallelic temperature-sensitive mutations, known as *rna* mutants, characterized originally as being conditionally defective in ribosome synthesis (Hartwell *et al.,* 1970).

The *rna* mutants have been the object of considerable study. Warner and Udem (1972) found that the transcription of rRNA is relatively unaffected in these mutants, but it fails to be processed and is degraded (thus the RNA⁻ phenotype). Further study revealed that upon temperature shift there is a specific and rapid decline in the amount of mRNA for ribosomal proteins, whereas the concentration of mRNA for nonribosomal proteins is relatively unaffected (Gorenstein and Warner, 1976; Warner and Gorenstein, 1977). This phenomenon was documented by measuring the synthesis of ribosomal proteins *in vivo* or the translational capacity of mRNA to synthesize ribosomal proteins *in vitro*. The results of Rosbash *et al.* (1981) confirmed that for three of the *rna* mutants, the accumulation of the mRNA for at least two ribosomal proteins is defective. These studies have been extended to the mRNA for five other ribosomal proteins by Kim and Warner (1983a). Further, Fried *et al.* (1981) found that for 8 of 11 ribosomal protein genes examined, a larger transcript appears in an *rna* mutant.

By comparing protected DNA fragments from S1 nuclease mapping of RNA obtained from the *rna*2 mutant incubated at the permissive and nonpermissive conditions, Rosbash *et al.* (1981) found that the higher molecular weight transcript of the gene for rp51 is an intron-containing precursor of the ribosomal protein mRNA. This precursor can be converted to mature mRNA if the *ts* cells are returned to a permissive temperature (Bromley *et al.,* 1982). Bollen *et al.* (1982) have observed introns in several ribosomal protein genes of *S. car-lsbergensis* by electron microscopic analysis of R-loops. DNA sequence analysis has now confirmed the presence of introns in the genes for ribosomal proteins rp51 (Teem and Rosbash, 1983), S10 (rp9) (Leer *et al.,* 1982), and L29 (rp44), the *cyh*2 gene (N. Kaufer *et al.,* 1983). Because most ribosomal protein genes were found to hybridize to large molecular weight transcripts, it appears that most ribosomal protein genes contain introns. Since these precursor molecules accumulate at the nonpermissive temperature in strains carrying mutations in at least three *rna* loci, it appears that the effect of mutations in these genes is to inactivate components required for splicing. The generality of the *rna* mutants is attested by the fact that they also prevent splicing of the actin gene (Teem and Rosbash, 1983).

The role of RNA processing in the coordinate synthesis of ribosomal proteins is open to speculation. Although most yeast ribosomal protein genes appear to contain introns, two at least do not, as shown by sequence analysis of the *tcm*1 gene (rp1) (Schultz and Frieser, 1983) and the gene for rp 39 (J. Woolford, personal communication). Thus, splicing cannot be a general control mechanism unless other factors regulate mRNA synthesis from non-intron-containing genes so as to keep in step with intron-containing genes. Introduction of the *cyh*2 gene on a multicopy plasmid in yeast results not only in an overabundance of mRNA for L29 (rp44) but in an increased amount of the precursor mRNA (H. M. Fried and J. R. Warner, unpublished). This observation may suggest that for those genes with introns the rate of processing is controlled in such a way as to balance the concentration of ribosomal protein mRNAs. Alternatively, the accumulation of precursor RNA may reflect a difference between transcripts from the genome and from a plasmid, perhaps in their location within the nucleus. The fact that the products of 10 genes or more (the *rna* genes) are somehow involved in splicing and that only ribosomal protein genes and the actin gene have retained this feature of genetic arrangement may imply that processing serves as an important regulatory mechanism for these essential proteins.

A number of questions pertaining to splicing remain unanswered. Why are the products of at least 10 genes necessary for splicing? Could they constitute a multiprotein complex which carries out the splicing reactions (Sharp, 1981), or are some of the gene products involved in steps that must precede splicing, such as formation of ribonucleoprotein particles? Are some of these proteins simply splicing enzymes or RNA-packaging proteins, or do some act in a regulatory capacity? Interestingly, Pearson *et al.* (1982b) have isolated a single, dominant, unlinked suppressor which overcomes the effect of the *rna* mutations, suggesting the involvement of at least one other component in RNA processing and showing that the products of at least these 6 genes must interact in a common process. On the other hand, the *rna* mutations have no effect on the rate of ribosomal protein synthesis when diploid cells are simultaneously induced to sporulate (although the mutations prevent completion of sporulation), suggesting that a different mode of regulation may be operating in this growth condition (Pearson and Haber, 1980). Perhaps the recent cloning of 3 of the *rna* genes (R. Last and J. Woolford, personal communication) will lead to answers to some of these questions.

E. DNA Sequences of Ribosomal Protein Genes

At this writing sequence analysis is near completion for approximately seven of the yeast ribosomal protein genes, although not all are yet published. It will soon be possible to examine the sequences to ask about characteristics of codon usage, to determine the size and amino acid sequences of the proteins, and to search for sequences that may play a role in coordination of ribosomal protein gene expression.

So far, the splice junction sequences in ribosomal protein genes conform to the consensus sequence common to all eukaryotic intron–exon boundaries (Mount, 1982):

$$5'\text{- - - - - - - - - -G/GTATGT- - - - - - - - - - -AGAG/ - - - - - - -3'}$$

as does the yeast actin gene. Of further interest is the fact that the single splice site occurs very near to the AUG initiation codon. Electron microscopic analysis of R-loops of most ribosomal protein genes has failed to reveal introns, although Northern analysis suggests the presence of precursor mRNAs for these proteins. Very short 5' exons would explain this discrepancy. The close proximity of the 5' splice junction to the 5' end of the mRNA could reflect a particular structural requirement of the enzyme(s) involved in processing.

The presence of intervening sequences in so many yeast ribosomal protein genes is striking. Because the 5' exon is so short it is difficult to argue that the two exons represent functional domains of the proteins (Gilbert, 1978). The 5' exons could, however, represent a regulatory domain of the genes. If one argues (Darnell, 1978) that intervening sequences were the rule in primordial genes, and that their absence in prokaryotes and paucity in lower eukaryotes is due to selection against them, then one must conclude that the retention of introns in yeast ribosomal protein genes implies that these genes are more conserved than most, or that the introns serve some regulatory function which is not, however, essential for all ribosomal protein genes.

IV. METABOLISM OF RIBOSOMAL PROTEINS

A. Economy of mRNAs for Ribosomal Proteins

By use of DNA-excess hybridization it is possible to measure the rate of synthesis, half-life, and steady state concentration of the mRNAs for individual ribosomal proteins. Kim and Warner (1983a) carried out such an analysis of the mRNAs for several yeast ribosomal proteins and two control proteins. Radioactive uracil was added to a culture from which aliquots were removed at frequent intervals. Determination of the specific activity of the UTP pool and the UMP in individual mRNAs over a period of time leads to a value for the half-life of each mRNA by analysis of the approach to equilibrium labeling (Greenberg, 1972). From the amount of radioactivity present at equilibrium and the size of the mRNA one can calculate the relative concentration of the mRNA for the individual proteins. These data suggest that there are approximately equimolar amounts of each mRNA for a ribosomal protein and that their half-lives are similar (Kim and Warner, 1983a). Therefore, there must be approximately equimolar transcription of the ribosomal protein genes in spite of their being scattered all over the genome and the fact that there are two active copies of some genes, e.g., S10 (rp9) and rp39, and one copy for others, e.g., L3 (rp1) and L32 (rp73).

These data are in significant contrast to those obtained using mouse cells, in which the mRNA from different genes appears to differ in concentration by as much as a factor of 10 (Meyuhas and Perry, 1980; Chapter 10, this volume).

B. Regulation of the Levels of mRNA for Ribosomal Proteins

As discussed in the Introduction, the rate of synthesis of ribosomes is a sensitive barometer for the mood of the cells. For example, when growing cells are deprived of a nitrogen source by being placed in sporulation medium, the rate of synthesis of ribosomal proteins (Pearson and Haber, 1980) and the level of mRNA for ribosomal proteins (Kraig et al., 1982) fall substantially. As another example, when wild-type yeast cells are subjected to an upshift in temperature, they undergo a transient decrease in the synthesis of all ribosomal proteins (Gorenstein and Warner, 1976) due to a decline in active mRNA for those proteins (Warner and Gorenstein, 1977). Using Northern gels probed with ribosomal protein genes, Rosbash et al. (1981) showed that there is in fact a substantial but transient decline in the concentration of mRNA sequences for four ribosomal proteins. These observations have been extended to several more ribosomal proteins (Kim and Warner, 1983b). It is now clear from hybridization of pulse-labeled RNA to cloned DNA that this decline is due to a precipitous drop in the rate of transcription of the genes for ribosomal proteins, whereas the transcription of other genes remains relatively unaffected (Kim and Warner, 1983b). Within 15 min of the temperature shift the transcription of ribosomal protein genes resumes.

The mechanisms that coordinate the expression of so many widely distributed genes are yet to be determined. A common transcriptional regulatory sequence, if it exists, will require definition of the promoter regions of ribosomal protein genes and extensive DNA sequence comparisons. Thus far, cross-hybridization between yeast ribosomal protein genes has not been observed even under conditions of very low stringency (H. M. Fried and J. R. Warner, unpublished observation), so a common regulatory sequence is not apparent at this level of analysis. The precise mechanism of such coordinate transcription will be an intriguing question for some time.

Analysis of the transcription of ribosomal protein genes in cells carrying temperature-sensitive mutations in the *rna* genes suggests these mutants are subject to the same transient inhibition of transcription as wild type after a shift from 23 to 36°C (Bromley *et al.*, 1982; Kim and Warner, 1983b). However, when they resume transcription, the mutant cells are unable to splice the intron from the transcript (Rosbash *et al.*, 1981) and thus are unable to synthesize ribosomes.

An analogous coordination in the transcription of ribosomal protein genes and/or the level of ribosomal protein mRNA has been demonstrated in regenerating rat liver (Faliks and Meyuhas, 1982) and developing embryos of *Xenopus* (Pierandrei-Amaldi *et al.*, 1982). These results are described in Chapter 10.

C. Regulation of the Production of Ribosomal Proteins

Many of the experiments on *E. coli* which led to the model of autogenous regulation of the synthesis of ribosomal proteins have been the result of introducing additional copies of their genes into a cell (reviewed by Lindahl and Zengel, 1982). The acquisition of biologically marked ribosomal protein genes has now made similar experiments possible in yeast. An autonomously replicating plasmid containing the resistance allele of the *tcm*1 gene introduced into a yeast cell conferred partial resistance to trichodermin, demonstrating that the plasmid-associated gene for protein L3 was active. Pearson *et al.* (1982a) found that such cells contained about seven times the usual number of *tcm*1 genes, all of which appeared to be transcribed normally. The steady state concentration of *tcm*1 mRNA, measured by DNA excess hybridization or by Northern gels, was only three to four times normal, suggesting that the lifetime of the transcript was reduced compared to normal cells. Translation in a heterologous extract demonstrated that all of the mRNA was potentially active. However, the rate of synthesis of protein L3, measured by a pulse of 3 min or even 30 sec (N. J. Pearson, personal communication), was only marginally greater than that in normal cells. Similar results have been obtained with a cell containing a plasmid carrying the *cyh*2 gene (H. M. Fried and J. R. Warner, unpublished).

These results led to the conclusion that the cell has a mechanism that maintains an equimolar synthesis of ribosomal proteins by regulating the efficiency of translation of their mRNAs. Using the data on mRNA levels (Section IV,A; Kim and Warner, 1983a), one can calculate that in cells growing normally on glucose the translation of mRNAs for ribosomal proteins is substantially less efficient than that of the average mRNA. Perhaps then the regulation that is observed in cells with an abnormal dosage of ribosomal protein genes is simply an exaggerated manifestation of the regulation that normally modulates the translation of ribosomal proteins to maintain the required equimolar ratio.

A similar regulation of the translation of ribosomal protein mRNA was observed by Pierandrei-Amaldi *et al.* (1982) in anucleolate *Xenopus* embryos, which accumulate mRNAs for ribosomal proteins but, in the absence of rRNA transcription, do not synthesize ribosomal proteins.

These results are formally analagous to the results in *E. coli* which led Nomura and colleagues (reviewed in Lindahl and Zengel, 1982) to propose autogenous regulation of the synthesis of ribosomal proteins. In this model, for each ribosomal protein operon one ribosomal protein can, if not bound immediately to newly formed rRNA, bind to its mRNA to prevent its translation. There is in effect a competition between rRNA and mRNA for the ribosomal protein. Although this suggestion is appealing, two factors caution against its direct extension to eukaryotic cells. (1) Since ribosomal protein genes are monocistronic, each ribosomal protein would have to autogenously regulate its own synthesis. (2) Since mRNA is in the cytoplasm and newly formed rRNA is in the nucleus, any competition between them may be hindered by the presence of the nuclear envelope. Further experiments are clearly in order.

V. FUTURE PROSPECTS

The cloning of ribosomal protein genes has provided a number of tools with which to dissect the regulation of ribosome synthesis. Cloned genes will be used to measure rates of transcription and processing, and half-life and steady state levels of accumulation of ribosomal protein mRNAs in cells growing under a variety of conditions, as well as during shifts between them. The ability to modify cloned genes by various methods of site-specific mutagenesis and to reintroduce these genes into yeast will make it possible to establish which nucleotide regions are responsible for transcriptional regulation seen after temperature shock, nitrogen starvation, etc. Similar methods will also make it possible to identify sites in transcribed sequences responsible for the translational modulation of ribosomal protein mRNAs which is observed when an imbalance of an individual mRNA exists. The role that mRNA processing plays in coordinating ribosomal protein synthesis should be amenable to analysis also. For instance, J. Teem and M. Rosbash (personal communication) have constructed a hybrid gene with lacZ sequences downstream from the intron of rp51. β-Galactosidase is observed only in the presence of a functional *RNA2* gene.

As mentioned in Section II,C, methods are now available for replacing a wild-type chromosomal allele of a cloned yeast gene with an allele incapable of producing a normal gene product. Application of this technique will permit the construction of diploid strains with a null mutation in one of two alleles of a ribosomal protein gene, as has already been accomplished for yeast histone genes (Rykowski *et al.*, 1981) and the yeast actin gene (Shortle *et al.*, 1982). One can ask then if transcriptional or translational modulation of the remaining functional allele can compensate for the deficiency, thus revealing further the mechanisms that balance synthesis of ribosomal proteins.

In addition to studying the coordinate regulation of ribosomal protein gene expression, the availability of cloned genes will permit analysis of ribosome structure and assembly. A number of ribosome-associated suppressor mutations have been identified (reviewed by Sherman, 1982) as well as mutations in other unlinked loci that cause loss of this suppression. These mutations probably occur in genes that specify the proteins comprising the mRNA-decoding region of the ribosome; cloning of these genes would provide a more detailed insight into this region of the eukaryotic ribosome.

The gene replacement methods will permit one to determine what regions of ribosomal proteins are important in their function and perhaps the function of a particular protein. A diploid strain in which a null mutation has been created in one of the two alleles of a gene can be transformed with a plasmid clone of the same gene. One can, in addition, introduce a small alteration in the coding sequence of the plasmid-associated gene prior to transformation. After transformation, sporulation of the diploid will produce two normal haploid spores and two spores whose viability depends on the degree to which the plasmid-associated gene can substitute for the null mutation. If viable spores are obtained, the segment of amino acids that was altered must not be essential for function. Grunstein and co-workers have pioneered this method by modifying the H2B histone gene to show that the first five amino acids are dispensable (Grunstein *et al.*, 1982).

The phosphorylation of protein S10 (analogous to S6 in mammalian cells) has been of intense interest because the role of this modification is unknown. The sequence of this protein [derived from the DNA sequence (Leer *et al.* 1982)] is now available and will provide the framework with which to identify the sites of phosphorylation. By replacing these sites with amino acids that cannot be phosphorylated, the function of phosphorylation can be determined *in vivo*. Similar approaches could be used to study the geometry of drug-binding sites, to verify RNA–protein and protein–protein contact points in the ribosome, and so forth. Finally, it may be possible to introduce mutations that cause either the assembly or functioning of a ribosomal protein to be temperature sensitive, thus providing conditional mutations to study these processes. Such temperature-sensitive mutations have in fact been produced in the yeast actin gene (D. Shortle, personal communication).

Thus, the possession of cloned genes will make possible analysis of the structure and function of eukaryotic ribosomes and the regulation of their synthesis.

As work in these areas proceeds it will be interesting to consider which elements of the ribosome have diverged through evolution and which have been rigidly conserved.

ACKNOWLEDGMENTS

We acknowledge fruitful discussions with John Woolford, Michael Rosbash, and Maria Pellegrini. We are grateful to Grace Sullivan for assistance in preparing the manuscript.

Work from the authors' laboratories was supported by grants from the NIH GM 25532 and CA 23330 and from the American Cancer Society NP 72L.

REFERENCES

Bakken, A., Morgan, G., Sollner-Webb, B., Roan, J., Busby, S., and Reeder, R. H. (1982). Mapping of transcription initiation and termination signals on *Xenopus laevis* ribosomal DNA. *Proc. Natl. Acad. Sci. U.S.A.* **79,** 56–60.

Benton, W., and Davis, R. (1977). Screening gt recombinant clones by hybridization to single plaques *in situ. Science* **196,** 180–182.

Berk, A. J., and Sharp, P. A. (1977). Sizing and mapping of early adenovirus mRNAs by gel electrophoresis of S1 endonuclease digested hybrids. *Cell* **12,** 721–732.

Bollen, G. H. P. M., Cohen, L. H., Mager, W. H., Klaassen, A. W., and Planta, R. J. (1981). Isolation of cloned ribosomal protein genes from the yeast *Saccharomyces carlsbergensis. Gene* **14,** 279–287.

Bollen, G. H. P. M., Molenaar, C. M. T., Cohen, L. H., van Raamsdonk-Duin, M. M. C., Mager, W. H., and Planta, R. J. (1982). Ribosomal protein genes of yeast contain intervening sequences. *Gene* **18,** 29–38.

Bozzoni, I., Beccari, E., Zhong, X. L., and Amaldi, F. (1981). *Xenopus laevis* ribosomal protein genes: Isolation of recombinant cDNA clones and study of the genomic organization. *Nucleic Acids Res.* **9,** 1069–1086.

Bozzoni, I., Tognoni, A., Pierandrei-Amaldi, P., Beccari, E., Buongiorno-Nardelli, M., and Amaldi, F. (1982). Isolation and structural analysis of ribosomal protein genes in *Xenopus laevis.* Homology between sequences present in the gene and in several different mRNA. *J. Mol. Biol.* (to be published).

Broach, J. R., and Hicks, J. B. (1980). Replication and recombination functions associated with the yeast plasmid, 2 circle. *Cell* **21,** 501–508.

Bromley, S., Hereford, L., and Rosbash, M. (1982). Further evidence that the rna2 mutation of yeast affects mRNA processing. *Mol. Cell. Biol.* **2,** 1205–1211.

Casey, J., and Davidson, N. (1977). Rates for formation and thermal stabilities of RNA:DNA and DNA:DNA duplexes at high concentrations of formamide. *Nucleic Acids Res.* **4,** 1539–1551.

Chambliss, G., Craven, G. R., Davies, J., Davis, K., Kahan, L., and Nomura, M., eds. (1980). "Ribosomes: Structure, Function and Genetics." University Park Press, Baltimore, Maryland.

Cooper, D., Banthorpe, D. V., and Wilkie, D. (1967). Modified ribosomes conferring resistance to cycloheximide in mutants of *Saccharomyces cerevisiae. J. Mol. Biol.* **26,** 347–350.

Darnell, J. E. (1978). Implications of RNA–RNA splicing in evolution of eukaryotic cells. *Science* **202,** 1257–1260.

Dennis, P. P. (1974). *In vivo* stability, maturation and relative differential synthesis rates of individual ribosomal proteins in *Escherichia coli* B/r. *J. Mol. Biol.* **88**, 25–41.

D'Eustachio, P., Meyuhas, O., Ruddle, F., and Perry R. P. (1981). Chromosomal distribution of ribosomal protein genes in the mouse. *Cell* **24**, 304–312.

Fabijanski, S., and Pellegrini, M. (1982a). A *Drosophila* ribosomal protein gene is located near repeated sequences including rDNA sequences. *Nucleic Acids Res.* **10**, 5979–5991.

Fabijanski, S., and Pellegrini, M. (1982b). Isolation of a cloned DNA segment containing a ribosomal protein gene of *Drosophila melanogaster*. *Gene* **18**, 267–276.

Faliks, D., and Meyuhas, O. (1982). Coordinate regulation of ribosomal protein mRNA level in regenerating rat liver. Study with the corresponding mouse cloned cDNAs. *Nucleic Acids Res.* **10**, 789–801.

Fangman, W. L., and Zakian, V. A. (1982). Genome structure and replication. *In* "Molecular Biology of the Yeast *Saccharomyces*" (J. Strathern, E. Jones, and J. Broach, eds.), pp. 27–58. Cold Spring Harbor Lab., Cold Spring Harbor, New York.

Fried, H. M., and Warner, J. R. (1981). Cloning of yeast gene for trichodermin resistance and ribosomal protein L3. *Proc. Natl. Acad. Sci. U.S.A.* **78**, 238–242.

Fried, H. M., and Warner, J. R. (1982). Molecular cloning and analysis of yeast gene for cycloheximide resistance and ribosomal protein L29. *Nucleic Acids Res.* **10**, 3133–3148.

Fried, H. M., Pearson, N. J., Kim, C. H., and Warner, J. R. (1981). The genes for fifteen ribosomal proteins of *Saccharomyces cerevisiae*. *J. Biol. Chem.* **251**, 10176–10183.

Gallwitz, D., and Sures, I. (1980). Structure of a split yeast gene: Complete nucleotide sequence of the actin gene in *Saccharomyces cerevisiae*. *Proc. Natl. Acad. Sci. U.S.A.* **77**, 2546–2550.

Gilbert, W. (1978). Why genes in pieces? *Nature (London)* **271**, 501–503.

Gorenstein, C., and Warner, J. R. (1976). Coordinate regulation of the synthesis of eukaryotic ribosomal proteins. *Proc. Natl. Acad. Sci. U.S.A.* **73**, 1547–1551.

Gorenstein, C., and Warner, J. R. (1979). The monocistronic nature of ribosomal protein genes in yeast. *Curr. Genet.* **1**, 9–12.

Grant, P. G., Sanchez, L., and Jimenez, A. (1974). Cryptopleurine resistance: Genetic locus for a 4OS ribosomal component in *Saccharomyces cerevisiae*. *J. Bacteriol.* **120**, 1308–1318.

Grant, P. G., Schindler, D., and Davies, J. E. (1976). Mapping of trichodermin resistance in *Saccharomyces cerevisiae:* A genetic locus for a component of the 60S ribosomal subunit. *Genetics* **83**, 667–682.

Greenberg, J. R. (1972). High stability of messenger RNA in growing cultured cells. *Nature (London)* **240**, 102–104.

Grunstein, M., Wallis, J., Kolodrubetz, D., Rykowski, M., and Choe, J. (1982). A genetic approach to histone function in yeast. *Abstr., Int. Conf. Yeast Genet. Mol. Biol., 11th*.

Hardy, S. J. S. (1975). The stoichiometry of the ribosomal proteins of *Escherichia coli*. *Mol. Gen. Genet.* **140**, 253–261.

Hartwell, L., McLaughlin, C., and Warner, J. R. (1970). Identification of ten genes that control ribosome formation in yeast. *Mol. Gen. Genet.* **109**, 42–56.

Hereford, L. M., and Rosbash M. (1977). Regulation of a set of abundant messenger RNA sequences. *Cell* **10**, 463.

Herford, L. M., Fahrner, K., Woolford, J., and Rosbash, M. (1979). Isolation of yeast histone genes H2A and H2B. *Cell* **18**, 1261–1271.

Hicks, J., Strathern, J. N., and Klar, A. J. S. (1979). Transposable mating type genes in *Saccharoyces cerevisiae*. *Nature* **282**, 478–483.

Hinnen, A., Hicks, J. B., and Fink, G. R. (1978). Transformation of yeast. *Proc. Natl. Acad. Sci. U.S.A.* **75**, 1929–1933.

Holland, M. J., Holland, J. P., Thill, G. P., and Jackson, K. A. (1981). The primary structures of two yeast enolase genes. *J. Biol. Chem.* **256**, 1385–1395.

Jimenez, A., Littlewood, A., and Davies, J. (1972). In "Molecular Mechanisms of Antibiotic Action on Protein Synthesis and Membranes" (E. Munoz, F. Garcia-Ferrandiz, and D. Vazquez, eds.), p. 292. Elsevier, Amsterdam.

Kaback, D. B., Angerer, L. M., and Davidson, N. D. (1979). Improved methods for the formation and stabilization of R-loops. *Nucleic Acids Res.* **6**, 2499–2515.

Kaltschmidt, E., and Wittman, H. G. (1970). Ribosomal proteins VII: Two dimensional polyacrylamide gel electrophoresis for fingerprinting of ribosomal proteins. *Anal. Biochem.* **36**, 401.

Käufer, N. F., Fried, H. M., Schwindinger, W. F., Jasin, M., and Warner, J. R. (1983). Cycloheximide resistance in yeast: The gene and its protein. *Nucleic Acids Res.* **11**, 3123–3135.

Kim, C. K. (1982). The transcriptional regulation of yeast ribosomal protein biosynthesis, Ph.D. Thesis, Albert Einstein College of Medicine.

Kim, C. K., and Warner, J. R. (1983a). Messenger RNA for ribosomal proteins in yeast. *J. Mol. Biol.* **165**, 79–89.

Kim, C. K., and Warner, J. R. (1983b). Mild temperature shock alters the transcription of a discrete class of *Saccharomyces cervisiae* genes. *Mol. Cell. Biol.* **3**, 457–465.

Kraig, E., Haber, J. E., and Rosbash, M. (1982). Sporulation and rna2 lower ribosomal protein mRNA levels by different mechanisms in *Saccharomyces cerevisiae*. *Mol. Cell. Biol.* **2**, 1199–1204.

Kruiswijk, T., and Planta, R. J. (1974). Analysis of the protein composition of yeast ribosomal subunits by two dimensional acrylamide gel electrophoresis. *Mol. Biol. Rep.* **1**, 409.

Larkin, J. C., and Woolford, J. (1983). Molecular cloning and analysis of the CRY1 gene: A yeast ribosomal protein gene. *Nucleic Acids Res.* (in press).

Leer, R. J., vanRaamsdonk-Dvin, M. M. C., Molenaar, C. M. T., Cohen, L. H., Mager, W. H., and Planta, R. J. (1982). The structure of the gene coding for the phosphorylated ribosomal protein S10 in yeast. *Nucleic Acids Res.* **10**, 5869–5878.

Lindahl, L., and Zengel, J. M. (1982). Expression of ribosomal genes in bacteria. *Adv. Genet.* **21**, 53–121.

Maaloe, O., and Kjeldgaard, N. O. (1966). "Control of Macromolecular Synthesis." Benjamin, New York.

McAlister, L., and Holland, M. J. (1982). Targeted deletion of a yeast enolase structural gene. *J. Biol. Chem.* **257**, 7181–7188.

Mets, L., and Bogorad, L. (1974). Two-dimensional polyacrylamide gel electrophoresis: An improved method for ribosomal proteins. *Anal. Biochem.* **57**, 200.

Meyuhas, O., and Perry, R. P. (1980). Construction and identification of cDNA clones for several mouse ribosomal proteins: Application for the study of r-protein gene expression. *Gene* **10**, 113–129.

Miller, O. L., Jr. (1981). The nucleolus, chromosomes and visualization of genetic activity. *J. Cell Biol.* **91**, No. 3, Pt. 2, 15s–27s.

Monk, R. J., Meyuhas, O., and Perry, R. P. (1981). Mammals have multiple genes for individual ribosomal proteins. *Cell* **24**, 301–306.

Mortimer, R., and Hawthorne, D. (1966). Genetic mapping in *Saccharomyces*. *Genetics* **53**, 165–172.

Mount, S. M. (1982). A catalogue of splice junction sequences. *Nucleic Acids Res.* **10**, 459–472.

Ng, R., and Abelson, J. (1980). Isolation and sequence of the gene for actin in *Saccharomyces cerevisiae*. *Proc. Natl. Acad. Sci. U.S.A.* **77**, 3912–3916.

Nomura, M., Tissières, A., and Lengyel, P., eds. (1974). "Ribosomes." Cold Spring Harbor Lab., Cold Spring Harbor, New York.

Otaka, E., and Osawa, S. (1981). Correlation of several nomenclatures for yeast ribosomal proteins: A proposal of standard nomenclature. *Mol. Gen. Genet.* **181**, 176.

Pearson, N. J., and Haber, J. (1980). Changes in regulation of ribosomal protein synthesis during vegetative growth and sporulation of *Saccharomyces cerevisiae*. *J. Bacteriol.* **143**, 1411–1420.

Pearson, N. J., Fried, H. M., and Warner, J. R. (1982a). Yeast use translational control to compensate for extra copies of a ribosomal protein gene. *Cell* **29**, 347–355.

Pearson, N. J., Thorburn, P. C., and Haber, J. E. (1982b). A suppressor of temperature sensitive *rna* mutations that affect general and specific messenger RNA processing in yeast. *Mol. Cell. Biol.* **2**, 571–577.

Pelham, H. R. B., and Brown, D. D. (1980). A specific transcription factor that can bind either the 5S RNA gene or the 5S RNA. *Proc. Natl. Acad. Sci. U.S.A.* **77**, 4170–4174.

Pierandrei-Amaldi, P., and Beccari, E. (1980). Messenger RNA for ribosomal proteins in *Xenopus laevis* oocytes. *Eur. J. Biochem.* **106**, 603–611.

Pierandrei-Amaldi, P., Camponi, N., Beccari, E., Bozzoni, I., and Amaldi, F. (1982). Expression of ribosomal protein genes in *Xenopus laevis* development. *Cell* **30**, 163–171.

Rao, S. S., and Grollman, A. P. (1967). Cycloheximide resistance in yeast: A property of the 60S ribosomal subunit. *Biochem. Biophys. Res. Commun.* **29**, 696–704.

Rosbash, M., Harris, P. K. W., Woolford, J. L., Jr., and Teem, J. L. (1981). The effect of temperature sensitive RNA mutants on the transcription products from cloned ribosomal protein genes of yeast. *Cell* **24**, 679–686.

Rykowski, M. C., Wallis, J. W., Choe, J., and Grunstein, M. (1981). Histone H2B subtypes are dispensable during the yeast cell cycle. *Cell* **25**, 477–487.

Schindler, D., Grant, P., and Davies, S. (1974). Trichodermin resistance-mutation affecting eukaryotic ribosomes. *Nature (London)* **248**, 535–536.

Schultz, L. D., and Friesen, J. D. (1983). Nucleotide sequence of the *tcml* gene (ribosomal protein L3) of *Saccharomyces cerevisiae J. Bacteriol.* **155**, 8–14.

Shalit, P., Coughney, K., Olsen, M. V., and Hall, B. D. (1981). Physical analysis of the CYC1-sup4 interval in *Saccharomyces cerevisiae*. *Mol. Cell. Biol.* **1**, 228–236.

Sharp, P. (1981). Speculations on RNA splicing. *Cell* **23**, 643–646.

Sherman, F. (1982). Suppression in the yeast *Saccharomyces cerevisiae*. *In* ''Molecular Biology of the Yeast *Saccharomyces*'' (J. Strathern, E. Jones, and J. Broach, eds.), Vol. 2. Cold Spring Harbor Lab., Cold Spring Harbor, New York.

Shortle, D., Haber, J., and Botstein, D. (1982). Lethal disruption of the yeast actin gene by integrative DNA transformation. *Science* **217**, 371–373.

Skogerson, L., McLaughlin, C., and Wakatama, E. (1973). Modification of ribosomes in cryptopleurine-resistant mutants of yeast. *J. Bacteriol.* **116**, 818.

Stocklein, W., and Pieperberg, W. (1980). Altered ribosomal protein L29 in cycloheximide-resistant strain of *Saccharomyces cerevisiae*. *Curr. Genet.* **1**, 177–181.

Strathern, J. N., Newlon, C. S., Heskowitz, I., and Hicks, J. B. (1979). Isolation of a circular derivative of yeast chromosome. III. Implications for the mechanism of mating type interconversion. *Cell* **18**, 309–319.

Struhl, K., Stinchcomb, D. T., Scherer, S., and Davis, R. W. (1979). High-frequency transformation of yeast: Autonomous replication of hybrid DNA molecules. *Proc. Natl. Acad. Sci. U.S.A.* **76**, 1035–1039.

Teem, J. L., and Rosbash, M. (1983). Expression of a β-galactosidase gene containing the ribosomal protein 51 intron is sensitive to the *rna2* mutation of yeast. *Proc. Natl. Acad. Sci. U.S.A.* **80**, 4403–4407.

Thomas, M., White, R. L., and Davis, R. W. (1976). Hybridization of RNA to double-stranded DNA: Formation of R-loop. *Proc. Natl. Acad. Sci. U.S.A.* **73**, 2294–2298.

Vaslet, C. A., O'Connell, P., Izquierdo, M., and Rosbash, M. (1980). Isolation and mapping of a cloned ribosomal protein gene of *Drosophila melanogaster*. *Nature (London)* **285**, 674–675.

Warner, J. R. (1977). In the absence of ribosomal RNA synthesis, the ribosomal proteins of HeLa cells are synthesized normally and degraded rapidly. *J. Mol. Biol.* **115**, 315–333.

Warner, J. R. (1982). The yeast ribosome: structure, function and synthesis. *In* "Molecular Biology of the Yeast *Saccharomyces*" (J. Strathern, E. Jones, and J. R. Broach, eds.), pp. 529–560. Cold Spring Harbor Lab., Cold Spring Harbor, New York.

Warner, J. R., and Gorenstein, C. (1977) The synthesis of eukaryotic ribosomal proteins *in vitro*. *Cell* **11**, 201–212.

Warner, J. R., and Udem, S. A. (1972). Temperature sensitive mutations affecting ribosome synthesis in *Saccharomyces cerevisiae*. *J. Mol. Biol.* **65**, 243–257.

Warner, J. R., Tushinski, R. J., and Wejksnora, P. J. (1980). Coordination of RNA and proteins in eukaryotic ribosome production. *In* "Ribosomes: Structure, Function, and Genetics" (G. Chambliss, G. R. Craven, J. Davies, K. Davis, L. Kahan, and M. Nomura, eds.), pp. 889–902. University Park Press, Baltimore, Maryland.

Williams, J. (1981). The preparation and screening of a cDNA clone bank. *In* "Genetic Engineering" (R. Williamson, ed.), Vol. 1, pp. 1–59. Academic Press, New York.

Wittman, H. G. (1982). Components of the bacterial ribosome. *Annu. Rev. Biochem.* **51**, 155–183.

Wool, I. G. (1979). The structure and function of eukaryotic ribosomes. *Annu. Rev. Biochem.* **48**, 719–754.

Woolford, J. L., Jr., and Rosbash, M. (1979). The use of R-looping for structural gene identification and mRNA purification. *Nucleic Acids Res.* **6**, 2483–2497.

Woolford, J. L., Jr., and Rosbash, M. (1981). Ribosomal protein genes rp39 (10-78), rp 39 (11-40), rp 51, and rp 52 are not contiguous to other ribosomal protein genes in the *Saccharomyces cerevisiae* genome. *Nucleic Acids Res.* **9**, 5021–5028.

Woolford, J. L., Jr., Hereford, L. M., and Rosbash, M. (1979). Isolation of cloned DNA sequences containing ribosomal protein genes from *Saccharomyces cerevisiae*. *Cell* **18**, 1247–1259.

Zylber, E. A., and Penman, S. (1971). Synthesis of 5S and 4S RNA in metaphase-arrested HeLa cells. *Science* **172**, 947–949.

II

Expression of Specific Genes
Associated with
Proliferation and
Differentiation

8

Cell Cycle Dependence of Globin Gene Expression in Friend Cells

DAVID CONKIE AND JOHN PAUL
The Beatson Institute for Cancer Research
Garscube Estate
Glasgow, Scotland

RECOMBINANT DNA AND
CELL PROLIFERATION

I. INTRODUCTION

The question of the relationship between cell proliferation resulting in expansion of a cell population and cell differentiation producing specialized cells has been a topic of interest to cell biologists for a number of years.

A major hypothesis of cell diversification is the concept of the "quantal" cell division (Holtzer, 1978; Holtzer *et al.*, 1981). This is envisaged as a critical cell division which produces daughter cells phenotypically different from the parent cell. Evidence for the concept is reviewed by Holtzer and colleagues (1975, 1981) in a variety of differentiating systems such as myogenesis, chondrogenesis, melanogenesis, and hematopoiesis.

The aim of this chapter is to present the evidence concerning the role of the cell division cycle in induced differentiation. Many of the complications of work in this field have arisen from the use of tissues with a complex mixture of differentiating cell types, when extrapolation from the results obtained to the metabolic options of single cells is often equivocal. To avoid this complication, exploitation of a differentiating homogeneous cell line or investigation of single differentiating cells is desirable. The Friend erythroleukemia cell line offers an opportunity for study of both single cells and homogeneous populations.

II. MURINE ERYTHROLEUKEMIA CELLS (FRIEND CELLS)

A. Friend Virus-Induced Leukemia

In 1957 Dr. Charlotte Friend found that Swiss mice inoculated with a cell-free extract of Ehrlich's ascites tumor cells developed an enlarged spleen from which a virus complex (Friend virus) was recovered. On inoculation into susceptible DBA/2 mice this virus produces a prominent erythroblastosis, localized principally in the spleen (Friend, 1957), which later develops into a leukemia.

The target cell for Friend virus appears to be a committed erythroid precursor (Tambourin and Wendling, 1975; Fredrickson *et al.*, 1975) more mature than the day 8 BFU-E and closely related to the CFU-E (Kost *et al.*, 1979). One of the main initial effects may be to render erythropoiesis independent of the normal hormone erythropoietin (reviewed by Harrison, 1977), resulting in the development of polycythemia.

Although Friend virus induces a generalized leukemia in DBA/2 mice and no tumor is found at the site of inoculation, solid tumors can be established from subcutaneous implants of mouse spleen or liver at the late stage of Friend disease (Friend and Haddad, 1960). Cells migrating from excised solid tumor fragments in organ culture replicate indefinitely in cell culture conditions (Friend *et al.*,

1966). Such cells are called Friend cells. Friend cell cultures release virus particles into the culture medium (de Harven and Friend, 1966), and approximately
1% of the cells spontaneously differentiate into hemoglobin-containing erythroblasts which stain with benzidine and incorporate ^{59}Fe into heme (Friend et al.,
1966, 1970). Otherwise, the cells appear undifferentiated and resemble pro-
erythroblasts. By contrast, Friend cells injected into irradiated mice form macroscopic colonies in which all stages of erythroid maturation can be observed
(Rossi and Friend, 1967), although no evidence of maturation is seen in the solid
tumors (Rossi and Friend, 1970). Hence, it is apparent that Friend cells have the
potential to differentiate along the erythroid pathway, but in culture they exhibit
maturational arrest prior to the stage at which hemoglobin is synthesized. Cells
of spleen and bone marrow from Friend virus-infected leukemic mice have
normal chromosomal numbers and karyotype (Majumdar and Bilenker, 1975)
whereas, in established cultures, the cells are aneuploid with chromosome rear-
rangements (Ostertag et al., 1972).

B. Induced Erythroid Differentiation of Friend Cells

In 1971 Friend et al. reported that a majority of cells in a cloned Friend cell
line can be induced by dimethyl sulfoxide to differentiate by accumulating heme
and hemoglobin, accompanied by morphological maturation to a cell stage resembling normal, orthochromatic erythroblasts.

Modification of the culture medium by adding bovine serum albumin permits
complete maturation of Friend cells to anucleate terminal cells resembling re-
ticulocytes (Volloch and Housman, 1982). Morphological differentiation is accompanied by an increase in iron uptake (Friend et al., 1971) and heme synthesis
(Ebert and Ikawa, 1974), accumulation of hemoglobin by 5 days of induction
(Boyer et al., 1972; Ostertag et al., 1972), α- and β-globins characteristic of the
DBA/2 mice from which Friend cells are derived, and cytoplasmic globin
mRNA reaching a maximum after 3 days of induction (Conkie et al., 1974),
appearance of erythrocyte-specific membrane antigens (Ikawa et al., 1973) including glycoproteins (Gazitt and Friend, 1981), and increased spectrin reaching
a peak on the first to second day after induction (Eisen et al., 1977). Other early
induced events include decreased membrane permeability (Dube et al., 1974)
and increased membrane microviscosity (Arndt-Jovin et al., 1976).

It is clear from this experimental evidence that induced Friend cells complete a
program of erythroid differentiation closely resembling that of normal eryth-
ropoietic cells.

In addition to dimethyl sulfoxide a number of other inducing agents are known
to be effective in reversing the maturational arrest of Friend cells. These are
classified as polar–planar compounds (Tanaka et al., 1975), short-chain fatty

acids (Leder and Leder, 1975; Takahashi *et al.*, 1974), purine derivatives (Gusella and Housman, 1976), hemin [which induces globin mRNA accumulation (Ross and Sautner, 1976) but not terminal differentiation (Gusella *et al.*, 1980)], inhibitors of cellular metabolism (Ebert *et al.*, 1976), and ouabain (Bernstein *et al.*, 1976). These compounds are effective at different concentrations and have different inducing properties. The most potent is hexamethylene bisacetamide (HMBA) which induces essentially the entire population of certain cloned Friend cells (Reuben *et al.*, 1976). The mechanism of action of inducers is not clear, although an alteration in membrane function may be the most likely effect. Genetic studies suggest that inducers do not all act by the same mechanism (reviewed by Harrison, 1977). Moreover, in a similar biological system, Graf *et al.* (1978a) describe a maturational arrest in avian erythroblasts transformed by temperature-sensitive avian erythroblastosis virus (*ts*AEV). This arrest can be reversed by culture at the nonpermissive temperature in anemic chick serum when the *ts* viral gene is repressed or by chemical inducers (Graf *et al.*, 1978b). Using clonal variants defective for either temperature-induced differentiation or chemical induction, Beug and colleagues (1982a) demonstrate that these two modes of differentiation induction seem to act by different mechanisms.

III. ROLE OF THE CELL CYCLE IN FRIEND CELL DIFFERENTIATION

A. Timing of Friend Cell Differentiation

The concept of a critical cell division being a prerequisite for differentiation of erythroid cells gains some support from the demonstration that prior DNA synthesis is required for erythropoietin-stimulated hemoglobin synthesis by mouse erythroid stem cells (Paul and Hunter, 1968, 1969; Paul and Conkie, 1973). The necessity for a critical or "quantal" cell division during differentiation of chick hematocytoblasts or presumptive chondroblasts or prior to fusion of myoblasts is reviewed by Holtzer *et al.* (1975).

More recent studies have explored cell cycle dependence of differentiation in Friend erythroleukemic cells. Evidence for a temporal relationship between the response to inducers of differentiation and the cell division cycle of Friend cells is apparent from the experiments of McClintock and Papaconstantinou (1974).

Exponentially growing Friend cells in asynchronous culture require two generations in inducer to accumulate hemoglobin detectable by spectrophotometry at 410 nm. Extension of the generation time by use of a low-serum medium or isoleucine deficiency results in a corresponding increase in the time required for hemoglobin synthesis.

Therefore, a circumstantial correlation exists between the artifically manipulated cell cycle time and the time of onset of hemoglobin production.

Cells synchronized by 44 hr of culture in a double, 2-mM thymidine block (Volpe and Eremenko, 1973) in the presence of dimethyl sulfoxide (DMSO) require a further 4 hr of DMSO treatment after release of the thymidine block for an enhanced (13–65%) induction of hemoglobin synthesis during subsequent culture without the inducer for 5 days; cells released from the block directly into medium without DMSO for 5 days produce only 2–5% benzidine-positive cells (Levy *et al.*, 1975). Synchronization of the cells involves progress through one S phase in the presence of DMSO and accumulation at the G_1/S boundary of the cell cycle. The results suggest that on release from this block, the presence of DMSO is still required during progress into the second S phase for induction of hemoglobin synthesis. Similarly, the work of Gambari *et al.* (1978) implies a critical requirement of inducer during the S phase.

By contrast, Levenson *et al.* (1980b) conclude that DNA synthesis is not required for the induced differentiation of Friend cells treated with fluorodeoxyuridine.

B. Hemoglobin Synthesis in Isolated Friend Cells

We have investigated the same question by studying the induction of single Friend cells isolated in semisolid medium (Harrison *et al.*, 1978). Friend cells plated in Methocel medium in the presence of inducer form a small colony of up to 32 cells during a 6-day incubation period. In these induced clones hemoglobin can be detected in every cell by staining with benzidine. By adding metabolic inhibitors such as cytosine arabinoside, hydroxyurea, or fluorodeoxyuridine, or by depriving the cells of isoleucine, it is possible to arrest up to 80% of the freshing plated cells at the single-cell stage. The remainder of the cells divide once in the presence of inducer prior to growth arrest. Only 3% of the arrested single cells accumulate hemoglobin over 5 days of culture in inducer, whereas 65% of the paired cells are benzidine-positive, suggesting a requirement for cell division in the initial induction of Friend cells. Refeeding deprived cells after 5 days with isoleucine and inducer results in 80% of the single cells forming clones of benzidine-positive cells. This rules out any question of the growth-arrested cells not being viable.

By contrast, cells of an alternative Friend cell clone continue to synthesize hemoglobin in the presence of the inducer butyric acid, even when DNA synthesis is inhibited 88% by cytosine arabinoside or hydroxyurea (Leder *et al.*, 1975). However, most of these single benzidine-positive cells may be binucleate (Harrison *et al.*, 1978), and so this result is consistent with a requirement for nuclear division if not cytokinesis in the induction of differentiation in Friend cells.

IV. FRIEND CELL VARIANTS TEMPERATURE SENSITIVE FOR GROWTH

A. An Alternative Approach

In experiments designed to measure the synthesis of a specific protein (in this case globin), the use of metabolic inhibitors of nucleic acid or protein synthesis to synchronize cell populations may invalidate the end result. With this in mind we decided to adopt an alternative approach exploiting conditional lethal mutants of Friend cells which are thermosensitive for growth.

Using a novel selection procedure two categories of temperature-sensitive (ts) variants can be selected either as cells growth arrested at nonpermissive temperatures or as cells surviving the selection because of an extended generation time. In principle, using ts Friend cells growth arrested at the nonpermissive temperature, it is possible to study the effect of inducers of hemoglobin synthesis, thereby directly investigating the cell cycle dependence of Friend cell differentiation.

B. Isolation of Temperature-Sensitive Friend Cells

The experimental strategy for isolation of Friend cells temperature-sensitive for growth is modified from that described by Puck and Kao (1967) for isolation of auxotrophic mutants and adapted by Naha (1969) for isolating conditional lethal ts mutants of BSC-1 cells. The method has been described in detail (Conkie et al., 1980) and can be summarized as a four-step procedure.

1. Friend cells at 33.5°C are mutagenized with 0.5 μg/ml N-methyl-N-nitro-N'-nitrosoguanidine (MNNG) until the plating efficiency is reduced to 70%. Incubation at 33.5°C is continued for 5 days to permit expression of mutant phenotype.

2. The incubation temperature is raised to 39.0°C for 18 hr to arrest the growth of ts cells.

3. Wild-type replicating cells are eliminated from the culture by incubation for 48 hr at 39.0°C in 10^{-4} M hypoxanthine, 4×10^{-7} M aminopterin, and 10^{-5} M bromodeoxyridine (HAB). Incorporation by replicating cells of the bromodeoxyuridine (BrdUrd) followed by exposure to near-visible light for 1 hr results in lethal chromosome damage which is further enhanced by the presence of the Hoechst fluorescent dye 33258 at 1 μg/ml (Stetten et al., 1977). The role of the aminopterin, besides promoting BrdUrd uptake by wild-type cells, is to eradicate replicating TK$^-$ cells unable to utilize BrdUrd. TK$^-$ cells arise spontaneously in Friend cell populations at a relatively high rate of 2.1×10^{-6}/cell/generation (McKenna and Hickey, 1981) whereas in L5178Y populations,

for example, they occur at a rate of 5×10^{-11}/cell/generation (Clive et al., 1972).

4. Surviving cells are returned to permissive conditions of culture free from selective reagents. The selection is repeated four times, after which clonal replica plating permits simultaneous assessment of growth at 33.5 and 39°C.

Thirty *ts* variants from three independent cultures have now been classified in some detail (Conkie et al., 1980).

C. Characterization of Friend Cells Temperature Sensitive for Growth

The majority of the *ts* clones are strictly temperature sensitive for growth. They become arrested in G_1 of the cell cycle of 39°C as determined by flow cytofluorometry but proliferate with a generation time of 20 hr at 33.5°C, similar to that of wild-type cells at 33.5°C. The cells of these *ts* clones grow in suspension and resemble the parent cell line in morphology, karyotype, and response to inducers of hemoglobin synthesis at 33.5°C. The reversion frequency of representative mutants from each of the three independent selections varies from 1.0 to 3.5×10^{-7}.

As for most other *ts* cells, the original Friend cell *ts* mutants remain viable for only a short period in nonpermissive conditions. Two further procedures help to alleviate this problem. Using a water bath for accurate temperature control, a careful study of the growth curves of cells at various temperatures revealed that the minimum nonpermissive temperature is 37.25°C. At this temperature, 75–80% of the cells accumulate in G_1 within 12 hr. Second, a regimen of repeated clonal selection of *ts* cells subjected to 4.5-day incubations at 39°C produces subclones that retain the ts phenotype and hemoglobin inducibility. However, after three selection cycles these subclones are 10-fold more viable at the nonpermissive temperature than the original *ts* clone.

V. CELL CYCLE-DEPENDENT HEMOGLOBIN SYNTHESIS IN TEMPERATURE-SENSITIVE FRIEND CELLS

A. Induction of Differentiation in Proliferating and G_1- Arrested Friend Cells

The experimental design used to observe the effect of inducer on Friend cells arrested in G_1 is outlined in Fig. 1. In this protocol the *ts* cells are collected in G_1 of the cell cycle by a preincubation of 12 hr at 37.25°C and then exposed to inducer for 30 hr either at 34.25°C (Fig. 1A) or at the nonpermissive 37.25°C (Fig. 1B). Thereafter, for the remainder of the 6-day period normally allowed for

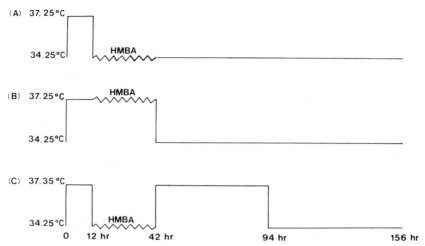

Fig. 1. Diagrammatic representation of the experimental rationale for investigating the effect of inducer on *ts* cells either replicating or growth arrested. See text for details.

assay of cellular hemoglobin accumulation, the cells are cultured in liquid medium without inducer or, in parallel, in semisolid Methocel medium minus inducer. The results are shown in Table I.

G$_1$-arrested *ts* cells maintained in permissive conditions by culture at 34.25°C for 144 hr in the presence of HMBA produce 82% benzidine-positive cells. When exposure to the inducer is reduced to 30 hr (Fig. 1A) followed by incubation without inducer for 114 hr at 34.25°C, 50% of the cells still accumulate hemoglobin. By contrast, no induction of hemoglobin synthesis is detected when G$_1$-arrested cells are exposed to inducer for 30 hr at 37.25°C (Fig. 1B). Globin mRNA levels in these cultures increase only when exposed to inducer at the permissive temperature.

In addition to inducing hemoglobin synthesis in *ts* Friend cells, 30 hr of treatment with HMBA at 34.25°C imposes a specific limitation in proliferative capacity of differentiating cells such that they cease replication after about five cell divisions. During the 114-hr incubation in Methocel medium without inducer, all of the cells of these microcolonies accumulate hemoglobin and can be stained *in situ* as an assay for the proportion of cells initially committed to differentiation. Further, from the proportion of undifferentiated macroscopic colonies evident several days later, it is possible to determine the fraction of the original population that responds to inducer by differentiating terminally. The results of these assays (Table I) show that cells treated with inducer at the nonpermissive temperature do not become restricted in proliferative function.

Taken together, all of these induction experiments with *ts* Friend cells suggest a requirement for some cell cycle-dependent event for induction of differentiation.

B. Critical Evaluation of Experimental Results

In experiments of the type described it is essential to demonstrate that during the G_1 arrest the cells remain viable. That this is substantially so is demonstrated by clonal analysis. The cloning efficiency of G_1-arrested cells treated with inducer at 37.25°C is 70% that of similar cells treated with inducer at 34.25°C. Moreover, when G_1 cells treated with inducer for 30 hr at the nonpermissive temperature are further incubated for 114 hr at 34.25°C in inducer, hemoglobin accumulates in 62% of the cells, showing that the G_1 arrest can be reversed to permit both replication and differentiation.

A further technical objection might be that 42 hr of culture at the nonpermissive temperature inhibits accumulation of hemoglobin quite apart from any requirement for cell proliferation. To test this possibility, G_1-arrested *ts* cells are exposed to inducer for 30 hr at the permissive temperature followed by incubation at 37.25°C for 52 and then 62 hr (total 114 hr) at 34.25°C without inducer (Fig. 1C, Table I). The accumulation of hemoglobin in 54% of the cells implies that incubation at the nonpermissive temperature for up to 52 hr does not per se prevent the accumulation of hemoglobin. Furthermore, it is also clear from this experiment that postcommitment expression of the erythroid phenotype is not cell cycle dependent.

Another possible doubt is that the temperature-sensitive lesion may be in protein or RNA synthesis. The rates of DNA, total RNA, and protein synthesis can be determined in *ts* cells at permissive and nonpermissive temperatures

TABLE I

Induction of Hemoglobin Synthesis in Proliferating and in G_1-Arrested Friend Cells[a]

Culture with 5 mM HMBA (hr)		Subsequent culture without HMBA (hr)		Hemoglobin synthesis		Commitment	
34.25°C	37.25°C	34.25°C	37.25°C	Benzidine-positive cells (%)	Cytoplasmic globin mRNA (ppm)	Benzidine-positive clones (%)	Terminal differentiation (%)
30	0	114	0	50	103	45	47
0	30	114	0	1	28	4	2
30	0	52	62	54	NT	NT	NT

[a] *ts* Friend cells are collected in G_1 by preincubation at 37.25°C for 12 hr. They are then treated with inducer at permissive or nonpermissive temperature as shown. Following removal of inducer by washing, the cells are further incubated for 114 hr as liquid cultures or, in parallel experiments, semisolid Methocel cultures. Benzidine-positive cells are scored on slides from liquid cultures and benzidine-positive clones are scored in Methocel cultures. Cytoplasmic globin mRNA is extracted from *ts* cells and hybridized to a fixed amount of α,β-globin cDNA under conditions in which all complementary globin sequences are known to hybridize. Percentage of terminal differentiation is calculated as 100% macroscopic clones at 11 days. NT, not tested; ppm, parts per million.

(Conkie *et al.*, 1981) using radioactive precursors. G_1-arrested cells cultured 30 hr with inducer at 37.25°C have a cellular content of labeled DNA 2% that of proliferating cells cultured 30 hr with inducer at 34.25°C. By contrast, RNA and protein synthesis continue at 45–50% of the rate found for randomly proliferating induced cells, about the values expected from other studies of actively growing G_1 cells synchronized by mitotic selection (Stambrook and Sisken, 1972; Ronning *et al.*, 1979). The reduced rate of RNA and protein synthesis in G_1-arrested *ts* cells does not therefore imply that the cells are metabolically defective. Indeed, cycloheximide inhibition of protein synthesis by 50% does not prevent differentiation (Levenson and Housman, 1979a).

Finally, in relating the induction of differentiation to the cell division cycle it is necessary to assume that the variant cell line used does not have a double *ts* mutation affecting both growth and induction of hemoglobin synthesis. Furthermore, it is unlikely that a single (temperature-sensitive) event will affect both cell replication and differentiation directly in all the independently isolated *ts* Friend cells, giving results similar to those aforementioned.

VI. COMMITMENT: THE DECISION TO DIFFERENTIATE

A. Evidence for Commitment as a Stochastic Event

In early work with Friend cells it became apparent that the process of induction and irreversible differentiation is preceded by a "lag" phase (Friend *et al.*, 1971). This 12- to 18-hr delay approximately coincides with the time required for one cell division (12–15 hr). Moreover, no increase in globin-specific mRNA can be detected prior to 11 hr of induction (Gambari *et al.*, 1978), whereas Friend virus RNA (which is not a prerequisite for differentiation) may be induced earlier (Pragnell *et al.*, 1980). Stimulation of Friend cells to synthesize hemoglobin may result from only 18–24 hr exposure to inducer. Thereafter, further incubation for up to 5 days in the absence of inducer permits expression of the complete erythroid program (Conkie *et al.*, 1974; Gusella *et al.*, 1976).

These results imply that an early induced cellular event leads to coordinated erythroid differentiation. This event, known as commitment, is implicated in a number of developmental systems (Levenson and Housman, 1981). Recently, an early irreversible commitment step in the hormone-dependent differentiation of *ts*AEV transformed erythroblasts has been described by Beug *et al.* (1982b). These *ts*AEV cells normally differentiate only at the nonpermissive temperature. However, they differentiate into erythrocytes even at the permissive temperature following a 30-hr pulse shift to the nonpermissive temperature.

Gusella *et al.* (1976) have reported a further, more detailed investigation of Friend cell differentiation resulting from a brief (24-hr) exposure to the inducer DMSO. Clonal analysis permits precise determination of the fate of every cell. A committed cell produces a small terminal colony of hemoglobin-containing cells despite subsequent absence of inducer. Uncommitted cells produce colonies without hemoglobin and with unrestricted proliferative potential. Commitment is dependent on the concentration of inducer and duration of exposure to inducer. The alternative inducer HMBA similarly recruits Friend cells to the committed state (Fibach *et al.*, 1977). The data from such experiments fit a mathematical model based on a stochastic commitment decision (Gusella *et al.*, 1976). Furthermore, a stochastic basis for the commitment of normal hematopoietic stem cells *in vivo* is suggested from the experimental evidence obtained by Till and McCulloch (1961) and Till *et al.* (1964).

B. Nature of Commitment at the Molecular Level

Additional experiments designed to investigate the lag phase of induction (Levenson and Housman, 1979b) establish that at least 9 hr of exposure to inducer are required before cells become committed. Cells withdrawn from inducer during this latent period can reinitiate commitment without a lag when exposed to inducer once again. Memory of the previous exposure to inducer appears to be effective for up to 18 hr, suggesting accumulation of a stable cellular component. Moreover, two different inducers used consecutively require the same lag period as a single inducer used alone (Housman *et al.*, 1978). This demonstrates that, at least for certain inducers, a common mechanism of commitment exists.

The molecular events leading to commitment are not fully understood. Yet it is clear from the following experiments that two processes are involved. A reversible inhibition of commitment occurs with concentrations of cycloheximide or puromycin that do not retard normal cell proliferation (Levenson and Housman, 1979a). The delay is equal to the duration of inhibitor treatment, suggesting a continuous requirement for protein synthesis as a rate-limiting step before commitment can be initiated. Similarly, protein synthesis is a prerequisite, as is RNA synthesis, for erythropoietin stimulation of normal erythroid stem cells (Paul and Hunter, 1969).

Experimental evidence for a second process is apparent from the effects of the nucleotide analogue cordycepin (3′-deoxyadenosine). At concentrations that are nontoxic for cell proliferation, cordycepin rapidly and reversibly inhibits commitment of DMSO-treated Friend cells (Levenson *et al.*, 1979). Removal of cordycepin leads to a rapid and synchronous commitment, suggesting that this inhibitor blocks a process (perhaps mRNA polyadenylation) essential, but not normally rate-limiting, for commitment, in contrast to the rate-limiting process sensitive to inhibitors of protein synthesis. The kinetics of recovery from cordycepin treatment also suggest that Friend cells can be synchronized at a point just prior to commitment. The continued presence of inducer is required for about 2 hr after reversal of the cordycepin effect for commitment to occur.

Further insight into the nature of commitment is provided by experiments using the local anesthetic, procaine. Procaine is known to inhibit the DMSO-induced differentiation of Friend cells, mediated by an effect on the cell membrane (Lyman *et al.*, 1976). Recently, the result of pretreating the Friend cells before induction with procaine reveals that commitment is reversibly inhibited by more than 90%. More importantly, reversal of procaine inhibition (Tsiftsoglou *et al.*, 1981) and amiloride inhibition (Levenson *et al.*, 1980a) is accompanied by accumulations of calcium ions. Observations on the effects of EGTA and calcium ionophore reinforce the evidence that cell membrane-controlled calcium flux may be important in the process leading to commitment. Moreover, it is now apparent that accumulation of calcium ions is accompanied by elimination of the lag phase of induction, and so calcium transport may be the rate-limiting event required for commitment to initiate a program of terminal erythroid differentiation (Bridges *et al.*, 1981). Nevertheless, inducer treatment is still essential for the non-rate-limiting step sensitive to low concentrations of cordycepin, since the provision of calcium alone in the absence of inducer does not initiate commitment to differentiation.

C. Transient Growth Arrest of Induced Friend Cells

The induction of asynchronous Friend cells with DMSO results in a transient growth arrest in G_1 of the cell cycle as judged by thymidine labeling index and flow microfluorometry (Terada *et al.*, 1977). Furthermore, induced globin mRNA is first detected during the transient G_1 arrest in Friend cells synchronized by excess thymidine and hydroxyurea (Gambari *et al.*, 1978) or by centrifugal elutriation (Pragnell *et al.*, 1980). Therefore, it is likely that completion of the latent stage of commitment to differentiation is marked by a transient growth arrest in G_1 of the cell cycle.

However, a lengthened G_1 phase is not found with all inducers of Friend cell differentiation (Friedman and Schildkraut, 1978). This implies that the prolonged G_1 phase observed with many inducers is not a prerequisite for terminal differentiation.

VII. SIGNIFICANCE OF THE CELL CYCLE DURING COMMITMENT TO DIFFERENTIATION

A. Growth Kinetics of Minimally Perturbed Temperature-Sensitive Friend Cells Synchronized by Centrifugal Elutriation

Many studies on the relationship between the cell division cycle and commitment to differentiation require totally viable synchronized cells. The technique of

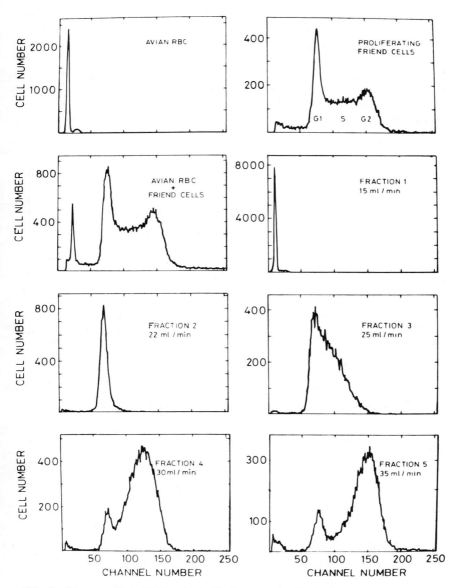

Fig. 2. Flow cytometry of cells stained with chromomycin (Sigma). Profiles are shown of avian peripheral blood cells (RBC), randomly proliferating Friend cells, and a mixture of both cell types. Further profiles record the DNA content of cell fractions elutriated at the flow rates shown.

centrifugal elutriation permits separation of Friend cells on the basis of their position in the cell cycle without the use of drugs or *ts* mutations and so with minimal perturbation of growth kinetics. The centrifugal force acting on the cells in the centrifuge rotor is exactly balanced by the flow of culture medium through the rotor chamber. By altering this equilibrium it is possible to collect cells into the various compartments of the cell division cycle. Moreover, their position in the cell cycle can be monitored by flow cytometry.

Figure 2 shows the flow cytometry profiles of avian red blood cells, randomly proliferating Friend cells, and a mixture of both cell types. Following elutriation

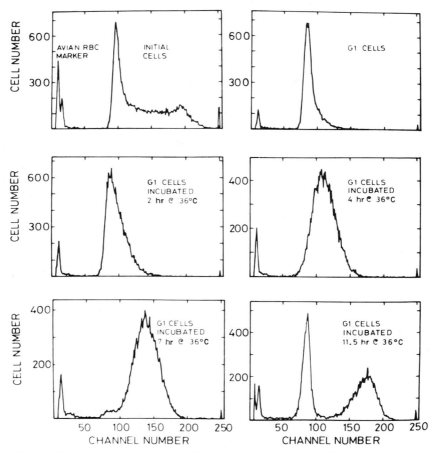

Fig. 3. Flow cytometry of randomly proliferating Friend cells (initial cells), elutriated G_1 Friend cells, and G_1 cells further cultured for the times shown at 36°C. Avian RBCs were added to each sample as a distance marker.

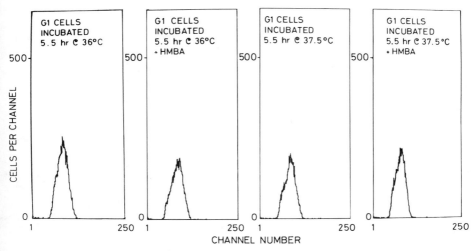

Fig. 4.   Flow cytometry profiles of G_1 cells further cultured for 5.5 hr at 36 or 37.5°C with or without the inducer HMBA.

at 15 ml/min only avian red blood cells are collected. At 22 ml/min G_1 Friend cells are collected, followed by G_1 + early S-phase cells at 25 ml/min. At 30 and 35 ml/min, mid-S and late S + G_2 cells are obtained, although these both contain a small fraction of G_1 cells of high sedimentation velocity. In Fig. 3 the result of reculturing the G_1 cells is demonstrated. By 2 hr of culture at the permissive temperature the cells move into early S phase, and by 4 hr are entirely in S phase while retaining a substantial measure of synchrony. After 7 hr of culture the cells are in mid- to late S and by 11.5 hr (almost one cell cycle time) the cells are in G_2 and reentering the second G_1 phase. The result of repeating and extending this experiment with *ts* Friend cells is shown in Fig. 4 where G_1 elutriated cells are recultured for 5.5 hr at either the permissive or nonpermissive temperatures with or without inducer. At 36°C the cells move synchronously into S phase. The same is true of cells cultured at 37.5°C; hence, the *ts* defect is not immediately operative and is possibly dependent on the depletion of a pool. Following 24 hr of culture at 36°C the cells are in G_2 + G_1 but seem to be losing synchrony (Fig. 5). In the presence of inducer the cells are transiently growth arrested in G_1, confirming results discussed earlier. G_1 cells cultured for 24 hr at 37.5°C are arrested in the second G_1 due to the *ts* defect whether or not inducer is present. By 30 hr of culture at 36°C (Fig. 6) the cells have lost synchrony. Cells transiently arrested in G_1 due to the effect of inducer at 36°C are now released into G_2 + G_1 of the cell cycle, whereas cells at the nonpermissive temperature are retained in G_1 because of the *ts* effect.

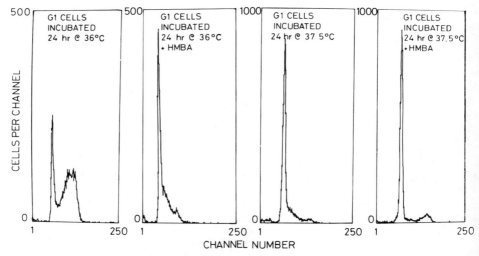

Fig. 5.   As for Fig. 4 except that the G_1 cells have been further cultured for 24 hr.

B. Is the Latent Phase Due Wholly or in Part to the Time Required for Acquisition of a Critical Intracellular Inducer Concentration?

An alteration in the rate of calcium transport appears to be the rate-limiting event for the commitment of murine erythroleukemia cells to a program of terminal erythroid differentiation. Provision of calcium in the presence, but not

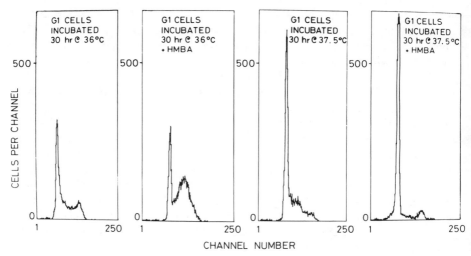

Fig. 6.   As for Fig. 4 except that the G_1 cells have been further cultured for 30 hr.

absence, of inducer initiates commitment without a significant lag phase (Bridges *et al.*, 1981). Thus, the lag phase is presumed to be the time required for inducer to alter the calcium transport properties of the cell. The question remains whether or not for such a mechanism the cell must acquire a critical intracellular inducer concentration. It is known that the intracellular concentration of the inducer HMBA is only 65–75% of the maximum after a 6-hr exposure (Gambari *et al.*, 1978).

In attempting to answer this question, elutriated G_1 *ts* Friend cells were held at 20 or 4°C for 16 hr. As shown in Fig. 7, the G_1 cells progress into S phase at 20°C whereas at 4°C they remain in G_1. The viability of cells held at 4°C for 16 hr is about 95% that of randomly proliferating cells at 36.5°C and about the same as G_1 cells cultured for 16 hr at 36.5°C. Further validation of viability is suggested by the observation that the G_1 cells held at 4°C for 16 hr progress synchronously into S phase during 5.5 hr of incubation at 36°C (Fig. 8). Similar results are obtained for G_1 cells treated for 16 hr at 4°C with the inducer HMBA (5 mM). That is, the elutriated G_1 cells remain in G_1 at 4°C yet progress into S phase at 36°C even in the presence of inducer. This demonstrates that G_1 cells equilibrated with inducer are not transiently growth arrested before the second G_1 phase.

Commitment to differentiation in the cells of these cultures is readily assayed by removing cells at various times and determining their ability to form a benzidine-positive clone in semisolid medium in the absence of inducer. The results in Table II reveal that there is no significant difference in the kinetics of commitment between cells treated with inducer at the start of incubation at 36°C and cells pretreated with inducer at 4°C for 16 hr prior to incubation at 36°C.

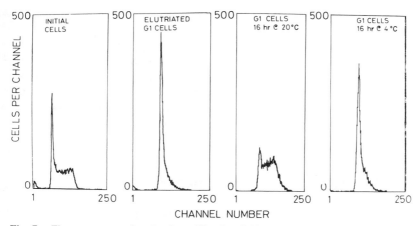

Fig. 7. Flow cytometry of randomly proliferating (initial cells) and G_1-elutriated Friend cells. The two profiles on the right are DNA profiles of the G_1 cells held for 16 hr at 20 or at 4°C.

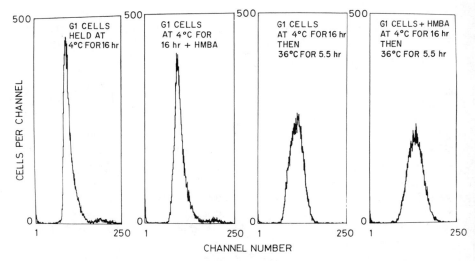

Fig. 8. Flow cytometry profiles of elutriated G_1 Friend cells held at 4°C for 16 hr either with or without HMBA, and then recultured at 36°C for 5.5 hr.

C. Is Exposure to Inducer during One Cell Division Cycle Sufficient for Commitment of Friend Cells to Differentiation?

G_1 *ts* Friend cells treated with inducer for 16 hr at 4°C progress through the cell cycle when reincubated with inducer at 37.5°C and become growth arrested at the second G_1 phase due to the *ts* effect. Although only one cell cycle traverse is accomplished, cells from these cultures have kinetics of commitment over 24 hr of incubation similar to cells cultured at permissive temperature (Table II).

Therefore, exposure to inducer during a single cell cycle is sufficient to initiate commitment to differentiation. This result is consistent with the report that G_1 Friend cells produce a greater proportion of committed cells in response to inducer than an unfractionated population (Geller *et al.*, 1978). Moreover, Friend cells synchronized by centrifugal elutriation accumulate induced globin mRNA during the transient G_1 arrest only after a complete traversal of S phase in inducer (Gambari *et al.*, 1979).

VIII. CONCLUSIONS AND FUTURE PROSPECTS

Multipotential hematopoietic stem cells can be cultured and assayed *in vitro* (Dexter and Testa, 1976) as can various committed erythroid pecursor cells (Axelrod *et al.*, 1973). However, difficulties in obtaining cell cycle synchrony have precluded exploitation of these cells in the present studies. Friend erythro-leukemia cells permit investigation of the cell cycle dependence of induced differentiation in synchronous cultures in the absence of unwanted perturbations which may arise in studies using drugs to arrest cells in the cell cycle.

It is clear from the results obtained that induction of Friend cell differentiation is dependent on some cell cycle-related event and that this event is complete following exposure to inducer during a single cell cycle traverse. It should now be possible to identify more precisely which part of this single, cell cycle traverse is important for induction to be initiated. How this crucial cell cycle event relates to the two stages of the commitment process is still obscure. However, using elutriated G_1 temperature-sensitive cells it should be feasible to investigate this further. A complementary line of investigation questions whether all of the induced erythroid program is cell cycle dependent. This study may be facilitated by the availability of DNA recombinants containing genes coding for erythroid-specific, nonglobin mRNAs (Harrison, 1982).

A major subject of interest is the molecular nature of the cell cycle-dependent events described. By monitoring DNase I sensitivity Groudine and Weintraub (1981) have shown that the chromatin structure of chicken globin genes under-goes a transition during differentiation and that this is dependent on DNA rep-lication. Furthermore, in *ts*AEV transformed chick erythroblasts which have an "inactive" globin chromatin structure, a switch from the permissive to the nonpermissive temperature is accompanied by acquisition of an "active" chro-matin structure (Weintraub *et al.*, 1982). This "active" globin chromatin struc-ture is retained by daughter cells even when permissive conditions of temperature

TABLE II

Commitment of *ts* Friend Cells Pretreated with Inducer[a]

	Time in culture at 36°C plus 5 m*M* HMBA (hr)					
Pretreatment	0	3	6	9	12	24
16 hr at 4°C	0	0	0	2	5	10
16 hr at 4°C + 5 m*M* HMBA	0	0	0	1	4	9

	Time in culture at 37.5°C plus 5 m*M* HMBA (hr)					
Pretreatment	0	3	6	9	12	24
16 hr at 4°C + 5 m*M* HMBA	0	0	0	1	4	12

[a] After the times shown for culture of pretreated cells at either the permissive or nonpermissive temperature in inducer, the cultures were sampled by resuspending the cells in Methocel medium in the absence of inducer. Commitment is the percentage of clones in which all cells are benzidine-positive 6 days later.

are reintroduced (Groudine and Weintraub, 1982). Similar experiments with Friend cell lines have established that globin genes are already in DNase I-sensitive configuration even when uninduced (Miller *et al.*, 1978). Nevertheless, induction results in a considerable enhancement of globin gene transcription (Orkin and Swerdlow, 1977). The mechanisms controlling this induced transcriptional activity remain to be resolved.

ACKNOWLEDGMENTS

The authors are grateful to Dr. Bryan Young for provision of flow cytometry facilities. Original work described is supported by grants from the Cancer Research Campaign.

REFERENCES

Arndt-Jovin, D. H., Ostertag, W., Eisen, H., and Jovin, T. M. (1976). *Haematol. Bluttransfus.* **19**, 137–150.

Axelrod, A. A., McLeod, D. L., Shreeve, M. M., and Heath, D. S. (1973). Properties of cells that produce erythrocytic colonies *in vitro*. *In* Proc. Second Int. Workshop on Hemopoiesis in Culture (W. A. Robinson, ed.), pp. 226–237 (DHEW Publication No. NIH 74-205).

Bernstein, A., Hunt, D. M., Crickley, V., and Mak, T. W. (1976). Induction by ouabain of haemoglobin synthesis in cultured Friend erythroleukaemic cells. *Cell* **9**, 375–381.

Beug, H., Doederlein, G., Freudenstein, C., and Graf, T. (1982a). Erythroblast cell lines transformed by a temperature-sensitive mutant of avian erythroblastosis virus: A model system to study erythroid differentiation *in vitro*. *J. Cell. Physiol., Suppl.* **1**, 195–207.

Beug, H., Palmieri, S., Freudenstein, C., Zentgraf, H. W., and Graf, T. (1982b). Hormone-dependent terminal differentiation *in vitro* of chicken erythroleukaemia cells transformed by ts mutants of avian erythroblastosis virus. *Cell* **28**, 907–919.

Boyer, S. H., Wuv, K. D., Noyes, A. N., Young, R., Scher, W., Friend, C., Preisler, H. D., and Bank, A. (1972). Haemoglobin biosynthesis in murine virus-induced leukaemic cells *in vitro:* Structure and amounts of globin chains produced. *Blood* **40**, 823–835.

Bridges, K., Levenson, R., Housman, D., and Cantley, L. (1981). Calcium regulates the commitment of murine erythroleukaemia cells to terminal erythroid differentiation. *J. Cell Biol.* **90**, 542–544.

Clive, D., Flam, W. G., Machesko, M. R., and Bernheim, N. J. (1972). A mutational assay system using the thymidine kinase locus in lymphoma cells. *Mutat. Res.* **16**, 77–87.

Conkie, D., Affara, N., Harrison, P. R., Paul, J., and Jones, K. (1974). *In situ* localisation of globin messenger RNA formation. *J. Cell Biol.* **63**, 414–419.

Conkie, D., Young, B. D., and Paul, J. (1980). Friend cell variants temperature-sensitive for growth. *Exp. Cell Res.* **126**, 439–444.

Conkie, D., Harrison, P. R., and Paul, J. (1981). Cell-cycle dependence of induced haemoglobin synthesis in Friend erythroleukaemia cells temperature-sensitive for growth. *Proc. Natl. Acad. Sci. U.S.A.* **78**, 3644–3648.

de Harven, E., and Friend, C. (1966). Origin of the viraemia in murine leukaemia. *Natl. Cancer Inst. Monogr.* **22**, 79–86.

Dexter, T. M., and Testa, M. G. (1976). Differentiation and proliferation of haemopoitic cells in culture. *Methods Cell Biol.* **14**, 387–405.

Dube, S. K., Gaedicke, G., Kluge, N., Weimann, B. J., Melderis, H., Steinheider, G., Crozier, T., Beckman, H., and Ostertag, W. (1974). *In* "Differentiation and Control of Malignancy of Tumour Cells" (W. Nakaharo, T. Ono, T. Sugimoto, and H. Sugano, eds.), pp. 99–135. Univ. of Tokyo Press, Tokyo.

Ebert, P. S., and Ikawa, Y. (1974). Induction of ALA synthetase during erythroid differentiation of cultured leukaemic cells. *Proc. Soc. Exp. Biol. Med.* **601**, 601–604.

Ebert, P. S., Wars, I., and Buell, D. N. (1976). Erythroid differentiation in cultured Friend leukaemia cells treated with metabolic inhibitors. *Cancer Res.* **36**, 1809–1813.

Eisen, H., Bach, R., and Emery, R. (1977). Induction of spectrin in Friend erythroleukaemic cells. *Proc. Natl. Acad. Sci. U.S.A.* **74**, 3898–3902.

Fibach, E., Reuben, R. C., Rifkind, R. A., and Marks, P. A. (1977). Effect of hexamethylene bisacetamide on the commitment to differentiation of murine erythroleukaemia cells. *Cancer Res.* **37**, 440–444.

Fredrickson, T., Tambourin, P., Jasmin, C., and Smadja, F. (1975). Target cells of the polycythaemia-inducing Friend virus: Studies with myleran. *JNCI, J. Natl. Cancer Inst.* **55**, 443–446.

Friedman, E. A., and Schildkraut, C. C. (1978). Lengthening of the G1 phase is not strictly correlated with differentiation in Friend erythroleukaemia cells. *Proc. Natl. Acad. Sci. U.S.A.* **75**, 3813–3817.

Friend, C. (1957). Cell-free transmission in adult Swiss mice of a disease having the character of a leukaemia. *J. Exp. Med.* **105**, 307–318.

Friend, C., and Haddad, J. R. (1960). Tumour formation with transplants of spleen or liver from mice with virus-induced leukaemia. *JNCI, J. Natl. Cancer Inst.* **25**, 1279–1285.

Friend, C., Patuleia, M. C., and de Harven, E. (1966). Erythrocytic maturation *in vitro* of murine (Friend) virus-induced leukaemic cells. *Natl. Cancer Inst. Monogr.* **22**, 505–514.

Friend, C., Scher, W., and Rossi, G. B. (1970). The biosynthesis of haem in Friend virus-induced leukaemic cell lines cloned *in vitro*. *In* "The Biology of Large RNA Viruses" (B. W. J. Mahy and R. D. Barry, eds.), pp. 267–275. Academic Press, New York.

Friend, C., Scher, W., Holland, J. G., and Sato, T. (1971). Haemoglobin synthesis in murine virus-induced leukaemic cells *in vitro:* Stimulation of erythroid differentiation by dimethyl sulphoxide. *Proc. Natl. Acad. Sci. U.S.A.* **68**, 378–382.

Gambari, R., Terada, M., Bank, A., Rifkind, R. A., and Marks, P. A. (1978). Synthesis of globin mRNA in relation to the cell cycle during induced murine erythroleukaemia differentiation. *Proc. Natl. Acad. Sci. U.S.A.* **75**, 3801–3804.

Gambari, R., Marks, P. A., and Rifkind, R. A. (1979). Murine erythroleukaemia cell differentiation: Relationship of globin gene expression and of prolongation of G1 to inducer effects during G1/early S. *Proc. Natl. Acad. Sci. U.S.A.* **76**, 4511–4515.

Gazitt, Y., and Friend, C. (1981). Early and late changes in the glycoproteins of Friend erythroleukaemia cells induced to differentiation. *Cancer Res.* **41**, 1070–1075.

Geller, R., Levenson, R., and Housman, D. (1978). Significance of the cell cycle in commitment of murine erythroleukaemia cells to erythroid differentiation. *J. Cell. Physiol.* **95**, 213–222.

Graf, T., Ade, N., and Beug, H. (1978a). Temperature-sensitive mutant of avian erythroblastosis virus suggests a block of differentiation as mechanism of leukaemogenesis. *Nature (London)* **275**, 496–501.

Graf, T., Beug, H., Royer-Pokora, B., and Meyer-Glauner, W. (1978b). *In vitro* transformation of haematopoietic cells by avian erythroid and myeloid leukaemia viruses: A model system for the differentiation of normal and neoplastic cells. *In* "Differentiation of Normal and Neoplastic Haemopoietic Cells" (B. Clarkson, P. A. Marks, and J. E. Till, eds.), pp. 625–639. Cold Spring Harbor Lab., Cold Spring Harbor, New York.

Groudine, M., and Weintraub, H. (1981). Activation of globin genes during chicken development. *Cell* **24**, 393–401.

Groudine, M., and Weintraub, H. (1982). Propagation of globin DNAase I-hypersensitive sites in absence of factors required for induction: A possible mechanism for determination. *Cell* **30**, 131–139.

Gusella, J. F., and Housman, D. (1976). Induction of erythroid differentiation *in vitro* by purines and purines analogues. *Cell* **8**, 263–269.

Gusella, J. F., Geller, R., Clarke, B., Weeks, V., and Housman, D. (1976). Commitment of erythroid differentiation by Friend erythroleukaemia cells: A stochastic analysis. *Cell* **9**, 221–229.

Gusella, J. F., Weil, S. C., Tsiftsoglou, A. S., Volloch, V., Neumann, J. R., Keys, C., and Housman, D. E. (1980). Haemin does not cause commitment of murine erythroleukaemia (MEL) cells to terminal differentiation. *Blood* **56**, 481–487.

Harrison, P. R. (1977). The biology of the Friend cell. *Int. Rev. Biochem.* **15**, 227–267.

Harrison, P. R. (1982). Regulation of differentiation in retrovirus-induced murine erythroleukemias: A personal view from the fringe. *Cancer Surv.* **1**, 231–277.

Harrison, P. R., Conkie, D., Rutherford, T., and Yeoh, G. (1978). Molecular aspects of erythroid cell regulation. *In* "Stem Cells and Tissue Homeostasis" (B. I. Lord, C. S. Patten, and R. J. Cole, eds.), pp. 241–257. Cambridge Univ. Press, London and New York.

Holtzer, H. (1978). Cell lineages, stem cells and the "quantal" cell cycle concept. *In* "Stem Cells and Tissue Homeostasis" (B. I. Lord, C. S. Patten, and R. J. Cole, eds.), pp. 1–27. Cambridge Univ. Press, London and New York.

Holtzer, H., Rubinstein, N., Fellini, S., Yeoh, G., Chi, J., Birnbaum, J., and Okayama, A. (1975). *Q. Rev. Biophys.* **8**, 523–557.

Holtzer, H., Pacifici, M., Croop, J., Boettinger, D., Toyama, Y., Payette, R., Biehl, J., Dlugosz, A., and Holtzer, S. (1981). Properties of cell lineages as indicated by the effects of ts-RSV and TPA on the generation of cell diversity. *Fortschr. Zool.* **26**, 207–225.

Housman, D., Gusella, J., Geller, R., Levenson, R., and Weil, S. (1978). Differentiation of murine erythroleukaemia cells: The central role of the commitment event. *In* "Differentiation of Normal and Neoplastic Haemopoietic Cells" (B. Clarkson, P. A. Marks, and J. Till, eds.), pp. 193–207. Cold Spring Harbor Lab., Cold Spring

Ikawa, Y., Furusawa, M., and Sugano, H. (1973). Erythrocyte membrane-specific antigens in Friend virus-induced leukaemic cells. *Bibl. Haematol.* **39**, 955–967.

Kost, T. A., Koury, M. J., Hankins, W. D., and Krantz, S. B. (1979). Target cells for Friend virus-induced erythroid bursts *in vitro*. *Cell* **18**, 145–152.

Leder, A., and Leder, P. (1975). Butyric acid, a potent inducer of erythroid differentiation in cultured erythroleukaemic cells. *Cell* **5**, 319–322.

Leder, A., Orkin, S., and Leder, P. (1975). Differentiation of erythroleukaemic cells in the presence of inhibitors of DNA synthesis. *Science* **190**, 893–894.

Levenson, R., and Housman, D. (1979a). Developmental programme of murine erythroleukaemia cells: Effect of the inhibition of protein synthesis. *J. Cell Biol.* **82**, 715–725.

Levenson, R., and Housman, D. (1979b). Memory of MEL cells to a previous exposure to inducer. *Cell* **17**, 485–490.

Levenson, R., and Housman, D. (1981). Commitment: How do cells make the decision to differentiate? *Cell* **25**, 5–6.

Levenson, R., Kernen, J., and Housman, D. (1979). Synchronisation of MEL cell commitment with cordycepin. *Cell* **18**, 1073–1078.

Levenson, R., Housman, D., and Cantley, L. (1980a). Amiloride inhibits murine erythroleukaemia cell differentiation: Evidence for a Ca^2 requirement for commitment. *Proc. Natl. Acad. Sci. U.S.A.* **77**, 5948–5952.

Levenson, R., Kerner, J., Matrani, A., and Housman, D. (1980b). DNA synthesis is not required for the commitment of murine erythroleukaemia cells. *Dev. Biol.* **74**, 224–230.

Levy, J., Terada, M., Rifkind, R. A., and Marks, P. A. (1975). Induction of erythroid differentiation by dimethylsulphoxide in cells infected with Friend virus: Relationship to the cell cycle. *Proc. Natl. Acad. Sci. U.S.A.* **72**, 28–32.

Lyman, G. H., Preisler, H. D., and Papahadjopoulos, D. (1976). Membrane action of DMSO and other chemical inducers of Friend leukaemic cell differentiation. *Nature (London)* **262**, 360–363.

McClintock, P. R., and Papaconstantinou, J. (1974). Regulation of haemoglobin synthesis in a murine erythroblastic leukaemic cell: The requirement for replication to induce haemoglobin synthesis. *Proc. Natl. Acad. Sci. U.S.A.* **71**, 4551–4555.

McKenna, P. J., and Hickey, I. (1981). The thymidine kinase locus of Friend erythroleukemic cells. I. Mutation rates and properties of mutants. *Mutat. Res.* **80**, 187–199.

Majumdar, S. K., and Bilenker, J. D. (1975). Failure to detect chromosome damage *in vivo* in Friend virus-infected leukaemic mice. *JNCI, J. Natl. Cancer Inst.* **54**, 503–505.

Miller, D. M., Turner, P., Nienhuis, A. W., Axelrad, D. E., and Gopalakrishnan, T. V. (1978). Active conformation of the globin genes in uninduced and induced mouse erythroleukemia cells. *Cell* **14**, 511–522.

Naha, P. M. (1969). Temperature sensitive conditional mutants of monkey kidney cells. *Nature (London)* **223**, 1380–1381.

Orkin, S. H., and Swerdlow, P. S. (1977). Globin RNA synthesis *in vitro* by isolated erythroleukemic cell nuclei: Direct evidence for increased transcription during erythroid differentiation. *Proc. Natl. Acad. Sci. U.S.A.* **74**, 2475–2479.

Ostertag, W., Melderis, H., Steinheider, G., Kluge, N., and Dube, S. (1972). Synthesis of mouse haemoglobin and globin mRNA in leukaemic cell cultures. *Nature (London), New Biol.* **239**, 231–234.

Paul, J., and Conkie, D. (1973). Effects of inhibitors of DNA synthesis on the stimulation of mouse erythroid cells by erythropoietin. *Exp. Cell Res.* **77**, 105–110.

Paul, J., and Hunter, J. A. (1968). DNA synthesis is essential for increased haemoglobin synthesis in response to erythropoietin. *Nature (London)* **219**, 1362–1363.

Paul, J., and Hunter, J. A. (1969). Synthesis of macromolecules during induction of haemoglobin synthesis by erythropoietin. *J. Mol. Biol.* **42**, 31–41.

Pragnell, I. B., Arndt-Jovin, D. J., Jovin, T. M., Fagg, B., and Ostertag, W. (1980). Commitment to differentiation in Friend cells and initiation of globin mRNA synthesis occurs during the G1 phase of the cell cycle. *Exp. Cell. Res.* **125**, 459–470.

Puck, T. T., and Kao, F. (1967). Genetics of somatic mammalian cells. V. Treatment with 5-bromodeoxyuridine and visible light for isolation of nutritionally deficient mutants. *Proc. Natl. Acad. Sci. U.S.A.* **58**, 1227–1234.

Reuben, R. C., Wife, R. L., Breslow, R., Rifkind, R. A., and Marks, P. A. (1976). A new group of potent inducers of differentiation in murine erythroleukaemia cells. *Proc. Natl. Acad. Sci. U.S.A.* **73**, 862–866.

Ronning, O. W., Pettersen, E. O., and Seglen, P. O. (1979). Protein synthesis and protein degradation through the cell cycle of human NHIK3025 cells *in vitro*. *Exp. Cell. Res.* **123**, 63–72.

Ross, J., and Sautner, D. (1976). Induction of globin mRNA accumulation by haemin in cultured erythroleukaemic cells. *Cell* **8**, 513–520.

Rossi, G. B., and Friend, C. (1967). Erythrocytic maturation of (Friend) virus-induced le·̇kaemic cells in spleen clones. *Proc. Natl. Acad. Sci. U.S.A.* **58**, 1373–1380.

Rossi, G. B., and Friend, C. (1970). Further studies on the biological properties of Friend virus-induced leukaemic cells differentiating along the erythrocytic pathway. *J. Cell. Physiol.* **76**, 159–166.

Stambrook, P. J., and Sisken, J. E. (1972). Induced changes in the rate of uridine-³H uptake and incorporation during the G1 and S periods of synchronised Chinese hamster cells. *J. Cell Biol.* **52,** 514–525.

Stetten, G., Davidson, R. L., and Latt, S. A. (1977). Hoechst enhances the selectivity of the BrdU-light methods of isolating conditional lethal mutants. *Exp. Cell Res.* **108,** 447–452.

Takahashi, E., Nagasawa, T., Sato, G., Matsushima, T., Sugimura, T., and Ohashi, A. (1974). Differentiation of cultured Friend leukemia cells induced by short-chain fatty acids. *Gann* **65,** 261–268.

Tambourin, P., and Wendling, F. (1975). Target cell for oncogenic action of polycythaemia-inducing Friend virus. *Nature (London)* **256,** 320–322.

Tanaka, M., Levy, J., Terada, M., Breslow, R., Rifkind, R. A., and Marks, P. A. (1975). Induction of erythroid differentiation in murine virus infected erythroleukaemia cells by highly polar compounds. *Proc. Natl. Acad. Sci. U.S.A.* **72,** 1003–1006.

Terada, M., Fried, J., Nudel, U., Rifkind, R. A., and Marks, P. A. (1977). Transient inhibition of initiation of S-phase associated with dimethyl sulphoxide induction of murine erythroleukaemia cells to erythroid differentiation. *Proc. Natl. Acad. Sci. U.S.A.* **74,** 248–252.

Till, J. E., and McCulloch, E. A. (1961). A direct measurement of the radiation sensitivity of normal mouse bone marrow cells. *Radiat. Res.* **14,** 213–222.

Till, J. E., McCulloch, E. A., and Siminovitch, L. (1964). A stochastic model of stem cell proliferation based on the growth of spleen colony-forming cells. *Proc. Natl. Acad. Sci. U.S.A.* **51,** 29–36.

Tsiftsoglou, A. S., Mitrani, A. A., and Housman, D. E. (1981). Procaine inhibits the erythroid differentiation of MEL cells by blocking commitment: Possible involvement of calcium metabolism. *J. Cell. Physiol.* **108,** 327–335.

Volloch, V., and Housman, D. (1982). Terminal differentiation of murine erythroleukaemia cells: Physical stabilisation of end-stage cells. *J. Cell Biol.* **93,** 390–394.

Volpe, P., and Eremenko, T. (1973). A method for measuring cell cycle phases in suspension cultures. *Methods Cell Biol.* **6,** 113–126.

Weintraub, H., Beug, H., Groudine, M., and Graf, T. (1982). Temperature-sensitive changes in the structure of globin chromatin in lines of red cell precursors transformed by ts-AEV. *Cell* **28,** 931–940.

9

Proliferation, Differentiation, and Gene Regulation in Skeletal Muscle Myogenesis: Recombinant DNA Approaches

KENNETH E. M. HASTINGS AND
CHARLES P. EMERSON, JR.
Department of Biology
University of Virginia
Charlottesville, Virginia

I. INTRODUCTION

Most of the proliferation controls operating in animal cells have probably evolved within the context of the whole organism. Presumably, their main function is to allow the animal to assign to each tissue-specific task an appropriate

219

RECOMBINANT DNA AND
CELL PROLIFERATION

number of cells. Accordingly, there may exist fundamental relationships between those mechanisms that control cell proliferation and those that control cell differentiation. By studying one of these phenomena, it may be possible to learn about the other. For example, the study of gene regulation during cell differentiation may yield useful insight into the less clearly perceived role of gene regulation in the mechanics and control of cell proliferation.

Recombinant DNA approaches provide uniquely powerful tools for the direct analysis of gene structure and regulation at the molecular level. These analytical tools have been used in the study of cell differentiation in a wide variety of experimental systems. The various cell types employed in these studies offer greater or lesser advantages for addressing questions concerning the relationships among gene regulation, cell differentiation, and cell proliferation. One of the most suitable cell types in this regard is vertebrate skeletal muscle.

A great deal of knowledge obtained through biochemical and amino acid sequence analysis of muscle-specific proteins, particularly contractile proteins, provides a solid background for studies of muscle genes and their expression. In addition, muscle differentiation or myogenesis can be examined in essentially pure cell culture *in vitro*. This is not only a great experimental advantage in terms of investigating the regulation of muscle protein synthesis, but several aspects of the cell biology of myogenesis in culture pertain directly to cell proliferation. First, myogenesis entails an abrupt change in the proliferative status of the cells involved. Second, there is evidence that myogenesis is regulated by cell growth factors, or mitogens. Thus, myogenesis provides a system in which to explore not only the regulation of genes encoding tissue-specific proteins, but also gene regulation related to changes in proliferative status and the effects of mitogens on gene expression.

In this chapter, we discuss the principal features of gene regulation in muscle cells. After supplying some background, we focus on results provided by recombinant DNA approaches, particularly work relating to contractile protein synthesis and proliferation-related gene regulation. Our aim was to develop the idea that through studying myogenesis we can learn something about the role of gene regulation in cell proliferation. It will become clear that although recombinant DNA approaches have already provided useful information of a specific nature, their most important contributions consist of establishing a general framework for further investigation and providing the tools to proceed.

II. BACKGROUND

A. Features of Skeletal Muscle

At the cellular level, vertebrate skeletal muscle is composed of parallel fibers, each of which is a multinucleate syncytial cell of relatively enormous length.

Each fiber exhibits microscopic cross-striations which are due to the almost complete dominance of muscle fiber cytoplasm by closely packed, structurally periodic myofibrils running parallel to the fiber axis. The myofibrils are the actual contractile apparatus of muscle and they consist of serially repeated structure–function units, the sarcomeres. It is the in-phase registry of sarcomeres in the parallel myofibrils that gives the muscle fiber its cross-striations.

The molecular structure of the sarcomere has been to a considerable extent determined and interpreted in terms of contractile function (see, for example, Squire, 1981). For our purposes the most salient point is that many of the sarcomere structural proteins—particularly the contractile proteins of the thick and thin filaments, myosin, α-actin, tropomyosin, and troponin (see Mannherz and Goody, 1976, for a review)—are well characterized and not known to be accumulated to any extent in tissues other than muscle. Although nonmuscle homologues of some of these proteins have been discovered, they are clearly encoded in distinct members of contractile protein multigene families (Vandekerckhove and Weber, 1978; Fine and Blitz, 1975). Thus, one important aspect of muscle differentiation concerns the tissue-specific expression and regulation of genes encoding the muscle contractile proteins.

Two other important features of skeletal muscle relate to the condition of the nuclei. Nuclei within skeletal muscle fibers replicate and divide extremely rarely, if ever, and this is often expressed by saying that they are postmitotic, or withdrawn from the proliferative cycle. This postmitotic state appears to be permanent because following injury to a muscle the damaged regions are repopulated, not by proliferation of nuclei within fibers, but instead by proliferation of mononucleated cells which subsequently fuse to form fibers (Lash *et al.*, 1957).

The DNA content of most, if not all, muscle fiber nuclei is much the same as that of diploid liver cells (Lash *et al.*, 1957). This quantity of DNA (2n) is also characteristic of proliferating diploid cells in the G_1 phase of the cell cycle, the significance of which is discussed later.

Thus, the chief features that must be accounted for when considering muscle development are (1) the activation of muscle gene expression and the accumulation of muscle proteins, (2) the formation of multinucleate fibers in which (3) all the nuclei contain the G_1 quantity of DNA, and (4) that these nuclei are postmitotic, or permanently withdrawn from the cell cycle. All of these features of skeletal muscle are established right at the outset of muscle fiber formation in myogenic cell cultures.

B. Myogenesis in Culture

1. *Myoblasts, Myofibers*

Cultures of myogenic cells—myoblasts—can be established from dispersed muscle from a variety of sources, usually embryonic, and the process of muscle differentiation can be studied *in vitro*. Myoblasts are mononucleated cells capable of active proliferation. Under certain initial conditions the history of an undisturbed myoblast culture shows two distinct phases (see Konigsberg, 1977). In the first phase the cells simply proliferate, rapidly increasing the total number of mononucleated cells in the culture. In the second phase the rate of proliferation declines and differentiation begins to occur, eventually involving essentially the entire culture. Differentiation, as seen in the microscope, consists of the fusion of individual myoblasts to form long, multinucleate syncytia called myotubes or myofibers (see Fig. 1). These soon contain hundreds of nuclei and seem to be homologous with the skeletal muscle fibers that form in the animal. In some culture systems, the myofibers develop cross-striations and contractility.

Analysis of individual nuclei of myofibers formed in culture shows that the distribution of measured DNA contents is much the same as that determined in late telophase pairs of dividing cells (Strehler et $al.$, 1963). Thus, most or all of these nuclei contain the diploid G_1 quantity of DNA as do the nuclei in a muscle fiber in the animal. Thymidine incorporation studies indicate that nuclei incorporated into myofibers do not appear ever to enter S phase, although mononucleated cells in the same culture can do so (Stockdale and Holtzer, 1961). Long-term microcinematography of muscle cultures does not reveal the appearance of mitotic figures within the myofibers (Capers, 1960). These findings imply that myoblasts fuse during the G_1 phase of the cell cycle and that their nuclei do not thereafter replicate or divide (Bischoff and Holtzer, 1969).

From the foregoing, it seems clear that the program of muscle differentiation entails the rather immediate establishment of a permanent nonproliferative status in the nuclei involved. There is some disagreement as to whether establishment of nonproliferative status occurs just prior to, simultaneously with, or immediately subsequent to, myoblast fusion (Dienstman and Holtzer, 1977; Nadal-Ginard, 1978; Hauschka et $al.$, 1982; Buckley and Konigsberg, 1974). Neither the mechanism nor the biological $raison$ $d'être$ of the postmitotic status of myofiber nuclei are known. Nonetheless, it is clear that myogenesis in culture provides an experimental system in which to study a change in proliferative status that mirrors events occurring in the normal development of the animal.

2. Activation of Contractile Protein Synthesis

Biochemical analysis of myoblast cultures in the proliferative phase shows very little or no detectable synthesis or accumulation of muscle-specific contractile proteins. However, during the fusion phase of the cultures, the synthesis of these tissue-specific gene products is activated to relatively high levels (reviewed by Buckingham, 1977). The extent of activation of synthesis of muscle-specific myosin heavy chain during myoblast differentiation has been estimated as high as several thousandfold (Emerson and Beckner, 1975).

In cultures manipulated to heighten the synchrony of fusion, the relative synthesis of α-actin and subunits of myosin, tropomyosin, and troponin show

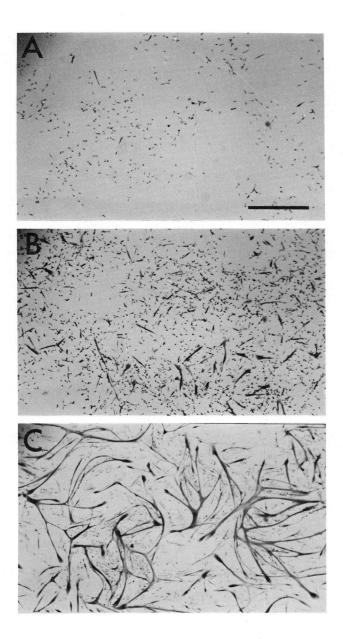

Fig. 1. Myogenesis in culture. Low magnification views of cultured myoblasts from the breast musculature of Japanese quail (*Coturnix coturnix*) embryos. (A) A culture in the proliferative stage when it consists almost exclusively of small, mononucleated myoblasts. (B) Fusion is just beginning, and in (C) most of the myoblasts have fused into myofibers. The scale bar in (A) represents 0.1 mm. The stain was Giemsa.

more or less coordinated increases along with the extent of fusion in the culture (Devlin and Emerson, 1978). However, fusion is not required for the activation of muscle contractile protein synthesis (Stockdale and Holtzer, 1961; Emerson, 1977). Neither is the permanent withdrawal from the cell cycle that is characteristic of myofiber nuclei (Devlin and Konigsberg, 1983; Devlin et al., 1982). These two striking attributes of myogenesis do not appear to play any essential role in the expression of muscle protein genes.

   Studies of the molecular basis of the activation of contractile protein synthesis have focused on the mRNA populations in differentiating muscle cultures. Analyses by RNA-excess cDNA hybridization kinetics have indicated that as myogenesis proceeds in the cultures, new species of poly(A)$^+$ RNA become abundant in the cytoplasm and on polyribosomes (Paterson and Bishop, 1977; Affara et al., 1980a; Zevin-Sonkin and Yaffe, 1980). Moreover, evidence was obtained that some of the new, abundant RNA species were mRNAs encoding myosin heavy chain, actin, tropomyosin, myosin light chains, and troponin C (Paterson and Bishop, 1977; Affara et al., 1980b). Analyses of mRNA populations by in vitro translation have lead to conflicting results in various culture systems [reviewed briefly in Devlin and Emerson (1979) and Shani et al., (1981)]. However, it is clear in some culture systems that in vitro-translatable mRNAs encoding muscle contractile proteins are not accumulated to any significant extent in proliferating myoblasts, but do accumulate rapidly in fused myofibers (Devlin and Emerson, 1979). Thus, both hybridization kinetics and in vitro translation results provided evidence that contractile protein synthesis during myogenesis is chiefly regulated via the accumulation of contractile protein mRNAs. This provides a focal point and foothold for analysis of gene regulation during myogenesis by recombinant DNA techniques.

III. GENE REGULATION DURING MYOGENESIS

   Given the general aforementioned features of myogenesis in culture there seem to be two chief areas in which questions of gene regulation arise.

   First, there is the question of the molecular basis of the activation of muscle contractile protein synthesis during myoblast differentiation. The evidence reviewed indicates that this activation is based on an accumulation of contractile protein mRNAs during myoblast differentiation.

   Another, perhaps less obvious aspect concerns the simple fact that whereas myoblasts are rapidly proliferating cells, differentiated myofibers are postmitotic. Thus, the transition from myoblasts to myofibers might involve changes in gene expression related to the change of proliferative status per se. This could reflect a variety of potentially significant gene regulatory phenomena. For example, any RNA specifically accumulated in the S and/or G_2 phases of the myoblast cell cycle would be more abundant in RNA extracted from a culture of proliferating myoblasts than in RNA extracted from a culture of differentiated myofibers. Also, depending on the exact relationship between the G_1 state of the cycling myoblast and the G_1/G_0 state of the differentiated myofiber, one might expect that RNAs preferentially accumulated in the G_1 phase of the cell cycle would accumulate to different extents in myoblast and myofiber cultures. Thus, a study of changes in mRNA populations during myogenesis might be relevant not only to contractile protein synthesis but also to proliferation and/or progress through the cell division cycle.

A. cDNA Cloning Strategies

What the foregoing considerations emphasize is that there are reasons to be interested in any RNA whose abundance changes during myogenesis, regardless of whether or not it encodes a known muscle protein. It is possible, using an appropriate cloning and screening strategy, to isolate cloned DNA copies of regulated RNAs on the basis of their differing abundance in myoblast and myofiber RNA preparations.

One such strategy would be to clone randomly the entire population of poly(A)$^+$ RNA sequences from myofiber cultures. Such a shotgun cDNA clone library could then be hybridized with highly ^{32}P-labeled cDNA preparations representing the sequences present in either myoblast poly(A)$^+$ RNA or myofiber poly(A)$^+$ RNA. One would expect to be able to classify the hybridization pattern of each cDNA clone in the library as being either unregulated, myoblast-specific, or myofiber-specific. In this way, cloned cDNA copies of regulated (and unregulated) RNAs would be identified in the shotgun cDNA clone library.

An example of this type of differential hybridization screen is shown in Fig. 2. A set of colonies picked randomly from a shotgun cDNA clone library of quail myofiber poly(A)$^+$ RNA sequences was replica plated and the replicas were hybridized with either myoblast ^{32}P-labeled cDNA (Fig. 2A) or myofiber ^{32}P-labeled cDNA (Fig. 2B). Two colonies showing a myofiber-specific pattern of hybridization are indicated in the figure. These colonies presumably harbor cloned cDNA copies of RNA sequences that are considerably more abundant in myofiber poly(A)$^+$ RNA than in myoblast poly(A)$^+$ RNA.

The example shown in Fig. 2 was drawn from a study in which a total of 890 randomly chosen cDNA clones were analyzed by the differential colony hybridization screen (Hastings and Emerson, 1982). Of the 890 clones, 28 showed a myofiber-specific pattern of hybridization, 3 showed a myoblast-specific pattern, and the remainder ($\sim$95%) showed an unregulated pattern, as is the case in Fig. 2. These cDNA clones provide specific probes for RNAs whose abundances increase, decrease, or remain the same during myogenesis. It became important to determine whether any of the regulated RNAs are involved in contractile

A

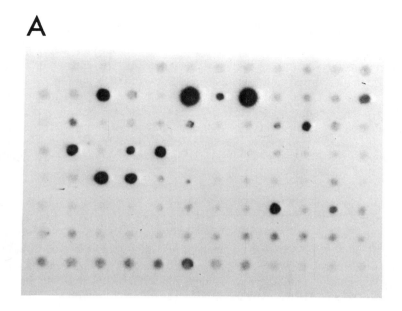

B

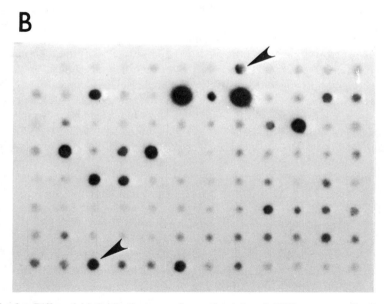

Fig. 2. Differential hybridization screen for regulated cloned cDNA sequences. Duplicate sets of transformant colonies from a shotgun cDNA clone library of quail myofiber poly(A)$^+$ RNA sequences were hybridized (Grunstein and Hogness, 1975) with highly ^{32}P-labeled cDNA made on either (A) myoblast or (B) myofiber poly(A)$^+$ RNA. Two colonies showing a myofiber-specific pattern of hybridization are indicated by arrowheads. Other differences between (A) and (B) were not reproduced in a duplicate set of hybridizations.

protein synthesis and whether any might represent proliferation-related gene regulation.

B. Contractile Protein Gene Regulation

Analysis of the myofiber-specific set of 28 clones indicates that they represent approximately 17 different RNA species, several of them being represented by 2 or 3 independently cloned cDNAs. Six of these RNAs have been identified by sequence analysis and hybridization–translation techniques as mRNAs encoding the muscle contractile proteins α-actin, α-tropomyosin, myosin heavy chain, myosin light chain 2, troponin C, and troponin I (Hastings and Emerson, 1982). The finding of contractile protein cDNA clones among the set of sequences showing myofiber-specific regulation indicated the validity of the cloning and screening protocols, as well as confirming the expected pattern of expression of contractile protein genes.

In other laboratories, other approaches have been successfully used to isolate cloned cDNA copies of contractile protein mRNAs. In most of these cases, cDNA clone libraries were made from RNA extracted from either adult or embryonic whole muscle, rather than from cultured myofibers. In some cases, RNA fractions enriched for particular mRNA species were used for cloning. Contractile protein cDNA clones were identified by a variety of methods including, in some cases, differential hybridization screens similar to that illustrated in Fig. 2. As a result of this work, muscle contractile protein cDNA clones have been isolated from several species of bird and mammal. Those reported to date are summarized in Table I. In addition, recombinant DNA probes for contractile protein genes have been isolated from organisms in other phyla (Durica *et al.*, 1980; MacLeod *et al.*, 1981; Storti *et al.*, 1981; Tobin *et al.*, 1980; Bernstein *et al.*, 1983) and kingdoms (Kindle and Firtel, 1978; Gallwitz and Sures, 1980).

cDNA clones have been used as probes to assay contractile protein-coding RNA levels during myogenesis in culture. These studies have clearly demonstrated that, whereas contractile protein mRNAs are readily detectable in differentiated myofibers, proliferating myoblasts do not contain significant quantities of contractile protein-coding RNAs—translatable or untranslatable (Katcoff *et al.*, 1980; Shani *et al.*, 1981; Hastings and Emerson, 1982). In one study the accumulation during myogenesis of mRNAs encoding myosin heavy chain, myosin light chain 2, and α-actin was found to be proportional to the extent of activation of synthesis of these proteins (Shani *et al.*, 1981). There can be no doubt that the activation of contractile protein synthesis is chiefly regulated by mechanisms that determine the abundance of contractile protein mRNAs.

Although it has not been directly demonstrated, there is fairly convincing indirect evidence that the increased accumulation of contractile protein mRNAs following myoblast fusion is due to a transcriptional activation of contractile

TABLE I

cDNA Clones Encoding Vertebrate Muscle Contractile Proteins

	Reference pertaining to organism[a]				
Protein	Quail	Chicken	Rat	Mouse	Rabbit
Myosin heavy chain	9	8	2,3,10		12
Myosin light chain 1/3		15	14	13	
Myosin light chain 2	9		1,14		12
α-Actin	9	4,5	1,14	7,11	12
α-Tropomyosin	9	6	14		12
Troponin C	9		14		12
Troponin I	9				12
Troponin T			14		12

[a] References: 1, Katcoff *et al.* (1980); 2, Medford *et al.* (1980); 3, Nudel *et al.* (1980); 4, Ordahl *et al.* (1980); 5, Schwartz *et al.* (1980); 6, MacLeod (1981); 7, Minty *et al.* (1981); 8, Umeda *et al.* (1981); 9, Hastings and Emerson (1982); 10, Mahdavi *et al.* (1982); 11, Minty *et al.* (1982); 12, Putney *et al.* (1983); 13, Robert *et al.* (1982); 14, Garfinkel *et al.* (1982); 15, Nabeshima *et al.* (1982).

protein genes. (The alternative would be that these genes could be similarly active in myoblasts and myofibers, but the mRNA accumulation differs because of different stabilities of transcripts.) First, the appearance of new abundant sequences in myofiber polyribosomal poly(A)$^+$ RNA is reflected in the new appearance of these same sequences in nuclear poly(A)$^+$ RNA (Affara *et al.*, 1980a). This indicates that the accumulation of this set of RNAs [which is known to include contractile protein mRNAs (Affara *et al.*, 1980b)] is regulated in the nucleus, not in the cytoplasm. Second, the genes encoding these new abundant RNAs are relatively more susceptible to DNase I digestion in myofiber chromatin than in myoblast chromatin (Affara *et al.*, 1980a). This result has been confirmed for two specific genes, encoding α-actin and myosin light chain 2, by the use of cloned DNA sequence probes (Carmon *et al.*, 1982). In view of the correlation between DNase I-sensitivity and transcriptional activity that has been established for other genes (Weintraub and Groudine, 1976; Wu *et al.*, 1979), these results indicate that muscle contractile protein synthesis is regulated at the level of gene transcription. This has yet to be directly demonstrated by pulse labeling nuclear RNA sequences produced in myoblasts and myofibers and assaying contractile protein gene transcripts by hybridization with cloned probes.

C. Proliferation-Related Gene Regulation

Contractile protein cDNA clones have, for obvious reasons, been the focus of much attention on the part of muscle biologists. However, as we indicated earlier, there are reasons for taking a more general view of gene regulation during myogenesis. In a further attempt to characterize the gene sets involved, we have initiated studies of the possible proliferation-related expression of genes represented in our cDNA clone collection of regulated sequences.

In these experiments, we compared gene expression in muscle cultures with that in a nonmuscle cell type. The rationale behind this approach will become clear when we consider the following specific examples. As a nonmuscle cell type, we used primary cultures of fibroblasts isolated from the skin of quail embryos. These cells were chosen because fibroblasts have been the subject of a great deal of study concerning the control of cell proliferation and proliferation-related gene regulation. Also, under the culture conditions used the fibroblasts, like myoblasts, exhibit a phase of rapid proliferation followed by a phase in which the cells replicate much more slowly. For example, between days 4 and 7 the DNA content (and cell number) of our fibroblast cultures increased sixfold, but between days 7 and 9 it increased only by 50%. Thus, we could contrast mitotically active versus more quiescent cells in both cell types.

We have thus far focused our studies on the cloned myoblast-specific cDNA sequences discussed in Section III,A. Two possible interpretations of the myoblast-specific pattern of regulation of these sequences could be proposed. On one hand, it could be supposed that the corresponding RNAs had some function in proliferating cells in general, but not in nonproliferating cells. Since myoblasts are rapidly dividing cells whereas myofibers are postmitotic, this would then explain the myoblast-specific pattern of expression. On the other hand, it could be supposed that the corresponding RNAs had some function specific to the muscle cell lineage. According to the latter idea, these RNAs should not be found in fibroblast cultures. The former idea implies not only that these RNAs should be found in fibroblast cultures but that they should be more abundant in rapidly proliferating fibroblast cultures.

The method of analysis in these studies was by Northern blot hybridization. RNA was extracted from proliferating myoblasts, differentiated myofibers, rapidly proliferating fibroblasts (day 4 of culture), and slowly growing fibroblasts (days 7 and 9 of culture). Equal amounts of total RNA were electrophoresed in an agarose gel, and the size-separated RNA species were transferred to a sheet of diazotized paper and hybridized with ^{32}P-labeled cDNA clones. This allows the visualization of specific RNAs complementary to the cDNA probes (Alwine *et al.*, 1977), presumably those RNA species from which the cDNAs were originally transcribed. The cDNA clones used as probes in these experiments had shown a variety of hybridization patterns in the colony hybridization screen described in Section III,A.

Figure 3 shows some of the results obtained. The same RNA blot was hybridized first with cDNA clone cC130 (Fig. 3A), then with cC131 (Fig. 3B), and finally with a mixture of cC127 and cC135 (Fig. 3C). The RNA species

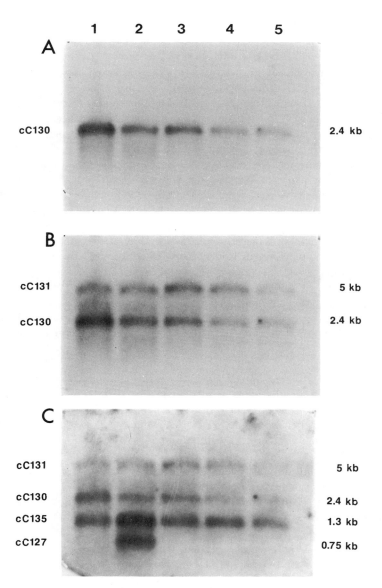

Fig. 3. Northern gel analysis of proliferation-related gene expression. Five 20-μg samples of RNA (numbered 1–5) were subjected to electrophoresis (from top to bottom) in a methylmercury–agarose gel (Bailey and Davidson, 1976) and transferred to a sheet of diazotized paper (Seed, 1982). The paper-bound RNA was hybridized (Wahl *et al.*, 1979) successively with several cDNA clones which had been ^{32}P-labeled by nick translation (Rigby *et al.*, 1977). (A) The result of the first hybridization (with cC130), (B) the second (with cC131), and (C) the third (with cC127 and cC135). The hybridization pattern is cumulative because previously formed hybrids were not stripped off between successive hybridizations. RNA bands hybridizing to specific cDNA clones are identified on the left by the clone with which they hybridize and on the right by their apparent size in kilobases (kb). The five RNA samples were (1) myoblast RNA, (2) myofiber RNA, (3) day 4 fibroblast RNA, (4) day 7 fibroblast RNA, and (5) day 9 fibroblast RNA.

hybridizing with each of the cDNA clones are identified by the cDNA clone name and by their apparent molecular sizes.

cC130, cC131, and cC135 RNAs were detectable in all five RNA samples examined. In contrast, cC127 RNA was abundant in myofiber RNA (well 2) but was not detectable in either myoblast RNA (well 1) or RNA from either day 4 (well 3), day 7 (well 4), or day 9 (well 5) fibroblast cultures.

cC127 RNA encodes a known muscle contractile protein, myosin light chain 2 (Hastings and Emerson, 1982). cC127 was one of the 28 "myofiber-specific" cDNA clones picked in the differential hybridization screen described in Section III,A. In fact, this particular clone is the myofiber-specific sequence in the top row of colonies in Fig. 2. The fact that cC127 RNA is abundant in myofiber RNA (well 2) but not in myoblast RNA (well 1) confirms the colony hybridization result. The fact that cC127 RNA is not detectable in RNA from fibroblast cultures (wells 3–5) is consistent with the expected tissue-specific distribution of muscle contractile proteins and indicates the absence of any significant myogenic contaminant in the fibroblast cultures.

cDNA clone cC135 was picked in the initial colony hybridization screen as an example of an unregulated sequence, i.e., it hybridized to similar extents with both myoblast and myofiber RNA sequence probes. Thus, the presence of cC135 RNA in both myoblast RNA (well 1) and myofiber RNA (well 2) is an expected result. The presence of cC135 RNA in RNA from fibroblast cultures (wells 3–5) shows that the gene encoding this RNA is not active exclusively in cells in the muscle lineage. (Although we cannot be certain that it is exactly the same gene expressed in myogenic cells and fibroblasts, our experience with contractile protein gene family members indicates that the hybridization and washing conditions used in Fig. 3 are such that an extremely high degree of sequence homology is required for detectable hybridization.) The important point to note is that the relative abundance of cC135 RNA does not decline dramatically as the fibroblast cultures age, and the cells change from a state of rapid proliferation (well 3) to a more quiescent state (wells 4 and 5).

cDNA clones cC130 and cC131 are two of the three myoblast-specific sequences picked out in the initial colony hybridization screen. (The third, cC129, gives results similar to those of cC130.) Figure 3 shows that these sequences are not truly myoblast-specific, because both cC130 and cC131 RNAs are present in fibroblast RNA (wells 3–5), as well as in myoblast RNA (well 1). However, Fig. 3 does confirm that, at least in the case of cC130 RNA, this sequence is more abundant in myoblast RNA (well 1) than in myofiber RNA (well 2). Moreover, both cC130 and cC131 RNAs decline substantially in relative abundance as the fibroblast cultures age and proliferation declines. The decline in abundance of cC130 and cC131 RNAs in fibroblast cultures does not appear to be attributable to a general decline in poly(A)$^+$ RNAs with respect to the ribosomal RNA which makes up the bulk of the samples analyzed in Fig. 3. cC135 RNA does not show a similar decline. The contrast is most striking by comparing the relative abundances of cC130 RNA and cC135 RNA in day 4 fibroblast cultures (well 3) with those in day 7 cultures (well 4).

From these results it seems likely that, at least in the case of cC130 RNA, we are observing a regulation of RNA abundance that is related to the change in proliferative status per se. Apparently, cC130 RNA declines in relative abundance during myogenesis, not because it has a unique myoblast-specific function, but because it is more abundantly expressed in rapidly proliferating cells in general.*

These results are presented to illustrate the approach we have used to examine possible proliferation-related gene regulation during myogenesis. These studies are still at an early stage, but already it seems clear from the results with cC130 that it is possible to isolate cDNA clones for genes whose expression during myogenesis is related to the change in proliferative status per se. The significance of this kind of result lies in the further questions that one can now ask.

Why is cC130 RNA more abundant in proliferating than in nonproliferating cells? Is it perhaps only accumulated by cells in the S or G_2 phases of the cell cycle? Is it accumulated in a proliferative G_1, but not in the G_1/G_0, state of quiescent fibroblasts or myofiber nuclei? Does it encode any of the proteins whose synthesis is increased when serum-starved fibroblasts are stimulated to resume proliferation (Thomas et al., 1981; Riddle et al., 1979; Riddle and Pardee, 1979) or decreased when myoblasts differentiate (Devlin and Emerson, 1978)? Is its decline during myoblast differentiation coordinated with, or related to, the activation of contractile protein gene expression? Some of these questions can be addressed by using the cDNA clone as a sequence probe in studies of specific RNA synthesis (rather than accumulation) in cells blocked in defined phases of the cell division cycle and cells travelling synchronously through the cycle.

These specific questions relate to the more general question of the role of gene expression and regulation in the mechanism and control of cell proliferation. One of the most important aspects of cell proliferation control that we know about concerns the molecules that act as growth factors, or mitogens. These molecules are able to influence or even absolutely determine the proliferative behavior of cells in culture. It is natural to consider whether their effects are mediated through gene regulation. Certainly, their activity appears to affect gene expression in some fashion, for we can safely say that any proliferation-related gene expression (such as previously discussed) demonstrates an effect, direct or

---

*An alternative explanation we have not ruled out is that the abundance of cC130 RNA may be directly regulated by some relatively labile substance—perhaps a hormone—which is present in fresh medium but lost after several days of culture. Such specific gene regulation could possibly be independent of proliferative status.

indirect, of mitogens on gene expression. In the present connection it is relevant to consider what is known of the role of mitogens in myogenesis.

IV. CONTROL OF MYOGENESIS: THE ROLE OF MITOGENS

Manipulations of muscle culture parameters, such as number of cells plated and volume of medium used, have led to two related conclusions: (1) There is an inverse relationship between myoblast proliferation and differentiation, and (2) proliferating myoblasts modify, or condition, the medium in which they grow. Fresh medium promotes proliferation and delays differentiation when compared with conditioned medium, which advances the onset of differentiation and does not support rapid proliferation (see Konigsberg, 1977). Since myoblast fusion and muscle gene expression can occur in completely defined medium never before exposed to cells (Emerson, 1977), it seems clear that the conditioning process does not consist of the secretion into the medium of differentiation-promoting substances that are lacking in fresh medium. Rather, conditioning must consist of the removal from fresh medium of substances (present in the embryo extract and/or serum) that promote proliferation and/or inhibit differentiation.

That conditioning consists of a removal of something from—rather than an addition of something to—fresh medium is further supported by the fact that addition of purified fibroblast growth factor to conditioned medium restores the main properties of fresh medium (Linkhart *et al.*, 1981). This suggests that the substance(s) depleted by conditioning may be a growth factor(s). Consistent with this idea is the observation that myoblasts refed fresh medium have a shorter average G_1 period than cells refed conditioned medium (Buckley and Konigsberg, 1974). This is an effect one would expect on the basis of growth factor depletion since the mitogenic action of growth factors most often resides in their ability to stimulate the entry of G_1 cells into S phase.

There seem to be two ways to account for the dual effect of mitogens on myoblast cultures. First, it is possible that the prevention of myoblast differentiation represents a distinct activity of mitogens, separate from their effect on cell proliferation. A simple example of such an activity would be if mitogens acted directly as repressors of muscle gene transcription. Second, it is possible that the differentiation-inhibiting activity of mitogens is a result of the increased proliferation they induce in the cells.

A model of how a mitogen-induced increase in proliferation might inhibit myoblast differentiation has been suggested by Konigsberg (1977). According to this hypothesis, myoblast differentiation requires that some G_1-restricted activity—in which all myoblasts engage—attains some critical threshold. Myoblasts with long G_1 periods have a greater probability of attaining the threshold than do myoblasts with shorter G_1 periods. This could explain why mitogen-stimulated myoblasts with their short G_1 periods do not differentiate, but myoblasts with long G_1 periods, such as those in conditioned medium or in undisturbed cultures undergoing fusion (Konigsberg, 1977), do differentiate.

There is a need for further work at the whole cell level in order to distinguish between a direct action of mitogens on muscle gene expression and an indirect action such as that in Konigsberg's hypothesis. In either case, however, there must ultimately be a molecular explanation. In the following section, we discuss possible recombinant DNA approaches to elucidating the molecular basis of proliferation-related gene regulation and the action of mitogens on gene expression.

V. FUTURE LINES OF INQUIRY

Further research directions in myogenesis involving recombinant DNA approaches fall into two categories. The first is the experimental, in which attempts are made to express and regulate cloned genes in their native or altered form in organisms, cells, or extracts. We will not discuss possible experimental approaches other than to say that any answers suggested by analytical approaches must ultimately be tested by experiment.

Second, there are analytical approaches in which one hopes to deduce some fundamental principles of gene regulation by examining it as it happens or by a structural analysis of the genes involved. Although there are many questions that have yet to be approached by even analytical recombinant DNA approaches, we would like to focus on two that seem to be accessible and have broad biological significance: (1) a "compare and contrast" method of analysis of gene structure and (2) an analysis of changes in contractile protein gene chromatin during myogenesis.

A. Structural Analysis of Gene Sets

The basic premise of gene structure analysis is that gene regulation ultimately depends on some aspect of primary structure either in the gene itself or in neighboring chromosomal DNA regions. Accordingly, this approach would be concerned with analyzing cloned genomic DNA fragments (which can be isolated from genomic libraries using cDNA clones as probes). The "compare and contrast" method depends on the analysis of a variety of gene sets showing both common and distinct patterns of regulation. The rationale behind the approach is that genes regulated by the same mechanism (and hence showing the same pattern of regulation) should have a common structural feature related to that control mechanism. This common feature, however, should not be present in genes regulated by a different mechanism or showing different patterns of regulation. Thus, by comparing and contrasting appropriately chosen genes, one hopes to be able to pinpoint key regulatory features in the structure. The usefulness of this approach will depend to a large extent on the appropriate choice of genes to be compared. We think that the results obtained in cDNA clone analysis of myogenesis provide the tools and a logical framework with which to proceed.

For example, consider the genes encoding the muscle contractile proteins actin, myosin, tropomyosin, and troponin. This is a diverse set of genes whose regulation seems to be fairly tightly coordinated. They are activated in myoblasts, but not in fibroblasts, when mitotic activity declines. Do the contractile protein genes share a common structural feature involved in their coordinate regulation? Is this common feature also shared by other genes activated by declining mitotic activity, but outside of the muscle cell lineage, e.g., in fibroblasts? These questions can be pursued by a comparative structural analysis of contractile protein genes as well as genes that are activated only in fibroblasts, or in both cell types.

At the moment we do not know of a cloned cDNA probe for a gene activated by declining mitotic activity in fibroblasts. However, we have not yet examined all of the myofiber-specific sequences described in Section III,A for their possible expression in fibroblast cultures. The search for such genes—activable in both myoblasts and fibroblasts—should be continued. Not only would they be valuable for the "compare and contrast" analysis, but it would also be useful information to know whether or not such genes do, in fact, exist.

Another set of genes useful for the "compare and contrast" method are the genes whose expression is reduced when mitotic activity declines. At the moment, we know definitely of only one gene (represented in cC130) likely to be in this category, and it is similarly regulated in both myoblasts and fibroblasts. Other examples are needed and might conceivably be found by screening a cDNA clone library of myoblast RNA sequences. The way our experiments were done, by cloning myofiber RNA, favors the cloning of RNAs whose abundances increase, rather than decrease, during myogenesis. By cloning myoblast RNA, this bias would be reversed. Screening a library of cloned myoblast RNA sequences would also give a better chance of isolating sequences representing a thus far undiscovered set of genes—those whose expression is down-regulated when myoblasts reduce mitotic activity but that are not expressed in fibroblasts—in other words, proliferation-dependent genes active only in the muscle lineage. A comparison of such genes with those proliferation-dependent genes expressed in both myoblasts and fibroblasts (such as the gene corresponding to cC130) should reveal similarities related to their similar modes of proliferation-related regulation, but differences related to their expression in particular cell lineages.

The point in all of this is to find and study a number of genes that exhibit a variety of regulatory patterns both with regard to cell proliferation and cell lineage. The cloning and screening protocols and myoblast/fibroblast comparisons we have discussed provide a methodology for the isolation of probes for a variety of such genes.

B. Changes in Muscle Gene Chromatin

In addition to the "compare and contrast" method of searching for gene sequences that might be involved in specific forms of regulation, there is another fairly accessible recombinant DNA approach to myogenesis. This would be to explore further the molecular basis of the activation of contractile protein gene expression during myoblast differentiation. The evidence reviewed in Section III,B points to a transcriptional activation of previously dormant contractile protein genes during myogenesis.

Exactly how transcriptionally silent genes are activated in eukaryotes is not known. Recent work on the transcriptional activation of globin genes indicates that the process may require multiple, independent events which modify the structure of globin genes and chromatin (Weintraub et al., 1982). These modifications include undermethylation of DNA bases in and around the gene region, the establishment of nuclease hypersensitive (single-stranded?) sites in the chromatin near or adjacent to the gene, and the binding to nucleosomes in the gene region of high mobility group (HMG) proteins [which apparently confers the high level of DNase I sensitivity of transcriptionally active chromatin (Weisbrod and Weintraub, 1979; Sandeen et al., 1980)]. In addition, active globin genes sit in a defined chromosomal domain of intermediate DNase I sensitivity. This intermediate level of sensitivity seems not to be due to binding of HMG proteins, but may reflect a more open, higher-order chromatin structure in the region of active genes as compared with inactive regions (Weisbrod and Weintraub, 1979; Stalder et al., 1980).

Although the actual significance of each of these correlates of transcriptional activity is not known, it does seem clear that the transcriptional activation of silent genes is likely to involve some structural modification of DNA or chromatin. In the case of contractile protein genes, we know already of a change in DNase I sensitivity (Carmon et al., 1982), presumably reflecting the binding of HMG proteins to nucleosomes in contractile protein genes in myofiber, but not in myoblast, chromatin. Further structural analysis of the type pioneered in the globin system and made possible by recombinant DNA sequence probes may reveal additional molecular alterations in muscle protein genes and chromatin during myoblast differentiation.

From knowledge of structural alterations in chromatin, it may be possible to begin to work backward through the series of molecular causes and effects that links muscle gene activation with its ultimate causes. We can expect that this chain of events will have components related both to cell proliferation and cell lineage determination. A connection with cell proliferation is expected because the onset of myogenesis and hence, muscle gene activation in myoblast cultures appears to be determined ultimately by mitogen levels. The question of how direct the effect of mitogens is on muscle gene expression remains to be addressed. However direct or indirect it is, the link between mitogens and muscle genes must also comprise cell lineage-specific elements, for we are given to expect that no amount of mitogen manipulation will activate muscle genes in fibroblasts. This must be taken into account by any comprehensive explanation of the role of mitogens in muscle differentiation.

VI. SUMMARY

An important aspect of the study of gene regulation in myogenesis is the analysis and interpretation of changes in the mRNA population following myoblast fusion. The fusion of myoblasts to form multinucleate myofibers is accompanied by a dramatic activation of muscle-specific protein synthesis and by a permanent withdrawal of the fused myoblasts from the cell division cycle. Recombinant DNA studies have shown that both biochemical differentiation and the change in proliferative status are reflected in changes in the mRNA population.

cDNA cloning and screening strategies have provided cloned sequence probes for RNAs whose abundances increase, decrease, or remain much the same during myogenesis. Analysis of cDNA clones has shown that the class of RNAs that increase in abundance includes mRNAs encoding muscle-specific contractile proteins, as well as at least 10 other as yet unidentified RNA species. The class that decreases in abundance is not as well studied and has fewer examples represented by cDNA clones. At least one member of this class of RNAs shows a level of expression that correlates with proliferative behavior both in myoblast cultures and in fibroblast cultures. The cDNA clone of this RNA may provide an exciting opportunity to study an example of proliferation-related gene regulation. Further use of the comparative approach using myoblast and fibroblast cultures will allow one to determine whether other examples of gene regulation during myogenesis are related to tissue-specific protein synthesis or whether they result from the change in proliferative status per se.

In addition to the analysis of mRNA populations, recombinant DNA approaches permit a direct study of gene and chromatin structure which will be invaluable in determining the molecular basis of gene regulation during myogenesis. Alterations in muscle gene chromatin detected by such techniques indicate that muscle-specific contractile protein genes are transcriptionally silent in proliferating myoblasts but become transcriptionally active during the period of myoblast fusion. Further analysis along these lines may allow a reconstruction of the molecular chain of events that results in muscle gene activation. Because the onset of myogenesis in culture appears to be regulated ultimately by mitogen levels, the results of this work may provide useful insight into the effects of mitogens on gene expression.

ACKNOWLEDGMENTS

We thank Druen Robinson for expert assistance in the cell culture work. Original work described was supported by a research grant from the National Institutes of Health (to C.P.E.) and a Muscular Dystrophy Association Postdoctoral Fellowship (to K.E.M.H.).

REFERENCES

Affara, N. A., Robert, B., Jacquet, M., Buckingham, M. E., and Gros, F. (1980a). Changes in gene expression during myogenic differentiation. I. Regulation of messenger RNA sequences expressed during myotube formation. *J. Mol. Biol.* **140,** 441–458.

Affara, N. A., Daubas, P., Weydert, A., and Gros, F. (1980b). Changes in gene expression during myogenic differentiation. II. Identification of the proteins encoded by myotube-specific complementary DNA sequences. *J. Mol. Biol.* **140,** 459–470.

Alwine, J. C., Kemp, D. J., and Stark, G. R. (1977). Method for detection of specific RNAs in agarose gels by transfer to diazobenzyloxymethyl-paper and hybridization with DNA probes. *Proc. Natl. Acad. Sci. U.S.A.* **74,** 5350–5354.

Bailey, J. M., and Davidson, N. (1976). Methylmercury as a reversible denaturing agent for agarose gel electrophoresis. *Anal. Biochem.* **70,** 75–85.

Bernstein, S. I., Mogani, K., Donady, J. J., and Emerson, C. P., Jr. (1983). *Drosophila* muscle myosin heavy chain encoded by a single gene in a cluster of muscle mutations. *Nature* **302,** 393–397.

Bischoff, R., and Holtzer, H. (1969). Mitosis and the process of differentiation of myogenic cells in vitro. *J. Cell Biol.* **41,** 188–200

Buckingham, M. E. (1977). Muscle protein synthesis and its control during the differentiation of skeletal muscle cells in vitro. *Int. Rev. Biochem.* **15,** 269–332.

Buckley, P. A., and Konigsberg, I. R. (1974). Myogenic fusion and the duration of the post-mitotic gap (G_1). *Dev. Biol.* **37,** 193–212.

Capers, C. R. (1960). Multinucleation of skeletal muscle in vitro. *J. Biophys. Biochem. Cytol.* **7,** 559–566.

Carmon, Y., Czosnek, H., Nudel, U., Shani, M., and Yaffe, D. (1982). DNAase I sensitivity of genes expressed during myogenesis. *Nucleic Acids Res.* **10,** 3085–3098.

Devlin, B. H., and Konigsberg, I. R. (1983). Reentry into the cell cycle of differentiated skeletal myocytes. *Dev. Biol.* **95,** 175–192.

Devlin, B. H., Merrifield, P. A., and Konigsberg, I. R. (1982). The activation of myosin synthesis and its reversal in synchronous, skeletal-muscle myocytes in culture. *In* "Muscle Development: Molecular and Cellular Control" (M. L. Pearson and H. F. Epstein, eds.), pp. 355–366. Cold Spring Harbor Lab., Cold Spring Harbor, New York.

Devlin, R. B., and Emerson, C. P., Jr. (1978). Coordinate regulation of contractile protein synthesis during myoblast differentiation. *Cell* **13,** 599–611.

Devlin, R. B., and Emerson, C. P., Jr. (1979). Coordinate accumulation of contractile protein mRNAs during myoblast differentiation. *Dev. Biol.* **69,** 202–216.

Dienstman, S. R., and Holtzer, H. (1977). Skeletal myogenesis: Control of proliferation in a normal cell lineage. *Exp. Cell Res.* **107,** 355–364.

Durica, D. S., Schloss, J. A., and Crain, W. R., Jr. (1980). Organization of actin gene sequences in the sea urchin: Molecular cloning of an intron-containing DNA sequence coding for a cytoplasmic action. *Proc. Natl. Acad. Sci. U.S.A.* **77,** 5683–5687.

Emerson, C. P., Jr. (1977). Control of myosin synthesis during myoblast differentiation. *In* "Pathogenesis of Human Muscular Dystrophies" (L. P. Rowland, ed.), pp. 799–811. Excerpta Medica, Amsterdam.

Emerson, C. P., Jr., and Beckner, S. K. (1975). Activation of myosin synthesis in fusing and mononucleated myoblasts. *J. Mol. Biol.* **93,** 431–447.

Fine, R. E., and Blitz, A. L. (1975). A chemical comparison of tropomyosin from muscle and non-muscle tissues. *J. Mol. Biol.* **95,** 447–454.

Gallwitz, D., and Sures, I. (1980). Structure of a split yeast gene: Complete nucleotide sequence of the actin gene in *Saccharomyces cerevisiae. Proc. Natl. Acad. Sci. U.S.A.* **77,** 2546–2550.

Garfinkel, L. I., Periasamy, M., and Nadal-Ginard, B. (1982). Cloning and characterization of cDNA sequences corresponding to myosin light chains 1, 2, and 3, troponin C, troponin T, α-tropomyosin, and α-actin. *J. Biol. Chem.* **257,** 11078–11086.

Grunstein, M., and Hogness, D. S. (1975). Colony hybridization: A method for the isolation of cloned DNAs that contain a specific gene. *Proc. Natl. Acad. Sci. U.S.A.* **72,** 3961–3965.

Hastings, K. E. M., and Emerson, C. P., Jr. (1982). cDNA clone analysis of six co-regulated mRNAs encoding skeletal muscle contractile proteins. *Proc. Natl. Acad. Sci. U.S.A.* **79,** 1553–1557.

Hauschka, S. D., Rutz, R., Linkhart, T. A., Clegg, C. H., Merrill, G. F., Haney, C. M., and Lim, R. W. (1982). Skeletal muscle development. *In* "Disorders of the Motor Unit" (D. L. Schotland, ed.), pp. 903–921. Wiley, New York.

Katcoff, D., Nudel, U., Zevin-Sonkin, D., Carmon, Y., Shani, M., Lehrach, H., Frischauf, A. M., and Yaffe, D. (1980). Construction of recombinant plasmids containing rat muscle actin and myosin light chain DNA sequences. *Proc. Natl. Acad. Sci. U.S.A.* **77,** 960–964.

Kindel, K. L., and Firtel, R. A. (1978). Identification and analysis of *Dictyostelium* actin genes, a family of moderately repeated genes. *Cell* **15,** 763–778.

Konigsberg, I. R. (1977). The role of the environment in the control of myogenesis in vitro. *In* "Pathogenesis of Human Muscular Dystrophies" (L. P. Rowland, ed.), pp. 779–798. Excerpta Medica, Amsterdam.

Lash, J. W., Holtzer, H., and Swift, H. (1957). Regeneration of mature skeletal muscle. *Anat. Rec.* **128,** 679–697.

Linkhart, T. A., Clegg, C. H., and Hauschka, S. D. (1981). Myogenic differentiation in permanent clonal mouse myoblast cell lines: Regulation by macromolecular growth factors in the culture medium. *Dev. Biol.* **86,** 19–30.

MacLeod, A. R. (1981). Construction of bacterial plasmids containing sequences complementary to chicken α-tropomyosin mRNA. *Nucleic Acids Res.* **9,** 2675–2689.

MacLeod, A. R., Karn, J., and Brenner, S. (1981). Molecular analysis of the unc-54 myosin heavy chain gene of *Caenorhabditis elegans. Nature (London)* **291,** 386–390.

Mahdavi, V., Periasamy, M., and Nadal-Ginard, B. (1982). Molecular characterization of two myosin heavy chain genes expressed in the adult heart. *Nature (London)* **297,** 659–664.

Mannherz, H. G., and Goody, R. S. (1976). Proteins of contractile systems. *Annu. Rev. Biochem.* **45,** 427–465.

Medford, R. M., Wydro, R. M., Nguyen, H. T., and Nadal-Ginard, B. (1980). Cytoplasmic processing of myosin heavy chain messenger RNA: Evidence provided by using a recombinant DNA plasmid. *Proc. Natl. Acad. Sci. U.S.A.* **77**, 5749–5733.

Minty, A. J., Caravatti, M., Robert, B., Cohen, A., Daubas, P., Weydert, A., Gros, F., and Buckingham, M. E. (1981). Mouse actin mRNAs. Construction and characterization of a recombinant plasmid molecular containing a complementary DNA transcript of mouse α-actin mRNA. *J. Biol. Chem.* **256**, 1008–1014.

Minty, A. J., Alonso, S., Caravatti, M., and Buckingham, M. E. (1982). A fetal skeletal muscle actin mRNA in the mouse and its identity with cardiac actin mRNA. *Cell* **30**, 185–192.

Nabeshima, Y., Fujii-Kuriyama, Y., Muramatsu, M., and Ogata, K. (1982). Molecular cloning and nucleotide sequences of the complementary DNAs to chicken skeletal muscle myosin two alkali light chain mRNAs. *Nucleic Acids Res.* **10**, 6099–6110.

Nadal-Ginard, B. (1978). Commitment, fusion and biochemical differentiation of a myogenic cell line in the absence of DNA synthesis. *Cell* **15**, 855–864.

Nudel, U., Katcoff, D., Carmon, Y., Zevin-Sonkin, D., Levi, Z., Shaul, Y., Shani, M., and Yaffe, D. (1980). Identification of recombinant phages containing sequences from different rat myosin heavy chain genes. *Nucleic Acids Res.* **8**, 2133–2146.

Ordahl, C. P., Tilghman, S. M., Ovitt, C., Fornwald, J., and Largen, M. T. (1980). Structure and developmental expression of the chick α-actin gene. *Nucleic Acids Res.* **8**, 4889–5005.

Paterson, B. M., and Bishop, J. O. (1977). Changes in the mRNA population of chick myoblasts during myogenesis in vitro. *Cell* **12**, 751–765.

Putney, S. D., Herlihy, W. C., and Schimmel, P. (1983). A new troponin T and cDNA clones for 13 different muscle proteins, found by shotgun sequencing. *Nature* **302**, 718–721.

Riddle, V. G. H., and Pardee, A. B. (1980). Quiescent cells but not cycling cells exhibit enhanced actin synthesis before they synthesize DNA. *J. Cell. Physiol.* **103**, 11–15.

Riddle, V. G. H., Dubrow, R., and Pardee, A. B. (1979). Changes in the synthesis of actin and other cell proteins after stimulation of serum-arrested cells. *Proc. Natl. Acad. Sci. U.S.A.* **76**, 1298–1302.

Rigby, P. W., Dieckmann, M., Rhodes, C., and Berg, P. (1977). Labeling deoxyribonucleic acid to high specific activity in vitro by nick translation with DNA polymerase I. *J. Mol. Biol.* **113**, 237–251.

Robert, B., Weydert, A., Caravatti, M., Minty, A., Cohen, A, Daubas, P., Gros, F., and Buckingham, M. (1982). cDNA recombinant plasmid complementary to mRNAs for light chains 1 and 3 of mouse skeletal muscle myosin. *Proc. Natl. Acad. Sci. U.S.A.* **79**, 2437–2441.

Sandeen, G., Wood, W., and Felsenfeld, G. (1980). The interaction of high mobility proteins HMG 14 and 17 with nucleosomes. *Nucleic Acids Res.* **8**, 3757–3778.

Schwartz, R. J., Haron, J. A., Rothblum, K. N., and Dugaicyk, A. (1980). Regulation of muscle differentiation: Cloning of sequences from actin mRNA. *Biochemistry* **19**, 5883–5890.

Seed, B. (1982). Diazotizable arylamine cellulose papers for the coupling and hybridization of nucleic acids. *Nucleic Acids Res.* **104**, 1799–1810.

Shani, M., Zevin-Sonkin, D., Saxel, O., Carmon, Y., Katcoff, D., Nudel, U., and Yaffe, D. (1981). The correlation between the synthesis of skeletal muscle actin, myosin heavy chain, and myosin light chain and the accumulation of corresponding mRNA sequences during myogenesis. *Dev. Biol.* **86**, 483–492.

Squire, J. (1981). "The Structural Basis of Muscular Contraction." Plenum, New York.

Stalder, J., Larsen, A., Engel, J. D., Dolan, M., Groudine, M., and Weintraub, H. (1980). Tissue-specific DNA cleavages in the globin chromatin domain introduction by DNase I. *Cell* **20**, 451–460.

Stockdale, F. E., and Holtzer, H. (1961). DNA synthesis and myogenesis. *Exp. Cell Res.* **24**, 508–520.

Storti, R. V., Bautch, V., Mischke, D., and Pardue, M. L. (1981). Isolation and characterization of the *Drosophila* tropomyosin gene. *J. Supramol. Struct. Cell. Biochem., Suppl.* **5**, 396.

Strehler, B. L., Konigsberg, I. R., and Kelley, J. E. T. (1963). Ploidy of myotube nuclei developing in vitro as determined with a recording double beam microspectrophotometer. *Exp. Cell Res.* **32**, 232–241.

Thomas, G., Thomas, G., and Luther, H. (1981). Transcriptional and translational control of cytoplasmic proteins after serum stimulation of quiescent Swiss 3T3 cells. *Proc. Natl. Acad. Sci. U.S.A.* **78**, 5712–1716.

Tobin, S. L., Zulauf, E., Sanchez, F., Craing, E. A., and McCarthy, B. J. (1980). Multiple actin-related sequences in the *Drosophila melanogaster* genome. *Cell* **19**, 121–131.

Umeda, P. K., Sinha, A. M., Jakovcic, S., Merten, S., Hsu, H. J., Subramanian, K. N., Zak, R., and Rabinowitz, M. (1981). Molecular cloning of two fast myosin heavy chain cDNAs from chicken embryo skeletal muscle. *Proc. Natl. Acad. Sci. U.S.A.* **78**, 2843–2847.

Vandekerckhove, J., and Weber, K. (1978). Mammalian cytoplasmic actins are the products of at least two genes and differ in primary structure in at least 25 identified positions from skeletal muscle actin. *Proc. Natl. Acad. Sci. U.S.A.* **75**, 1106–1110.

Wahl, G. M., Stern, M., and Stark, G. R. (1979). Efficient transfer of large DNA fragments from agarose gels to diazobenzyloxymethyl-paper and rapid hybridization by using dextran sulfate. *Proc. Natl. Acad. Sci. U.S.A.* **76**, 3683–3687.

Weintraub, H., and Groudine, M. (1976). Chromosomal subunits in active genes have an altered conformation. *Science* **193**, 848–856.

Weintraub, H., Beug, H., Groudine, M., and Graf, T. (1982). Temperature-sensitive changes in the structure of globin chromatin in lines of red cell precursors transformed by ts-AEV. *Cell* **28**, 931–940.

Weisbrod, S., and Weintraub, H. (1979). Isolation of a subclass of nuclear proteins responsible for conferring a DNase I-sensitive structure on globin chromatin. *Proc. Natl. Acad. Sci. U.S.A.* **76**, 630–634.

Wu, C., Wong, Y. C., and Elgin, S. C. R. (1979). The chromatin structure of specific genes. II. Disruption of chromatin structure during gene activity. *Cell* **16**, 807–814.

Zevin-Sonkin, D., and Yaffe, D. (1980). Accumulation of muscle-specific RNA sequences during myogenesis. *Dev. Biol.* **74**, 326–334.

10

Ribosomal Protein Gene Expression in Proliferating and Nonproliferating Cells

ODED MEYUHAS

Developmental Biochemistry Research Unit
Institute of Biochemistry
Hebrew University–Hadassah Medical School
Jerusalem, Israel

243

RECOMBINANT DNA AND
CELL PROLIFERATION

I. INTRODUCTION

Many complex processes occurring during growth, differentiation, or development in eukaryotes involve modulation of ribosome biosynthesis. This organelle contains four molecules of ribosomal RNA (rRNA) and over 70 different species of ribosomal protein (r-protein), all of which appear in equimolar amounts. Thus, to maintain the proper stoichiometry of the ribosomal components, there must be regulatory mechanisms ensuring (1) coordinate synthesis of all r-proteins; (2) coordinate synthesis of rRNA and r-proteins; and (3) balanced formation of ribosomes with growth status of the cell.

Since the publication of the last review on the control of eukaryotic ribosome biosynthesis (Warner *et al.*, 1980), major developments have provided means for the study of the *rp* genes and the regulation of their expression. These include the molecular cloning of cDNAs containing rp-mRNA sequences from mouse (Meyuhas and Perry, 1980) and *Xenopus laevis* (Bozzoni *et al.*, 1981) and of *rp* genes from *Saccharomyces cerevisiae* (Woolford *et al.*, 1979; Fried *et al.*, 1981; Fried and Warner, 1981, 1982; Bollen *et al.*, 1981a) and *Drosophila melanogaster* (Vaslet *et al.*, 1980).

This review is primarily concerned with the application of recombinant cDNA clones containing vertebrate *rp* sequences to the study of the organization of *rp* genes and the control of their expression under various growth conditions. Ribosomal protein genes in invertebrates are discussed in greater detail by Fried and Warner in Chapter 7.

II. CONSTRUCTION AND IDENTIFICATION OF cDNA FOR EUKARYOTIC RIBOSOMAL PROTEINS

A. General Strategy

The strategy used to obtain recombinant plasmids with inserts of rp-mRNA sequences is one that is particularly suited to situations in which the relevant mRNAs are not abundant. Such mRNAs, which individually constitute about 0.01–0.5% of the total mRNA mass, are impractical or impossible to isolate by purely physical or immunological means, although they can be readily translated by suitable cell-free, protein-synthesizing systems to produce identifiable protein products. For these mRNAs it is convenient to make a plasmid cDNA library using cDNAs made from a fraction of poly(A)$^+$ mRNA enriched for the mRNAs of interest. One can then screen the various recombinant DNAs for their ability to hybridize selectively with an mRNA that can be recognized via its translatable product. This so-called positive selection technique has been used effectively to identify cDNA clones containing mRNA sequences for r-protein from mouse L cells (Meyuhas and Perry, 1980) and from *Xenopus laevis* oocytes (Bozzoni *et al.*, 1981).

B. Enrichment of the mRNA Fraction for Ribosomal Protein-Coding Species

Eukaryotic r-proteins are relatively small, ranging in size from 11,000–41,000 daltons (Wool, 1982) and correspondingly, their mRNAs can be found among the smaller mRNA species (Hackett *et al.*, 1978; Nabeshima *et al.*, 1979; Pierandrei-Amaldi and Beccari, 1980). Since most, if not all, eukaryotic r-proteins are encoded in poly(A)-containing mRNA (Hackett *et al.*, 1978; Warner and Gorenstein, 1977; Nabeshima *et al.*, 1979; Pierandrei-Amaldi and Beccari, 1980) poly(A)$^+$ mRNA should be used as the starting point for the fractionation procedure. Further enrichment is achieved by size fractionation of poly(A)$^+$ mRNA on sucrose gradient and identification of the r-protein-translatable activity in each fraction by its ability to direct the synthesis of r-protein in a cell-free translational system. Due to lack of purified antibodies against vertebrate r-proteins, the translational products are analyzed by two-dimensional gel electrophoresis together with a set of marker proteins from 80 S ribosomes or their 60 and 40 S subunits. It is clear from such analysis that nearly all of the mRNAs coding for r-proteins in mouse L cells (Meyuhas and Perry, 1980) and *Xenopus laevis* oocytes (Pierandrei-Amaldi and Beccari, 1980) are smaller than 16 S in size.

Selective separation of basic translational product either by one-dimensional gel electrophoresis (O. Meyuhas and R. P. Perry, unpublished data) or by two-dimensional gel electrophoresis (Hackett *et al.*, 1978; Meyuhas and Perry, 1980) suggests that most of the basic proteins encoded by mammalian poly(A)$^+$ mRNA are r-proteins. This fact can be exploited to distinguish the basic proteins by their ability to bind to carboxymethyl cellulose (Egberts *et al.*, 1977). Although the chromatography tests do not separate individual proteins or indicate their sizes, such analysis is simple and fast and indicates the approximate level of r-protein synthesis *in vitro*. According to this criterion 20% of the proteins translated by the ≤12 S poly(A)$^+$ mRNA from mouse L cells are basic proteins and consequently are r-proteins (Meyuhas and Perry, 1980). In *Xenopus* oocytes, contrary to other systems, a major type of basic protein encoded by small poly(A)$^+$ mRNAs are histones (Levenson and Marcu, 1978; Ruderman and

Pardue, 1977). To eliminate these mRNAs, a narrower size range of 10–16 S was chosen, after fractionation of the poly(A)$^+$ mRNA on a sucrose gradient. In this size class, about 10–20% of the template activity is specific for r-proteins (Pierandrei-Amaldi and Beccari, 1980).

C. Construction and Identification of Cloned cDNAs Specific for Ribosomal Proteins

The mRNA fraction enriched for r-protein was copied into cDNA, which was then made double-stranded, essentially as described by Efstratiadis et al. (1976). Insertion of the ds-cDNA into the EcoRI site of a bacterial plasmid was carried out either by annealing the two DNA molecules through tails of complementary homopolymers (Meyuhas and Perry, 1980) or by ligation through synthetic Eco-RI linkers added to the ds-cDNA (Bozzoni et al., 1981). These recombinant DNAs were used to transform Escherichia coli. The transformed clones containing sequences complementary to the mRNAs in the fraction enriched for rp-mRNA were selected by colony hybridization (Grunstein and Hogness, 1975) with ^{125}I-labeled mRNA (Meyuhas and Perry, 1980) or ^{32}P-labeled cDNA (Bozzoni et al., 1981).

In order to identify the mouse r-protein cDNA clones, each of 50 plasmid DNA samples was covalently bound to DBM paper and used to select its corresponding mRNA from a preparation of total poly(A)$^+$ mRNA (Meyuhas and Perry, 1980). The selectively bound mRNA molecules were eluted individually from the paper and translated in the wheat germ system. The basic proteins among the translational products were selectively extracted with acetic acid (Hardy et al., 1969). For an initial screen the basic products were analyzed on one-dimensional Triton X-100 slab gels (Zweidler, 1978). The high resolution of r-proteins on this gel system and the convenience of screening 20–30 samples on a single slab gel (Fig. 1) enabled us to select with confidence 13 out of the 50 clones as candidates for the r-protein. Of these, there were two pairs of recombinant and one set of three recombinants that selected mRNA that apparently encode the same protein.

The identity of the nine unique candidate clones having the largest inserts was further examined by analysis of translational products on two-dimensional gels. The two-dimensional gel analysis enabled us to categorize our r-protein clones according to the conventional nomenclature for mammalian r-proteins (Mc-Conkey et al., 1979). An example of two-dimensional analysis for 40 and 60 S r-proteins and an autoradiogram for one of the r-protein clones are shown in Fig. 2. Of the nine clones tested, seven contained sequences encoding large subunit proteins (L7, L10, L13, L18, L19, L30, and L32/33), one contained a sequence encoding a small subunit protein (S16), and one appeared to contain a sequence encoding a nonribosomal basic protein (Meyuhas and Perry, 1980).

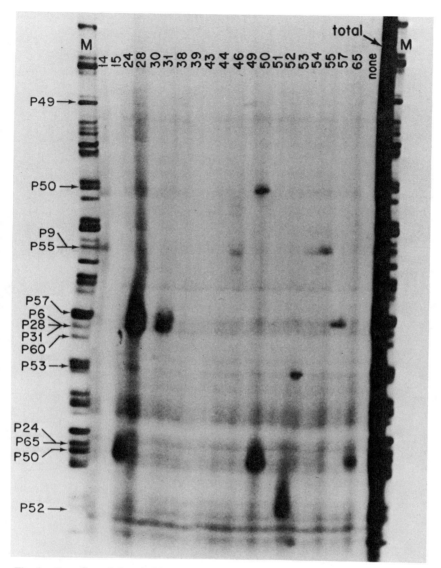

Fig. 1. Screening of cloned cDNAs by one-dimensional analysis of the translation products of selected mRNA. Poly(A)⁺ mRNA was hybridized to a set of 20 recombinant DNAs individually bound to DBM papers. The mRNAs that specifically hybridized to each recombinant (designated by the number in each lane) were recovered and translated in a wheat germ system. The basic translation products were resolved by electrophoresis on 10–20% polyacrylamide slab gel containing urea and Triton X-100. M, marker of ¹⁴C-labeled 80 S r-proteins. The lanes labeled "total" and "none" contain the basic translation products of mouse total poly(A)⁺ mRNA and wheat germ endogenous mRNA, respectively. The numbers of the plasmids that appear to have selected rp-mRNAs are indicated on the left. (From Meyuhas and Perry, 1980.)

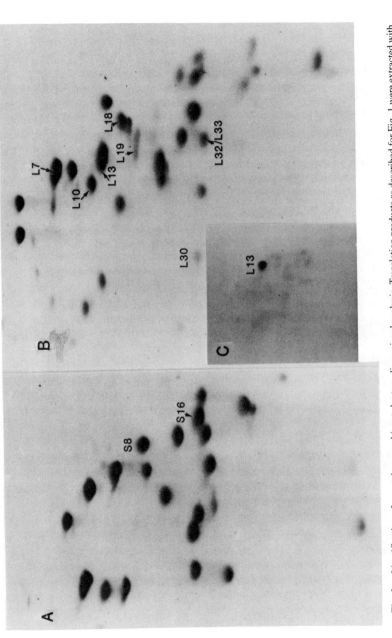

Fig. 2. Identification of r-protein recombinants by two-dimensional analysis. Translation products as described for Fig. 1 were extracted with acetic acid and divided into two portions, one of which was subjected to two-dimensional electrophoresis with mouse liver 40 S r-proteins and the other with 60 S r-proteins. (A and B) Examples of stained gels containing 40 and 60 S r-proteins, respectively. (C) Example of fluorogram (photographically reduced) that was used to identify p57 as containing the sequence of the mRNA coding for L13. The labeled r-proteins in (A) and (B) are those for which a comigration of stain and radioactive label was observed. (From Meyuhas and Perry, 1980.)

A similar approach was used in selection of cDNA clones containing sequences specific to *Xenopus* r-protein (Bozzoni *et al.*, 1981). Out of 127 recombinant clones screened by two-dimensional gel electrophoresis, 24 contained sequences corresponding to basic protein, but only 11 of them hybridized to mRNAs coding for r-proteins. Of these, there were three pairs of clones and one set of 3 clones that hybridized with the same mRNA. Of the six different recombinants three contained sequences encoding large subunit proteins (L1, L14, and L32) and three contained sequences of small subunit proteins (S1, S8, and S19). This numbering system is specific for *Xenopus laevis* and does not correspond to those used in yeast (Bollen *et al.*, 1981b) or mammals (McConkey *et al.*, 1979). Identification of 12 mouse rp clones out of the total of 50 that were screened and of 11 *Xenopus* rp-cDNA clones out of 127 examined is in good agreement with the estimates of the relative abundance of rp-mRNAs in the enriched fractions (20% and 10–20% respectively).

III. ORGANIZATION OF RIBOSOMAL PROTEIN GENES

A. Gene Dosage of Ribosomal Protein

The isolation of recombinant DNA clones that contain sequences for r-protein provides a powerful tool for examining the structure of *rp* genes and the regulation of their expression. Studies with such clones have shown that whereas in *E. coli* a single gene encodes for each r-protein (Nomura *et al.*, 1977), in eukaryotes the situation appears more complicated.

When mammalian DNAs are digested with *Eco*RI, blotted onto nitrocellulose (Southern, 1975), and hybridized with various rp-cDNA clones, complex fragment patterns are observed (Meyuhas and Perry, 1980; Monk *et al.*, 1981; D'Eustachio *et al.*, 1981; Faliks and Meyuhas, 1982). The number of discrete fragments hybridized with different r-protein probes varied from about 7 to 20, suggesting that there is some variation in multiplicity among the r-protein genes examined. Similar Southern blot analysis of *Eco*RI-digested *Xenopus* oocyte DNA (Bozzoni *et al.*, 1981) or yeast DNA (Fried *et al.*, 1981) reveals much simpler patterns. An example of the difference in the complexity of hybridization patterns between mammals and amphibians is shown in Fig. 3.

An estimate of 10 gene copies for mouse S16 r-protein was derived from reassociation kinetics analysis (Monk *et al.*, 1981). This value is consistent with the number of discrete restriction fragments observed in a Southern blot analysis of mouse DNA (Monk *et al.*, 1981). The reiteration of *rp* genes in *Xenopus* DNA was estimated to be 2–5 copies per haploid genome (Bozzoni *et al.*, 1981).

Our preliminary restriction enzyme analysis of mouse genomic clones suggests that in the mouse DNA there are at least four different genes for L18 and six for L7 (E. Yalif, A. Klein, and O. Meyuhas, unpublished results).

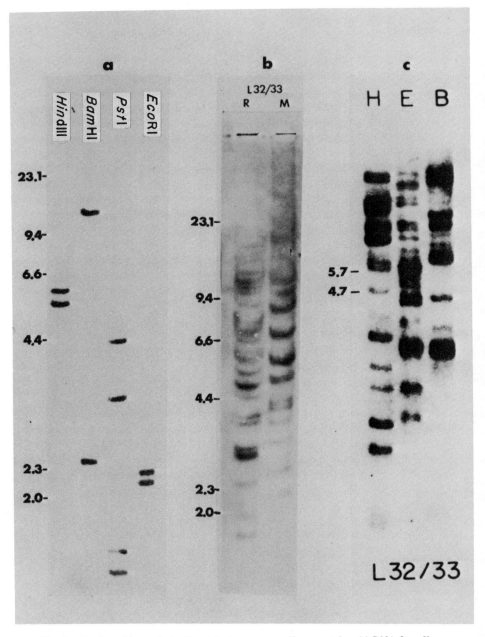

Fig. 3. Southern blot analysis of vertebrate genes encoding r-proteins. (a) DNA from *Xenopus laevis* digested with *Hind*III, *Bam*HI, *Pst*I, and *Eco*RI and hybridized to ^{32}P-labeled cloned S19 cDNA (from Bozzoni *et al.*, 1981). (b) Liver DNA from rat (R) and mouse (M) digested with *Eco*RI (from Faliks and Meyuhas, 1982). (c) Human DNA from HeLa cells digested with *Hind*III (H), *Eco*RI (E), and *Bam*HI (B). (From Monk *et al.*, 1981. Copyright © 1981 by M.I.T. Press.) All mammalian DNAs were hybridized to ^{32}P-labeled cloned L32/33 mouse DNA.

This unique multiplicity of r-protein genes in mammals is an important factor in considering their modes of regulation. The r-protein mRNAs are only moderately abundant (200–400 copies/cell) in rapidly growing cells (Meyuhas and Perry, 1980; Geyer *et al.*, 1982) and are relatively stable as compared with other mRNA species (Craig *et al.*, 1971; Geyer *et al.*, 1982). Thus, one functional r-protein gene could presumably satisfy the cell's requirement for r-protein mRNA production. One intriguing possibility is that the gene multiplicity serves some regulatory purpose, and that not all the competent genes are simultaneously expressed but rather, depend on developmental stage, differential state, or physiological stimulus. However, since analysis based on homology between rp probes and genomic sequences could detect but not distinguish incapacitated or pseudogenes, it remains to be established whether all of the *rp* genes are functional or complete.

Somatic cell hybrids constitute an experimental system with which to investigate the regulation of ribosome biosynthesis in mammalian cells under circumstances altering gene copy number. One such cell hybrid was formed by fusion of mouse and hamster cells, both resistant to emetine (Wejksnora and Warner, 1979), an antibiotic affecting the 40 S subunit. Emetine resistance is associated with an altered r-protein S14 (Wejksnora and Warner, 1981) as well as its mRNA (Madjar *et al.*, 1982). Despite the difference in gene dosage, similar synthesis and accumulation of the modified S14 have been shown in a subline in which both the mouse and hamster genes were active and in hybrids in which only the hamster gene was active (Wejksnora and Warner, 1981).

Alternatively, the technique of transfecting eukaryotic cells with purified DNA enables manipulation of the *rp* gene copy number. Pearson and associates (1982) have exploited similar methodology to transform yeast by isolated *rp* genes and to monitor the expression of the gene in excess at the transcriptional and the translational levels (see the following).

B.   Chromosomal Distribution of Ribosomal Protein Genes

The prokaryotic ribosome structure and biogenesis have been studied in great detail in *E. coli* (Chambliss *et al.*, 1980). The 53 r-proteins are synthesized coordinately and stoichiometrically under a variety of conditions of cellular growth rate and metabolism (Gausing, 1980). In *E. coli*, coordinate expression is facilitated by the presence of several polycistronic transcriptional units (Nomura and Post, 1980), controlled autogenously at both the transcriptional (Dennis and Fill, 1979; Little *et al.*, 1981) and translational levels (Fallon *et al.*, 1979; Nomura *et al.*, 1980).

Ribosome production in mammals requires the coordinate expression of genes encoding more than 70 different proteins as well as the genes specifying 18, 28, 5.8 and 5 S rRNA. Thus, to understand the coordinate regulation of these genes, it is important to know whether or not they are clustered. In order to study the chromosomal locations of r-protein gene families in the mouse, corresponding cDNA probes have been used to examine DNA from a panel of mouse–hamster hybrid cell lines that poses defined sets of different mouse chromosomes (D'Eustachio *et al.*, 1981).

The results indicate that (1) The *rp* genes are widely dispersed throughout the genome, i.e., extensive clustering of many *rp* gene families on a few chromosomes is unlikely; (2) although some members of a particular r-protein gene family may be clustered, generally all members of the family are not located on the same chromosome; and (3) there is no obligatory linkage between r-protein and rRNA genes.

A similar phenomenon has been demonstrated in yeast; its *rp* genes in general are not closely linked (Woolford *et al.*, 1979; Woolford and Rosbash, 1981; Fried *et al.*, 1981).

Although mammalian *rp* genes seem to belong to highly dispersed multigene families, some functionally significant clustering in certain subsets of *rp* genes cannot be ruled out completely.

C. Conservation of Ribosomal Protein Genes

A similarity in the electrophoretic properties of r-proteins among different mammalian species (McConkey *et al.*, 1979) suggests that there is some evolutionary conservation of r-protein genes. Indeed, hybridization of mouse rp-cDNA probes with blots of human DNA from HeLa cells, liver DNA from mouse and rat (Fig. 3), or hamster DNA (D'Eustachio *et al.*, 1981) demonstrated that there is substantially good cross-reactivity between the r-protein sequences of these species and those of the mouse. The degree of sequence homology between rp-mRNAs from mouse and rat livers was estimated by measuring the thermal stability of their hybrids with cloned rp-cDNA from mouse L cells. The melting curves obtained indicated 90–100% homology between mouse and rat rp-mRNAs (Faliks and Meyuhas, 1982). Southern blot analysis of *Drosophila* DNA hybridized with mouse r-protein sequences revealed only one or two relatively faint bands (R. Monk; O. Meyuhas and R. Perry, unpublished results), indicating that there is considerable divergence between the r-protein sequences of mammals and insects.

IV. RIBOSOMAL PROTEIN mRNA: SIZE, ABUNDANCE, AND STABILITY

A. Size of the rp-mRNAs and Their Presumptive Nuclear Precursors

The sizes of various mouse and *Xenopus* rp-mRNAs, determined by analysis on denaturing gels, correlate well with the sizes of the r-proteins encoded by the

TABLE I

Size of Several Mouse and *Xenopus laevis* rp-mRNA Sequences

Species	r-Protein	r-Protein mol. wt.[a] (× 10^{-3})	Size of coding sequence[b] (kb)[e]	mRNA size[c] (kb)	Size of nuclear-specific poly(A)+ RNA components[d] (kb)
Mouse	L7	29.2	0.64	1.03	3.5,3.1,2.7,2.4,1.7
	L13	26.3	0.58	0.80	4.2,3.5,2.6,1.3
	L19	25.3	0.56	0.90	3.9,3.5,3.1,2.7,2.2,1.8,1.6
	L18	24.5	0.54	0.83	2.4,2.1,1.3
	L10	24.2	0.53	0.92	4.2,3.3,2.6,1.3
	L32/33	17.2/15.6	0.38/0.34	0.58	3.6,3.1,2.9,2.6,2.3,1.75,1.65,1.3
	L30	14.5	0.34	0.61	3.2,3.0,2.6,2.3,1.9,1.1
	S16	17.1	0.38	0.66	>12.0,5.1,2.4,2.1,1.8
Xenopus laevis	L1	54.0	1.4	1.5	
	L14	20.8	0.54	0.75	
	L32	8.8	0.23	0.55	
	S1	32.0	0.85	1.0	
	S8	21.5	0.56	0.85	
	S19	12.5	0.32	0.65	

[a] Molecular weight of mouse r-proteins from Wool (1982) and those of *Xenopus* from Pierandrei-Amaldi *et al.* (1982).

[b] Data for mouse r-proteins calculated as described by Meyuhas and Perry (1980) and for *Xenopus* as described by Pierandrei-Amaldi *et al.* (1982).

[c] Poly(A)+ mRNA from various mouse cell lines (Meyuhas and Perry, 1980) and *Xenopus laevis* oocytes (Pierandrei-Amaldi *et al.*, 1982) was size fractionated by electrophoresis on denaturing gels, transferred to DBM paper, and hybridized to several ^{32}P-labeled rp-cDNA clones.

[d] Nuclear poly(A)+ RNA from mouse L cells was electrophoresed on agarose gels containing formaldehyde, blotted onto nitrocellulose filter (Thomas, 1980), and hybridized to various ^{32}P-labeled cloned mouse rp-cDNAs (O. Meyuhas, unpublished).

[e] Kb, kilobases.

mRNA (Table I). Assuming that the poly(A) tail of the mRNA constitutes about 100 nucleotides, we estimate that whereas mouse rp-mRNAs have a relatively uniform proportion (about 70%) of coding sequence; in *Xenopus* this proportion may be more variable.

Each species of mouse rp-mRNA is associated with a distinctive set of poly(A)-containing nuclear components, the largest transcript of each set being 3–18 times larger than the corresponding mRNA (Table I). Some or all of these components are presumably precursors of the mRNA. Their multiplicity may reflect successive processing stages. This interpretation is consistent with the presence of intervening sequences observed in some yeast *rp* genes (Bollen *et al.*, 1982); the accumulation of larger rp transcripts in certain yeast temperature-sensitive mutants, which fail to synthesize functional rp-mRNA (Rosbash *et al.*, 1981; Fried *et al.*, 1981); and the demonstration of split *rp* genes in *Xenopus* (Bozzoni *et al.*, 1982) and mouse (O. Meyuhas, unpublished results). However, due to the multiplicity of the *rp* genes in mammals, some of the nuclear components might represent transcripts of different genes specifying the same r-protein.

The relatively small size of the noncoding sequence in rp-mRNA tends to argue against the idea that the rp-mRNAs are polycistronic, as are some rp-mRNAs in bacterial cells (Nomura *et al.,* 1977). Similarly, the fact that each rp-mRNA exhibits a distinctive pattern of nuclear components makes unlikely the possibility that several rp-mRNAs are synthesized as part of a common primary transcript (for discussion, see Meyuhas and Perry, 1980).

B. Relative Abundance of rp-mRNAs and Nuclear Components

The relative abundance of the rp-mRNAs in exponentially growing mouse cells was estimated with the technique of DNA-excess filter hybridization. For these measurements, radioactive cDNA complementary to poly(A)$^+$ mRNA of mouse MPC-11 myeloma cells or L cells was annealed with a large excess of the eight cloned mouse rp sequences immobilized individually (Meyuhas and Perry, 1980). Alternatively, ^{32}P-labeled poly(A)$^+$ mRNA extracted from mouse 3T6 fibroblasts labeled to equilibrium with ^{32}PO$_2$ was hybridized to each of seven plasmids containing mouse rp-cDNA sequences (Geyer *et al.,* 1982). These analyses indicated that the average rp-mRNA in the particular group sampled comprises about 0.7% of the poly(A)$^+$ mRNA in L cells and 3T6 cells and about 0.18% in MPC-11 cells (Table II). These values are in good agreement with the expected average abundance (0.1%) based on the fraction of total poly(A)$^+$ mRNA translation product comprising basic proteins (8%) and the number of ribosomal proteins (approximately 80). Similar average abundance (0.1%) for rp-mRNA in *Drosophila* was suggested by indirect estimates (Hereford and Rosbash, 1977).

From a comparison of the intensities of RNA blots, the concentration of rp-mRNA sequences in nuclear poly(A)$^+$ mRNA was estimated to be 10% or less, comparable to the cytoplasmic poly(A)$^+$ mRNA (Meyuhas and Perry, 1980). Calculations of the copy number of an rp-mRNA sequence in L cells suggest 300–400 molecules of each rp-poly(A)$^+$ mRNA per cell, of which just 10–30 are found in the nucleus (mostly in molecules the size of the mature rp-mRNAs). The number of rp-mRNA sequences in larger components is probably on the order of 2–5 molecules of each species per cell (Meyuhas and Perry, 1980).

C. Stability of rp-mRNA

Early estimates of the relative turnover of the rp-mRNA were obtained indirectly by measuring the extent of synthesis of their specific proteins on polysomes isolated from mouse L cells at various times after actinomycin D treatment

TABLE II

Relative Abundance of rp-mRNAs in Various Mouse Cells[a]

Species of rp-mRNA	Percentage of rp-mRNA in poly(A)$^+$ mRNA		
	MPC-11 cells	L cells	3T6 cells[b]
L7	0.32	0.12	0.06
L10	0.13	0.12	—
L32/33	0.07	0.08	0.15
L18	0.38	0.06	0.10
S16	0.07	0.06	0.07
L19	0.38	0.06	0.05
L30	0.07	0.06	0.07
L13	0.02	0.01	0.015
Average	0.18	0.07	0.07

[a] The relative abundance of rp-mRNA was determined by the proportion of either ^{32}P-labeled cDNA complementary to mRNA from MPC-11 cells or ^{125}I-labeled DNA complementary to L cell mRNA, hybridizing to individually immobilized rp plasmid DNAs (Meyuhas and Perry, 1980). Similarly, ^{32}P-labeled mRNA from 3T6 cells was hybridized to various filter-bound rp plasmids and the relative content of the rp-mRNAs was calculated. (From Geyer et al., 1982.)

[b] Values for 3T6 cell rp-mRNAs are based on (1) the observation that under growth conditions, the mRNA for seven r-proteins in these cells constitutes 0.5% of the total mRNA and (2) the known proportion of each mRNA (Table I in Geyer et al., 1982).

(Craig et al., 1971). The half-life of rp-mRNAs was found to be similar to the average half-life of the total mRNA population. Because actinomycin D can influence polyribosomal decay via an effect on the initiation of protein synthesis (Singer and Penman, 1972), this method could not give a reliable value for the absolute half-life of rp-mRNAs.

The available cloned mouse rp-cDNAs permitted a direct measurement of the half-life of rp-mRNA, and thus avoided problems associated with drug utilization. Figure 4 shows that in growing 3T6 cells, the average half-life of rp-mRNAs (8 hr) is similar to that of total poly(A)$^+$ mRNA (9 hr), and is hardly affected by the growth status, since in resting cells these half-lives are 11 and 9 hr, respectively.

D. Coordinate Synthesis of Ribosomal Proteins

The synthesis of most r-proteins is coordinately controlled during transition between stationary and growing states in yeast (Gorenstein and Warner, 1976; Warner and Gorenstein, 1977, 1978; Kief and Warner, 1981a,b), mouse 3T3 cells (Tushinski and Warner, 1982), and regenerating rat liver (Nabeshima and Ogata, 1980).

With one exception (L13) the variation in abundance among the rp-mRNAs is

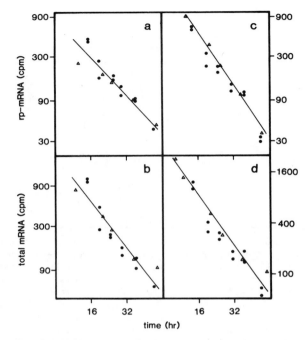

Fig. 4. Stability of rp-mRNA cultures of resting (a and b) or growing (c and d) cells labeled with [³H]uridine for 4 hr. At time 0, the medium was replaced with medium containing 5 mM uridine and 2.5 mM cytidine. At various times, cultures were harvested and the amount of radioactivity in counts per minute (cpm) in total poly(A)$^+$ mRNA (b and d) and rp-poly(A)$^+$ mRNA (a and c) was determined. (From Geyer *et al.,* 1982.)

small in L cells or 3T6 cells and moderate in MPC-11 cells (Table II). Such differences may result from variability in gene dosage, transcription rate, or stability of the rp-mRNA. Nevertheless, maintaining the equimolar amounts of all r-proteins in resting and growing cells, as well as during the transition between the two states, requires a compensation for the excess of certain rp-mRNAs. Several regulatory mechanisms can be proposed:

1. Posttranslational control. Ribosomal proteins are synthesized in non-equimolar amounts, but the unassembled r-proteins are rapidly degraded. Such regulation has been shown to maintain a balance between ribosomal RNA and ribosomal proteins in yeast; when ribosomal proteins are synthesized in the absence of rRNA they do not accumulate but are rapidly degraded (Gorenstein and Warner, 1977).

2. Differential translation efficiency of rp-mRNAs. Evidence from a number of sources demonstrates convincingly that the amount of protein synthesized in

eukaryotic systems is not necessarily directly related to the molar amount of mRNA present. The equimolar synthesis of the α- and β-globin peptides was suggested to result from a higher rate of peptide initiation on β-globin mRNA to compensate for its lower content (Lodish, 1971).

3. Translational feedback regulation of r-protein synthesis. Studies of *E. coli* aimed to elucidate the mechanisms that ensure the balanced and coordinate synthesis of all r-protein led to a model that features autogenous translation control (Fallon *et al.*, 1979). This model proposes that r-protein synthesis and ribosome assembly are coupled. If r-protein synthesis exceeds the rate of ribosome biosynthesis, particular r-proteins act as repressors that prevent the translation of their mRNA and of other r-proteins encoded in the same operon (for review, see Nomura *et al.*, 1982).

Similar translational control has been shown to operate in yeast (Pearson *et al.*, 1982): the dosage of the gene coding for r-protein L3 was altered by introducing into the yeast an autonomously replicating plasmid carrying this gene. The rate of transcription was increased in proportion to the gene dosage, whereas the synthesis of L3 was same as in control cells. Thus, balanced synthesis is established by increasing the turnover of L3 mRNA as well as decreasing the efficiency of its translation.

While coordinate control of r-protein synthesis is a fundamental feature of growing and resting cells, minor irregularities have been observed in the developing slime mold (Ramagopal and Ennis, 1981, 1982) and during early embryogenesis of *Drosophila* (Santon and Pellegrini, 1980), *Xenopus* (Pierandrei-Amaldi *et al.*, 1982), and the mouse (La Marca and Wassarman, 1979). Elucidation of the mechanisms regulating the predominantly balanced synthesis of r-proteins and the few exceptional cases, particularly during development, requires further investigation.

V. RIBOSOMAL PROTEIN SYNTHESIS IN REGENERATING LIVER

Most eukaryotic tissues are in a relatively homeostatic condition where major fluctuations in ribosome synthesis occurs rarely if at all. The regenerative response of the liver following removal of ⅔ of its mass provides a suitable *in vivo* system to monitor changes in the ribosomal content in cell-initiating growth. The early stages of this response in rats consist of a hypertrophia lasting approximately 12–16 hr, during which the rate of RNA and protein synthesis rises, and a subsequent hyperplasia characterized by a peak in DNA synthesis at about 24 hr and mitosis 6–8 hr later (Bucher and Malt, 1971; Lewan *et al.*, 1977).

Measurements of the concentration and rate of synthesis of the RNA and its 45 S precursor indicate 1.5 to 10-fold increase in the ribosomal content of regenerating hepatocytes (Chaudhuri and Lieberman, 1968; Rizzo and Webb, 1972; Tsurugi *et al.,* 1972; Dabeva and Dudov, 1982). Despite the discrepancy of estimates, it was evident in all cases that ribosomes accumulate prior to a substantial increase in cell division rate. In regenerating rat liver both the relative rate of biosynthesis of the r-proteins and the translational activity of their mRNA parallel the changes in rRNA synthesis (Tsurugi *et al.,* 1972; Nabeshima and Ogata, 1980).

Since the rp-mRNA sequences are highly homologous among mammals we could monitor the changes in the rp-mRNA in regenerating rat liver using eight cloned mouse rp-cDNAs (Faliks and Meyuhas, 1982). The relative abundance of rat L7, L13, L18, L30, L32/33, and S16 mRNAs increases after partial hepatectomy. Their maximal level is about twice that of normal rat liver and is achieved 12–18 hr after the operation. Comparison of the time courses suggests that these mRNAs belong to a family of mRNAs exhibiting coordinate control. L10 and L19 mRNAs seem to differ in that their relative initial elevation is considerably slower and does not subside even 48 hr after the operation (Fig. 5).

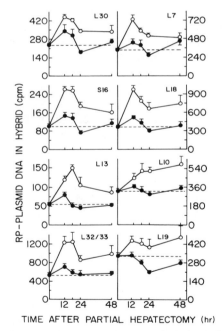

Fig. 5. Time course of the effect of partial hepatectomy on the level of rp-mRNAs in rate liver. Poly(A)$^+$ mRNA was extracted from livers of sham-operated (●———●) and partially hepatectomized (○———○) rats at different times after the operation and from intact rats (— —). One microgram of poly(A)$^+$ mRNA from each preparation was dot blotted onto nitrocellulose filters and hybridized to the indicated ^{32}P nick translated rp-plasmid DNA. The filters were washed, and retained radioactivity was counted. (From Faliks and Meyuhas, 1982.)

The increase in relative abundance of rp-mRNAs most likely reflects a net increase in the rp-mRNA content during liver regeneration because the amount of poly(A)$^+$ mRNA per gram of rat liver is similar for normal and regenerating rat liver at various times after partial hepatectomy (Nabeshima and Ogata, 1980). This increase follows a time course similar to that reported for the template activity for rp-mRNA (Nabeshima and Ogata, 1980). Thus, the stimulation of r-protein biosynthesis in regenerating liver parallels the increase in content of rp-mRNAs. This relationship between the accumulation of rp-mRNAs and the concurrent increase in the r-protein is not universal, however, since in other situations, as will be described, the appearance of newly made r-proteins is translationally controlled.

Whether changes in the rp-mRNA level in the regenerating liver are consequences of alterations in transcription rate, processing efficiency, or mRNA stability, or of activation of inactive members of the r-protein multigene family is yet to be determined.

VI. RIBOSOMAL PROTEIN SYNTHESIS IN RESTING AND GROWING CELLS

Cells in culture can exist in either one of two reversible growth states: (1) proliferating cells (exponential phase or growing cells) and (2) resting cells (stationary phase or nongrowing cells). Transitions from nonproliferative to proliferative states are accompanied by an increase in the synthesis and accumulation of ribosomes (measured as rRNA). Such alterations have been observed when: (1) resting cells are serum stimulated to enter the cell cycle (Becker *et al.*, 1971; Johnson *et al.*, 1974, 1976; Abelson *et al.*, 1974); (2) contact-inhibited cells are seeded into sparse cultures (Emerson, 1971; Weber, 1972); (3) lymphocytes are stimulated with a mitogen such as phytohemagglutinin (Cooper, 1969); and (4) neoplastic growth is induced in epidermis (De Young *et al.*, 1977).

The content of rRNA in resting mouse 3T6 fibroblasts is two to three times lower than in exponentially growing cells (Johnson *et al.*, 1974). The rate of transcription of pre-rRNA increases within 10 min of serum stimulation (Mauck and Green, 1973; Grummt *et al.*, 1977). Similar fluctuations in the synthesis of r-proteins have been observed in mouse 3T3 fibroblasts, (Tushinski and Warner, 1982). When these cells rest due to serum deprivation, the relative synthesis of r-protein is only half that of the growing cell. Upon growth stimulation by addition of serum, all ribosomal protein responds coordinately as demonstrated by rapid increase in the synthesis of each r-protein.

Of special interest is the regulation of ribosomal content in transformed cells which do not control their growth. Stanners and associates (1979) have shown that entry of transformed hamster cells into a nongrowing state leads to decreased rates of synthesis of protein and DNA. Though this reduction is similar to that seen in normal resting cells, the transformed cell maintains a ribosomal content characteristic of exponential growth. This observation implies a loss of control over ribosome complement by the transformed cell. The factors responsible for the high ribosome level in transformed cells are unknown.

A. Regulation of rp-mRNA Content in Growth-Stimulated Mouse Fibroblasts

To estimate the content of rp-mRNA under various growth conditions, resting, exponentially growing, and serum-stimulated 3T6 cells were labeled with $^{32}PO_2$. The amount of radioactivity in rp-mRNA was determined by DNA-excess filter hybridization using seven cloned mouse rp-cDNAs and compared to that in total poly(A)$^+$ mRNA (Geyer et al., 1982).

The relative abundance of rp-mRNA is similar in resting (0.65%) and growing (0.5%) cells. However, since the level of total mRNA in growing cells is greater than in resting cells (Johnson et al., 1974) the content of rp-mRNA per cell in growing cells is at least twice that in resting cells.

The rate of synthesis of r-protein in 3T3 cells increased threefold during the first 2 hr after serum stimulation (Tushinski and Warner, 1982). Somewhat lower response of the rate of rRNA synthesis (twofold) has been observed in 3T6 cells 2–3 hr after the stimulation (Mauck and Green, 1973). These changes lead to an increase in ribosome content, beginning about 6 hr after stimulation (Johnson et al., 1974). However, the total amount of rp-mRNA remains constant during this period (Fig. 6). In the light of these observations, the presence of a translational control mechanism regulating the r-protein synthesis during the resting–growing transition is strongly indicated.

B. Translational Control of Ribosomal Protein Synthesis during Growth Stimulation

Investigation of the processes by which, shortly after growth stimulation, the rp-mRNA may be more efficiently translated requires estimation of the extent of rp-mRNA associated with polysomes. Using the cloned mouse rp-cDNAs it was found that in exponentially growing cells about 85% of the rp-mRNA was in the polysome fraction, whereas in resting cells only one-half of the rp-mRNA was in the polysome fraction. The portion of rp-mRNA in polysomes increased to that of the growing cells within 3 hr after the serum stimulation (Fig. 7). This observation supported the hypothesis that the rate of synthesis of r-proteins is indeed controlled during the first 6 hr after stimulation by an alteration of the efficiency of translation of rp-mRNA. The translational control of rp-mRNA appears to be highly specific since the proportion of toal mRNA in polysomes

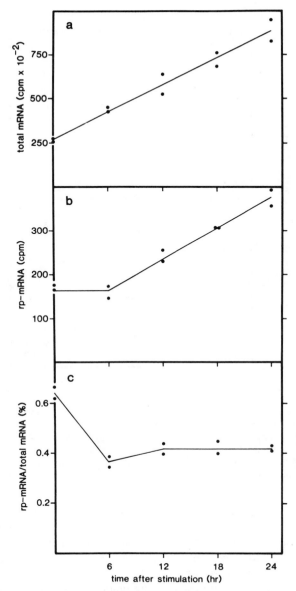

Fig. 6. Content of rp-mRNA in serum-stimulated cultures. Resting 3T6 cells were seeded in medium containing 0.5% serum and $^{32}PO_4$. Cultures were used for an experiment 7 days later. At time 0, cultures were serum stimulated with fresh medium containing 10% serum and $^{32}PO_4$ at the same initial specific activity. Cultures were harvested at the indicated times and assayed for the amount of radioactivity (proportional to content) in total poly(A)$^+$ mRNA (a), rp-poly(A)$^+$ mRNA (b), and the ratio of rp-mRNA to total poly(A)$^+$ mRNA (c). (From Geyer *et al.*, 1982.)

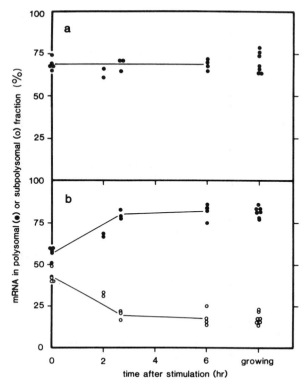

Fig. 7. Polysome distribution of rp-mRNA. Cultures of resting, exponentially growing, or serum-stimulated cells were labeled for 2 or 3 hr with [³H]uridine. After the labeling the cells were harvested and the cytoplasmic compartment was separated into polysomal and subpolysomal fractions. The percentage of labeled total poly(A)⁺ mRNA (a) and rp-poly(A)⁺ mRNA (b) isolated from the polysomal (●) and subpolysomal (○) fractions was determined and expressed as a function of time after stimulation (From Geyer *et al.*, 1982.)

(70%) does not change after stimulation. An increase in the number of ribosomes per rp-mRNA in growing cells as compared with resting cells might result from increase initiation of rp-mRNA translation during the resting–growing transition. At later times, rp-mRNA content increases, permitting still greater enhancement in the rate of synthesis of r-proteins.

Translational control of rp-mRNA is not unique to mammalian fibroblasts. It has also been postulated in resting cultures of chick embryo fibroblasts. The production of r-protein in these cells is increased by insulin stimulation (De Philip *et al.*, 1980) and decreased by insulin deprivation (Ignotz *et al.*, 1981). Analysis of the changes led to the conclusion that rp-mRNAs are less efficiently initiated in resting chick fibroblasts compared to most other cell mRNAs and that insulin acts directly at the level of translation to raise the production of r-proteins.

VII. RIBOSOMAL PROTEIN SYNTHESIS DURING DEVELOPMENT

Protein synthesis varied markedly as a function of development (Davidson, 1976). The content and rate of synthesis of ribosomes vary in accordance with changing requirements for protein synthetic machinery. Analysis of r-protein from various stages of development in *Dictyostelium discoideum* demonstrates that cell differentiation is accompanied by quantitative as well as qualitative changes in r-protein (Ramagopal and Ennis, 1981). Measurements of the translational activity of the corresponding mRNAs (Ramagopal and Ennis, 1982) suggest that the synthesis of most of r-protein during spore germination and vegetative growth is regulated at the level of the mRNA content. However, translational control was postulated for the spore-specific rp-mRNAs since these mRNAs are present in the amoeba while their corresponding proteins cannot be detected in these cells.

Synthesis of r-proteins in *Drosophila melanogaster* embryos starts at least as early as 90 min after fertilization (Santon and Pellegrini, 1980). This significantly precedes the synthesis of rRNA which does not begin before blastoderm stage (Anderson and Lengyel, 1979).

During embryogenesis of mammals, the synthesis of rRNA is initiated at least as early as the 16-cell stage in rabbit (Manes, 1971) or 4-cell stage in mouse (Clegg and Pikó, 1971). Estimation derived from electron microscopic morphometry indicates that in mouse the number of ribosomes per embryo rises sharply between the 2- and 8-cell stages as well as between the 8-cell and early blastocyst stages (Pikó and Clegg, 1982).

The schedule of increase of ribosomal content corresponds to that of rRNA (Clegg and Pikó, 1977) and r-protein synthesis (La Marca and Wassarman, 1979). Whereas in the unfertilized mouse egg the synthesis of r-protein accounts for 1.1% of total protein synthesis it rises to 8.1% at the 8-cell stage. It seems likely then that the number of ribosomes increases to keep pace with the increasing rate of total protein synthesis between the 2-cell stage and the blastocyst (Brinster *et al.*, 1976).

During oogenesis in *Xenopus laevis*, ribosomes are synthesized as a storage product for later use in the developing embryo. The genes for the 28 and 18 S RNA are selectively amplified about a thousandfold early in oogenesis (Brown and Dawid, 1968; Gall, 1968). The synthesis of r-protein is coordinately regulated with the intense accumulation of 28 and 18 S rRNA, constituting about 30% of the protein synthesized during mid-oogenesis (Hallberg and Smith,

1975). The following lines of evidence support the idea that in amphibian embryo, protein synthesis is dependent on a supply of maternal mRNA and ribosomes through much of embryogenesis.

1.   No detectable rRNA synthesis occurs in normal *Xenopus laevis* embryos until the onset of gastrulation (stage 10) when each embryo consists of tens of thousands of cells (Brown and Littna, 1964, 1966).

2.   Homozygous mutant *Xenopus laevis* embryos, which lack the nucleolar organizer region (0-nu), develop normally until early swimming tadpole stage, even though they do not synthesize rRNA (Brown and Gurdon, 1964).

Estimations of the rp-mRNA activity support the idea that the increased synthesis of r-proteins occurs between the blastula and tadpole stages and may be attributed to accumulations of rp-mRNA during embryogenesis (Weiss *et al.*, 1981).

A. Expression of Ribosomal Protein Genes during *Xenopus laevis* Development

An available set of cloned *Xenopus* rp-cDNAs (Bozzoni *et al.*, 1981) has enabled Pierandrei-Amaldi and associates (1982) to monitor fluctuations of rp-mRNA abundance during *Xenopus laevis* oogenesis and embryogenesis. In the oocyte the amount of rp-mRNA increased from the very early stages, reaching a plateau at the onset of vitellogenesis. This accumulation of rp-mRNA precedes the maximal rate of ribosome production. During embryogenesis the amount of rp-mRNA per cell varies considerably, starting with low levels, probably of maternal origin, at early cleavage (stages 2–3), almost disappearing before gastrulation (stages 10–11), then increasing until neurulation (stage 16), and finally remaining constant thereafter (Fig. 8). The appearance of both newly synthesized rRNA (Brown and Littna, 1964) and rp-mRNA seems to occur simultaneously at gastrulation and much earlier than synthesis of the r-protein. Measurement of r-protein synthesis during early embryogenesis (Pierandrei-Amaldi *et al.*, 1982) demonstrates that newly made r-proteins are not detectable before stage 26 except for the constantly synthesized S7, L17, and L31 and the early synthesized L5 (stages 7–9). The onset of r-protein synthesis occurs in tailbud embryos (stages 28–32) several hours after the appearance of their mRNAs. Such a time lag between the appearance of mRNA in the cytoplasm and its translation into proteins suggests a translational control of rp-mRNAs. Indeed, this hypothesis was further supported by study of the polysomal distribution of S1-mRNA during embryogenesis (Pierandrei-Amaldi *et al.*, 1982). This rp-mRNA initially appears only in the subpolysomal fraction (stage 15) and starts to be mobilized to polysomes around stage 26, and more than 50% of it is associated with polysomes by

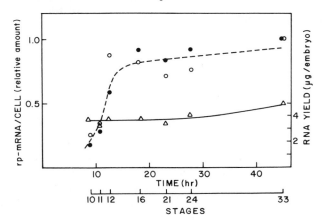

Fig. 8. Relative amount of rp-mRNA per cell during *Xenopus* embryogenesis. Cytoplasmic RNA derived from 10 embryos of each stage was denatured, run on agarose gels, and transferred to nitrocellulose filters. Filters were hybridized to ^{32}P-labeled plasmids containing r-protein L14 and S1 sequences. The extent of hybridization of L14 mRNA (●——●) and S1 mRNA (○——○) to the corresponding probes was estimated by densitometric tracing of the autoradiogram. The values have been normalized to the number of cells per embryo of the specific stage. The extraction yields of RNA from the various developmental stages are shown △——△) (From Pierandrei-Amaldi *et al.*, 1982. Copyright © 1982 by M.I.T. Press.)

stage 31. The intensive mobilization of rp-mRNA to polysomes coincides with its active translation and the substantial accumulation of rRNA (Brown and Littna, 1964).

B. Coupling of Ribosomal Proteins and Ribosomal RNA Synthesis

The parallel appearance of newly synthesized r-proteins (Hallberg and Smith, 1975) with the substantial increase of rRNA (Brown and Littna, 1964) in *Xenopus leavis* embryos suggests that r-protein synthesis is coordinately regulated with that of the rRNA. Further support for this idea comes from the anucleolate (0-nu) *Xenopus*. These mutants lack the capacity to synthesize 28 and 18 S rRNA and also fail to synthesize most of the r-protein at a time when normal embryos are synthesizing complete ribosomes (Hallberg and Brown, 1969; Pierandrei-Amaldi *et al.*, 1982). To determine whether the inability to synthesize r-proteins originates at the transcriptional level, the amount of mRNA for two r-proteins from normal and 0-nu embryos has been monitored (Pierandrei-Amaldi *et al.*, 1982). The results indicate that in 0-nu embryos the genes for these r-proteins are regularly transcribed, and at stage 26 the rp-mRNA is normally accumulated, although rRNA is not synthesized. Later (stage 35), the amount of rp-mRNA decreases in the 0-nu, compared to the normal, embyos.

Nevertheless, the accumulation of rp-mRNA is not followed by appearance of newly made r-protein. This may be attributed either to a translational control of the rp-mRNA or to a rapid degradation of the newly made r-protein.

Similar coupling of rRNA and r-protein synthesis has been proposed to occur in various systems like *Saccharomyces* after addition of glucose to a culture growing in ethanol (Kief and Warner, 1981a), regenerating rat liver (Tsurugi *et al.*, 1972) and during early embryogenesis in the mouse (La Marca and Wassarman, 1979). However, in other experimental systems the synthesis of the ribosomal components is clearly uncoordinated. The yeast rRNA is normally transcribed when r-protein synthesis is inhibited in temperature-sensitive mutants at the nonpermissive temperature (Shulman and Warner, 1978). During inhibition of rRNA synthesis in HeLa cells by low concentration of actinomycin D, r-protein synthesis continues (Craig and Perry, 1971; Warner, 1977). When rat myoblast L_6E_9 cell line is induced to differentiate into myotubes the synthesis of r-proteins remains constant, although that of rRNA decreases five- to tenfold (Krauter *et al.*, 1980). One might conclude, therefore, that coupling of rRNA and r-protein is characteristic of proliferating cells and growth stimulation when ribosome synthesis is increased to supply the requirement for enhanced protein synthesis. However, when growth is slowed down at a nonpermissive temperature by action of drugs or during differentiation the control of ribosomal components is probably dependent on the degradation of excess of rRNA or r-protein rather than finely tuned coordination of their synthesis.

VIII. CONCLUDING REMARKS

Eukaryotic r-proteins represent a group of "housekeeping" proteins whose requirement is ubiquitous to all cell types, developmental states, growth statuses, and environmental circumstances. Since these proteins are synthesized in equimolar amounts under a variety of different conditions and in coordination with rRNA, they afford a useful model for identifying the mechanisms controlling the expression of many genes. This may be applicable to understanding the regulation of the biosynthesis of other complex structures as well as shedding light on developmental processes requiring coordinate expression of multiple genes.

The recent construction and isolation of recombinant DNAs containing rp-cDNA or *rp* gene sequences have provided the proper experimental tool with which to approach these questions. Using these molecular clones we hope to gain information about the relationship between the balanced production of r-proteins and various cellular factors. These include the multiplicity of the *rp* genes and their chromosomal dispersion; possible regulatory sequences inside the genes, in their vicinity, and/or in the corresponding mRNA; and transcription, processing, stability, and translational activity of rp-mRNA molecules. Given the number and complexity of the reactions involved in the coordinated biosynthesis of r-proteins, we may expect to disclose a similar diversity of regulatory mechanisms as we deepen our understanding of them.

ACKNOWLEDGMENTS

The preparation of this chapter and research in my laboratory were supported by grants from the Israel Academy of Sciences and Humanities—Basic Research Foundation and from the United States–Israel Binational Science Foundation (BSF), Jerusalem, Israel.

The author wishes to thank Dr. R. P. Perry and Mr. D. Mencher for their critical comments on the manuscript.

REFERENCES

Abelson, H. T., Johnson, L. F., Penman, S., and Green, H. (1974). Changes in RNA in relation to growth of the fibroblast. II. The time life of mRNA, rRNA and tRNA in resting and growing cells. *Cell* **1**, 161–165.

Anderson, K. V., and Lengyel, J. (1979). Rates of synthesis of major classes of RNA in *Drosophila* embryos. *Dev. Biol.* **70**, 217–231.

Becker, H., Stanners, C. P., and Kudlow, J. E. (1971). Control of macromolecular synthesis in proliferating and resting Syrian hamster cells in monolayer culture. II. Ribosome complement in resting and early G1 cells. *J. Cell. Physiol.* **11**, 43–50.

Bollen, G. H. P. M., Cohen, L. H., Mager, W. H., Klaassen, A. W., and Planta, R. J. (1981a). Isolation of cloned ribosomal protein genes from the yeast *Saccharomyces carlsbergensis*. *Gene* **14**, 279–287.

Bollen, G. H. P. M., Mager, W. H., and Planta, R. J. (1981b). High resolution mini two dimensional gel electrophoresis of yeast ribosomal proteins. A standard nomenclature for yeast ribosomal proteins. *Mol. Biol. Rep.* **8**, 37–44.

Bollen, G. H. P. M., Molenaar, C. M. T., Cohen, L. H., van Raamsdonk-Duin, M. M. C., Mager, W. H., and Planta, R. J. (1982). Ribosomal protein genes of yeast contain intervening sequences. *Gene* **18**, 29–37.

Bozzoni, I., Beccari, E., Luo, Z.-X., Amaldi, F., Pierandrei-Amaldi, P., and Campioni, N. (1981). *Xenopus laevis* ribosomal protein genes: Isolation of recombinant cDNA clones and study of the genomic organization. *Nucleic Acids Res.* **9**, 1069–1086.

Bozzoni, I., Tognoni, A., Peirandrei-Amaldi, P., Beccari, E., Buongiorno-Nardelli, M., and Amaldi, F. (1982). Isolation and structural analysis of ribosomal protein genes in *Xenopus laevis*. Homology between sequences present in the gene and in several different messenger RNAs. *J. Mol. Biol.* **161**, 353–371.

Brinster, R. L., and Wiebold, J. L., and Brunner, S. (1976). Protein metabolism in preimplanted mouse ova. *Dev. Biol.* **51**, 215–224.

Brown, D. D., and Dawid, I. B. (1968). Specific gene amplification in oocytes. *Science* **160**, 272–280.

Brown, D. D., and Gurdon, J. B. (1964). Absence of ribosomal RNA synthesis in the anucleolate mutant of *Xenopus laevis*. *Proc. Natl. Acad. Sci. U.S.A.* **51**, 139–146.

Brown, D. D., and Littna, E. (1964). RNA synthesis during the development of *Xenopus laevis,* the South African clawed toad. *J. Mol. Biol.* **8**, 669–687.

Brown, D. D., and Littna, E. (1966). Synthesis and accumulation of DNA-like RNA during embryogenesis of *Xenopus laevis*. *J. Mol. Biol.* **20**, 81–91.

Bucher, N. L. R., and Malt, R. A. (1971). "Regeneration of Liver and Kidney." Little, Brown, Boston, Massachusetts.

Chambliss, G., Craven, G. R., Davies, J., Davis, K., Kahan, L., and Nomura, M., eds. (1980). "Ribosomes: Structure, Function and Genetics." University Park Press, Baltimore, Maryland.

Chaudhuri, S., and Lieberman, I. (1968). Control of ribosome synthesis in normal and regenerating liver. *J. Biol. Chem.* **243**, 29–33.

Clegg, K. B., and Pikó, L. (1977). Size and specific activity of the UTP, pool and overall rates of RNA synthesis in early mouse embryos. *Dev. Biol.* **58**, 76–95.

Cooper, H. L. (1969). Ribosomal ribonucleic acid wastage in resting and growing lymphocytes. *J. Biol. Chem.* **244**, 5590–5596.

Craig, N., and Perry, R. P. (1971). Persistent cytoplasmic synthesis of ribosomal proteins during the selective inhibition of ribosome RNA synthesis. *Nature (London), New Biol.* **229**, 75–80.

Craig, N., Kelley, D. E., and Perry, R. P. (1971). Lifetime of the messenger RNAs which code for ribosomal proteins in L-cells. *Biochim. Biophys. Acta* **246**, 493–498.

Dabeva, M. D., and Dudov, K. P. (1982). Transcriptional control of ribosome production in regenerating rat liver. *Biochem. J.* **208**, 101–108.

Davidson, E. H. (1976). "Gene Activity in Early Development." Academic Press, New York.

Dennis, P., and Fill, N. P. (1979). Transcriptional and post transcriptional control of RNA polymerase and ribosomal protein genes cloned on composite Col E_2 plasmids in the bacterium *Escherichia coli*. *J. Biol. Chem.* **254**, 7540–7547.

De Philip, R. M., Rudert, W. A., and Lieberman, I. (1980). Preferential stimulation of ribosomal protein synthesis by insulin and in the absence of ribosomal and messenger ribonucleic acid formation. *Biochemistry* **19**, 1662–1669.

D'Eustachio, P., Meyuhas, O., Ruddle, F., and Perry, R. P. (1981). Chromosomal distribution of ribosomal protein genes in the mouse. *Cell* **24**, 307–312.

De Young, L. M., Argyris, T. S., and Gordon, G. B. (1977). Epidermal ribosome accumulation during two stage skin tumorigenesis. *Cancer Res.* **37**, 388–393.

Efstratiadis, A., Kafatos, F. C., Maxam, A. M., and Maniatis, T. (1976). Enzymatic in vitro synthesis of globin genes. *Cell* **7**, 279–288.

Egberts, E., Hackett, P. B., and Traub, P. (1977). Shutoff of histone synthesis by mengovirus infection of Ehrlich ascites tumor cells. *Hoppe-Seyler's Z. Physiol. Chem.* **358**, 463–474.

Emerson, C. P., Jr. (1971). Regulation of the synthesis and the stability of ribosomal RNA during contact inhibition of growth. *Nature (London), New Biol.* **232**, 101–106.

Faliks, D., and Meyuhas, O. (1982). Coordinate regulation of ribosomal protein mRNA level in regenerating rat liver. Study with the corresponding mouse cloned cDNAs. *Nucleic Acids Res.* **10**, 789–801.

Fallon, A. M., Jinks, C. S., Strycharz, G. D., and Nomura, M. (1979). Regulation of ribosomal protein synthesis in *Escherichia coli* by selective mRNA inactivation. *Proc. Natl. Acad. Sci. U.S.A.* **76**, 4311–4315.

Fried, H. M., and Warner, J. R. (1981). Cloning of yeast gene for trichodermin resistance and ribosomal protein L3. *Proc. Natl. Acad. Sci. U.S.A.* **78**, 238–242.

Fried, H. M., and Warner, J. R. (1982). Molecular cloning and analysis of yeast gene for cycloheximide resistance and ribosomal protein L29. *Nucleic Acids Res.* **10**, 3133–3148.

Fried, H. M., Pearson, N. J., Kim, C. H., and Warner, J. R. (1981). The genes for fifteen ribosomal proteins of *Saccharomyces cerevisia* . *J. Biol. Chem.* **256**, 10176–10182.

Gall, J. R. (1968). Differential synthesis of the genes for ribosomal RNA during amphibian oogenesis. *Proc. Natl. Acad. Sci. U.S.A.* **60**, 553–560.

Gausing, K. (1980). Regulation of ribosome biosynthesis in *E. coli*. *In* "Ribosomes: Structure,

Function and Genetics'' (G. Chambliss, G. R. Craven, J. Davies, K. Davis, L. Kahan, and M. Nomura, eds.), pp. 693–718. Univ. Park Press, Baltimore, Maryland.

Geyer, P. K., Meyuhas, O., Perry, R. P., and Johnson, L. F. (1982). Regulation of ribosomal protein mRNA content and translation in growth-stimulated mouse fibroblasts. *Mol. Cell. Biol.* **2**, 685–693.

Gorenstein, C., and Warner, J. R. (1976). Co-ordinate regulation of the synthesis of eukaryotic ribosomal proteins. *Proc. Natl. Acad. Sci. U.S.A.* **73**, 1547–1551.

Gorenstein, C., and Warner, J. R. (1977). Synthesis and turnover of ribosomal proteins in the absence of 60S subunit assembly in *Saccharomyces cerevisiae*. *Mol. Gen. Genet.* **157**, 327–382.

Grummt, F., Paul, D., and Grummt, A. (1977). Regulation of ATP pools, rRNA and DNA synthesis in 3T3 cells in response to serum or hypoxanthine. *Eur. J. Biochem.* **76**, 7–12.

Grunstein, M., and Hogness, D. S. (1975). Colony hybridization: A method for the isolation of cloned DNAs that contain a specific gene. *Proc. Natl. Acad. Sci. U.S.A.* **72**, 3861–3965.

Hackett, P. B., Egberts, E., and Traub, P. (1978). Characterization of Ehrlich ascites tumor cell messenger RNA specifying ribosomal proteins by translation in vitro. *J. Mol. Biol.* **119**, 253–267.

Hallberg, R. L., and Brown, D. D. (1969). Co-ordinate synthesis of some ribosomal proteins and ribosomal RNA in embryo of *Xenopus laevis*. *J. Mol. Biol.* **46**, 393–411.

Hallberg, R. L., and Smith, D. C. (1975). Ribosomal protein synthesis in *Xenopus laevis* oocytes. *Dev. Biol.* **42**, 40–52.

Hardy, S. J. S., Kurland, C. G., Voynow, P., and Mora, G. (1969). The ribosomal proteins of *E. coli:* A purification of the 30S ribosomal proteins. *Biochemistry* **8**, 2897–2905.

Hereford, L. M., and Rosbash, M. (1977). Regulation of a set of abundant mRNA sequences. *Cell* **10**, 463–467.

Ignotz, G. G., Hokari, S., De Philip, R. M., Tsukada, K., and Lieberman, I. (1981). Lodish model and regulation of ribosomal protein synthesis by insulin-deficient chick embryo fibroblasts. *Biochemistry* **20**, 2550–2558.

Johnson, L. F., Abelson, H. T., Green, H., and Penman, S. (1974). Changes in RNA in relation to growth of the fibroblast. I. Amount of mRNA and tRNA in resting and growing cells. *Cell* **1**, 95–100.

Johnson, L. F., Levis, R., Abelson, H. T., Green, H., and Penman, S. (1976). Changes in RNA in relation to growth of the fibroblast. IV. Alterations in the production and processing of mRNA and rRNA in resting and growing cells. *J. Cell Biol.* **71**, 933–938.

Kief, D. R., and Warner, J. R. (1981a). Co-ordinated control of synthesis of ribosomal ribonucleic acid and ribosomal proteins during nutritional shift-up in *Saccharomyces cerevisiae*. *Mol. Cell. Biol.* **1**, 1007–1015.

Kief, D. R., and Warner, J. R. (1981b). Hierarchy of elements regulating synthesis of ribosomal proteins in *Saccharomyces cerevisiae*. *Mol. Cell. Biol.* **1**, 1016–1023.

Krauter, K. S., Soeiro, R., and Nadal-Ginard, R. (1980). Uncoordinate regulation of ribosomal RNA and ribosomal protein synthesis during L_6E_9 myoblast differentiation. *J. Mol. Biol.* **142**, 145–159.

La Marca, M. J., and Wassarman, P. M. (1979). Program of early development in the mammal: Changes in absolute rates of synthesis of ribosomal proteins during oogenesis and early embryogenesis in the mouse. *Dev. Biol.* **73**, 103–119.

Levenson, R. G., and Marcu, K. B. (1976). On the existence of polyadenylated histone mRNA in *Xenopus laevis* oocytes. *Cell* **9**, 311–322.

Lewan, L., Ynger, T., and Engelbrecht, C. (1977). The biochemistry of the regenerating liver. *Int. J. Biochem.* **8**, 477–487.

Little, R., Fill, N. P., and Dennis, P. (1981). Transcriptional control of ribosomal proteins ribonucleic acid polymerase genes. *J. Bacteriol.* **147,** 25–35.

Lodish, H. F. (1971). Alpha and beta globin in messenger ribonucleic acid. Different amounts and rates of initiation of translation. *J. Biol. Chem.* **24,** 7131–7138.

McConkey, E. M., Bielka, H., Gordon, J., Lastick, S. M., Lin, A., Ogata, K., Riboud, J. P., Traugh, J. A., Traut, R. R., Warner, J. R., Welfe, H., and Wool, I. G. (1979). Proposed uniform nomenclature for mammalian ribosomal proteins. *Mol. Gen. Genet.* **169,** 1–6.

Madjar, J.-J., Nielsen-Smith, K., Frahm, M., and Roufa, D. J. (1982). Emetine resistance in Chinese hamster ovary cells is associated with an altered ribosomal protein S14 mRNA. *Proc. Natl. Acad. Sci. U.S.A.* **79,** 1003–1007.

Manes, C. (1971). Nucleic acid synthesis in preimplantation rabbit embryos. II. Delayed synthesis of ribosomal RNA. *J. Exp. Zool.* **176,** 87–95.

Mauck, J. C., and Green, H. (1973). Regulation of RNA synthesis in fibroblasts during transition from resting to growing state. *Proc. Natl. Acad. Sci. U.S.A.* **70,** 2819–2822.

Meyuhas, O., and Perry, R. P. (1980). Construction and identification of cDNA clones for several mouse ribosomal proteins. Application for the study of r-protein gene expression. *Gene* **10,** 113–129.

Monk, R., Meyuhas, O., and Perry, R. P. (1981). Mammals have multiple genes for individual ribosomal proteins. *Cell* **24,** 301–306.

Nabeshima, Y.-I., and Ogata, K. (1980). Stimulation of the synthesis of ribosomal proteins in regenerating rat liver with special reference to the increase in the amounts of effective mRNAs for ribosomal protein. *Eur. J. Biochem.* **107,** 323–329.

Nabeshima, Y.-I., Imai, K., and Ogata, K. (1979). Biosynthesis of ribosomal proteins by poly(A)-containing mRNAs from rat liver in a wheat germ cell free system and sizes of mRNAs coding ribosomal proteins. *Biochim. Biophys. Acta* **564,** 105–121.

Nomura, M., and Post, L. E. (1980). Organization of ribosomal genes and regulation of their expression in *Escherichia coli. In* "Ribosomes: Structure, Function and Genetics" (G. Chambliss, G. R. Craven, J. Davies, K. Davis, L. Kahan, and M. Nomura, eds.), pp. 671–691. University Park Press, Baltimore, Maryland.

Nomura, M., Morgan, E. A. and Jaskunas, S. P. (1977). Genetics of bacterial ribosomes. *Annu. Rev. Genet.* **11,** 287–347.

Nomura, M., Yates, J. L., Dean, D., and Post, L. E. (1980). Feedback regulation of ribosomal protein gene expression in *Escherichia coli:* Structural homology of ribosomal RNA and ribosomal protein mRNA. *Proc. Natl. Acad. Sci. U.S.A.* **77,** 7084–7088.

Nomura, M., Dean, D., and Yates, L. (1982). Feedback regulation of ribosomal protein synthesis in *Escherichia coli. Trends Biochem. Sci.* **7,** 92–95.

Pearson, N. J., Fried, H. M., and Warner, J. R. (1982). Yeast use translational control to compensate for extra copies of a ribosomal protein gene. *Cell* **29,** 347–355.

Pierandrei-Amaldi, P., and Beccari, E. (1980). Messenger RNA for ribosomal proteins in *Xenopus laevis* oocytes. *Eur. J. Biochem.* **106,** 603–611.

Pierandrei-Amaldi, P., Campioni, N., Beccari, E., Bozzoni, I., and Amaldi, F. (1982). Expression of ribosomal protein genes in *Xenopus laevis* development. *Cell* **30,** 163–171.

Pikó, L., and Clegg, K. B. (1982). Quantitative changes in total RNA, total poly(A) and ribosomes in early mouse embryos. *Dev. Biol.* **89,** 362–378.

Ramagopal, S., and Ennis, H. L. (1981). Regulation of synthesis of cell specific ribosomal proteins during differentiation of *Dictyostelium discoideum. Proc. Natl. Acad. Sci. U.S.A.* **78,** 3083–3087.

Ramagopal, S., and Ennis, H. L. (1982). Ribosomal protein synthesis during spore germination and vegatitative growth in *Dictyostelium discoideum. J. Biol. Chem.* **257,** 1025–1031.

Rizzo, A. J., and Webb, T. E. (1972). Regulation of ribosome formation in regenerating rat liver. *Eur. J. Biochem.* **27,** 136–144.

Rosbash, M., Harris, P. K., Woolford, J. L. Jr., and Teem, J. L. (1981). The effect of temperature-sensitive RNA mutants on the transcription products from cloned ribosomal protein genes of yeast. *Cell* **24,** 679–686.

Ruderman, J. V., and Pardue, M. L. (1977). Cell-free translation analysis of messenger RNA in Echinoderm and Amphibian early development. *Dev. Biol.* **60,** 48–68.

Santon, J. B., and Pellegrini, M. (1980). Expression of ribosomal proteins during *Drosophila* early development. *Proc. Natl. Acad. Sci. U.S.A.* **77,** 5649–5653.

Shulman, R. W., and Warner, J. R. (1978). Ribosomal RNA transcription in mutant of *Saccharomyces cerevisiae* defective in ribosomal protein synthesis. *Mol. Gen. Genet.* **161,** 221–223.

Singer, R. H., and Penman, S. (1972). Stability of HeLa cell mRNA in actinomycin. *Nature (London)* **240,** 100–102.

Southern, E. M. (1975). Detection of specific sequences among DNA fragments separated by gel electrophoresis. *J. Mol. Biol.* **98,** 503–517.

Stanners, C. P., Adams, M. E., Harkins, J. L., and Pollard, J. W. (1979). Transformed cells have lost control of ribosome number through their growth cycle. *J. Cell. Physiol.* **100,** 127–138.

Thomas, P. S. (1980). Hybridization of denatured RNA and small DNA fragments transferred to nitrocellulose. *Proc. Natl. Acad. Sci. U.S.A.* **77,** 5201–5205.

Tsurugi, K., Morita, J., and Ogata, K. (1972). Studies on the metabolism of ribosomal structural proteins of regenerating rat liver. *Eur. J. Biochem.* **25,** 117–128.

Tushinski, R.J., and Warner, J. R. (1982). Ribosomal proteins are synthesized preferentially in cell commencing growth. *J. Cell. Physiol.* **112,** 128–135.

Vaslet, C. A., O'Connel, P., Izquierdo, M., and Rosbash, M. (1980). Isolation and mapping of cloned ribosomal protein gene of *Drosophila melanogaster. Nature (London)* **285,** 674–676.

Warner, J. R. (1977). In the absence of ribosomal RNA synthesis, the ribosomal proteins of HeLa cells are synthesized normally and degraded rapidly. *J. Mol. Biol.* **115,** 315–333.

Warner, J. R., and Gorenstein, C. (1977). The synthesis of eucaryotic ribosomal proteins in vitro. *Cell* **11,** 201–212.

Warner, J. R., and Gorenstein, C. (1978). Yeast has a true stringent response. *Nature (London),* **275,** 338–339.

Warner, J. R., Tushinski, R. J., and Wejksnora, P. J. (1980). Coordination of RNA and proteins in eukaryotic ribosome production. *In* "Ribosomes: Structure, Functions and Genetics" (G. Chambliss, G. R. Craven, J. Davies, K. Davis, K. Kahan, and M. Nomura, eds.), pp. 889–902. University Park Press, Baltimore, Maryland.

Weber, M. J. (1972). Ribosomal RNA turnover in contact inhibited cells. *Nature (London), New Biol.* **235,** 58–61.

Weiss, Y. C., Vaslet, C. A., and Rosbash, M. (1981). Ribosomal protein mRNA increases dramatically during *Xenopus* development. *Dev. Biol.* **87,** 330–339.

Wejksnora, P. J., and Warner, J. R. (1979). Hybrid mammalian cells assemble hybrid ribosomes. *Proc. Natl. Acad. Sci. U.S.A.* **76,** 5554–5558.

Wejksnora, P. J., and Warner, J. R. (1981). Regulation of ribosomal RNA and proteins in mouse–hamster hybrid cells. *J. Biol. Chem.* **256,** 9406–9413.

Wool, I. G. (1982). The structure of eukaryotic ribosomes. *In* "Protein Biosynthesis in Eukaryotes" (A. Perez-Bercoff, ed.), pp. 69–95. Plenum, New York.

Woolford, J. L., Jr., and Rosbash, M. (1981). Ribosomal protein genes rp39 (10-78), rp39 (11-40), rp51 and rp52 are not contiguous to other ribosomal protein genes in the *Saccharomyces cerevisiae* genome. *Nucleic Acids Res.* **9,** 5021–5036.

Woolford, J. L., Jr., Hereford, L. M., and Rosbash, M. (1979). Isolation of cloned DNA sequences containing ribosomal protein genes from *Saccharomyces cerevisiae. Cell* **18,** 1247–1259.

Zweidler, A. (1978). Resolution of histones by polyacrylamide gel electrophoresis in presence of non-ionic detergents. *Methods Cell Biol.* **17,** 223–293.

11

Regulation of Nonmuscle Actin Gene Expression during Early Development

LEWIS J. KLEINSMITH, N. KENT PETERS, AND MARY E. ZEIGLER[1]

Division of Biological Sciences
The University of Michigan
Ann Arbor, Michigan

I. INTRODUCTION

Actin is an abundant protein constituent of all eukaryotic organisms. This protein is involved in the construction of the cytoskeleton and carries out contractile functions in events such as cell division, cytokinesis, ameboid movement,

[1]Present address: Department of Plant Pathology, Cornell University, Ithaca, New York.

RECOMBINANT DNA AND
CELL PROLIFERATION

and intracellular transport (Allen, 1981; Pollard, 1981; Pollard and Weihing, 1974). Actin is also an essential component of the contractile apparatus of skeletal, cardiac, and smooth muscle tissues (Franzini-Armstrong and Peachey, 1981; Goldman *et al.*, 1976).

Three different forms of actin have been identified by two-dimensional gel electrophoresis. They are designated α, β, and γ in order of increasing iso-electric points. Actins of the β and γ type are generally referred to as "nonmuscle" actins because of their widespread occurrence in nonmuscle cells, although β-actin is present in smooth muscle as well (Vandekerckhove and Weber, 1978a). α-Actins are present only in muscle. Amino acid sequence analysis of amino-terminal tryptic peptides has shown that at least four different types of α-actin are present in mammals, one in skeletal muscle, one in cardiac muscle, and two in smooth muscle (Vandekerckhove and Weber, 1978b).

From such protein analysis it has become evident that actin represents a family of highly conserved, closely related proteins that is differentially expressed during organismal development. The existence of such a complex population of related proteins raises the question of how the family of genes coding for these proteins is organized and regulated. The development of recombinant DNA techniques has allowed a great deal of progress to be made in this area.

II. ORGANIZATION OF ACTIN GENE FAMILY

Knowledge of the number of actin genes, their linkage relationships, and their sequence organization is an important prerequisite for investigating the regulation of actin gene expression. Investigation of this area was first made possible when gene sequences coding for actin were identified in cloned cDNA libraries made from *Dictyostelium* and sea urchin mRNA (Kindle and Firtel, 1978; Merlino *et al.*, 1980). Recombinant cDNA plasmids that contained actin DNA sequences were identified by their ability to hybridize to mRNA which produced actin when translated *in vitro*. Because actin is such a highly conserved protein, these cloned cDNAs were subsequently used to identify actin genes in other organisms.

A. Actin Gene Number

Most estimates of the number of actin genes per haploid genome have been obtained by using either DNA–DNA filter hybridization or solution hybridization. DNA–DNA filter hybridizations are of two varieties, Southern blot analysis and dot blots. For Southern blot analysis (Southern, 1975) genomic DNA is generally cleaved with enzymes that have six-basepair (bp) recognition sites to limit the number of fragments obtained. The resulting DNA fragments are then fractionated by agarose gel electrophoresis. The DNA fragments are denatured, transferred to a support medium such as nitrocellulose filter paper, and hybridized to a radioactive actin gene probe which does not contain any highly repeated sequences. Autoradiography reveals the presence of restriction fragments containing actin gene sequences, which can then be counted to yield a rough estimate of the number of different actin genes. Unfortunately, this method can lead to an overestimate of the gene number if restriction sites reside within the actin gene. Underestimation of the gene number can also occur if more than one actin gene is contained on a single restriction fragment or if different actin genes are contained on restriction fragments of the same size. It is therefore necessary to determine how many actin genes or what fraction of an actin gene is represented in a single band on an autoradiogram. This information can be obtained by electrophoresing DNA standards that contain known amounts of actin gene sequences (Durica et al., 1980).

The second type of DNA filter hybridization is called a dot blot (Thomas, 1980). This procedure is simpler than the Southern blot because the DNA is spotted directly onto the support medium. As in the Southern blot, the genomic DNA is probed for actin sequences with a radioactive probe. Dot intensities are related to DNA standards containing known amounts of actin gene sequences. A major advantage of this method over Southern blot analysis is that less error is incurred in the evaluation of the intensity of a single dot as opposed to several bands. In addition, possible variation in the efficiency of the transfer of DNA from the gel to the support media is circumvented by the dot blot method.

The use of solution hybridization to determine the reiteration frequency of actin sequences within the genome involves hybridization of a labeled DNA fragment containing actin sequences (the "tracer") with excess amounts of sheared genomic DNA (the "driver"). The more copies of the actin DNA sequence that exist in solution, the faster the tracer will hybridize with its complementary strand (Fyrberg et al., 1980). The extent of hybridization can be determined in two ways. The hybridization mixture can be treated with S1 nuclease to digest all unreacted single-stranded tracer DNA before precipitating reacted tracer with trichloroacetic acid. Alternatively, the reaction mixture can be passed over a hydroxyapatite column, which will bind the reacted double-stranded tracer and not the single-stranded tracer. It is possible to estimate the copy number of the actin sequence from the kinetics of hybridization (McKeown et al., 1978). Unfortunately, the reiteration frequency may be underestimated by this method if sequence divergence exists between multiple copies of the reiterated sequence. At best, cot analysis indicates the reiteration frequency within a factor of two.

Use of the aforementioned methods to determine the number of actin genes in various organisms has led to the conclusion that with the single exception of yeast, all organisms have multiple actin genes (Table I). There seems to be no

TABLE I

Number of Actin Genes per Haploid Genome of Various Organisms

Organism	Number of actin genes	Reference
Sea urchin		
Strongylocentrotus purpuratus	5–20	Overbeek *et al.* (1981); Durica *et al.* (1980); Schuler and Keller (1981); Cleveland *et al.* (1980); Scheller *et al.* (1981); Johnson *et al.* (1983)
S. franciscanis	15–20	Johnson *et al.* (1983)
L. pictus	15	Johnson *et al.* (1983)
Chicken	4–7	Cleveland *et al.* (1980)
Rat	12	Nudel *et al.* (1982)
Human	25–30	Hamada *et al.* (1982)
Fruit fly (*Drosophila melanogaster*)	6	Fyrberg *et al.* (1980)
Slime mold (*Dictyostelium discoideum*)	17	McKeown *et al.* (1978)
Yeast (*Saccharomyces cerevisiae*)	1	Gallwitz and Seidel (1980); Ng and Abelson (1980)
Oxytricha fallax	3	Kaine and Spear (1980)

obvious correlation between the level of complexity of an organism and its number of actin genes. The greatest number of actin genes has been identified in the slime mold *Dictyostelium*, where cot analysis suggests that the actin-coding sequences are repeated 10–20 times in the genome (McKeown *et al.*, 1978). Southern blot analysis reveals between 15 and 20 bands, depending on which restriction enzyme is used to cut the genomic DNA (McKeown *et al.*, 1978). To control for multiple cut sites within the actin-coding sequences, blots of genomic DNA have been hybridized separately with labeled probes complementary to the 3' or 5' ends of the coding sequence. The banding patterns obtained from these two types of blots were found to be the same, indicating that each band represents one actin gene. The actin gene copy number in *Dictyostelium* is therefore at least 15 (Kindle and Firtel, 1978).

Like *Dictyostelium*, the number of actin genes in the sea urchin *Strongylocentrotus purpuratus* is between 5 and 20. Cot analysis has indicated a reiteration frequency of only 5 copies per haploid genome (Durica *et al.*, 1980), but Southern blots suggest that the number could be at least 10 (Durica *et al.*, 1980; Overbeek *et al.*, 1981). Dot blot analysis of genomic DNAs from *S. purpuratus*, *S. franciscanis*, and *Lytechinus pictus* set the actin gene reiteration frequency in each of these species between 15 and 20 (Johnson *et al.*, 1983).

The most accurate analysis of the number of actin genes present in an organism has been carried out in *Drosophila*. The number of actin genes in *Drosophila* has been estimated in three different ways. Southern blot analysis yielded an estimate of 3–5 actin genes, while cot analysis yielded an estimate of 2.5 copies (Fyrberg *et al.*, 1980). The most definitive analysis, however, has come from *in situ* hybridization of a labeled actin gene probe to the polytene chromosomes of the fruit fly's salivary gland. In this approach, the salivary tissue was prepared to expose the polytene chromosomes of these cells, and the ^{3}H-labeled actin gene probe was then hybridized directly to the chromosomes. When the chromosomes were examined by autoradiography, six radioactive regions were detected, setting the actin gene number in *Drosophila* at 6 (Fyrberg *et al.*, 1980).

Not all organisms have large numbers of actin genes. In yeast, only one actin gene is present (Gallwitz and Seidel, 1980; Ng and Abelson, 1980; Water *et al.*, 1980), whereas the ciliated protozoan *Oxytricha fallax* appears to have but three actin genes (Kaine and Spear, 1980). The number of actin genes in other organisms has not been rigorously quantified as of yet, but Southern blots have revealed the presence of multiple actin genes in chickens, rats, and humans (Cleveland *et al.*, 1980; Nudel *et al.*, 1982). Nine unique actin clones have been isolated from a human genomic DNA library, suggesting the presence of at least nine actin genes in humans (Engel *et al.*, 1982). Likewise, eight unique actin clones have been isolated from a rat genomic library (Nudel *et al.*, 1982).

B. Actin Gene Linkage

The chromosomal arrangement of the members of gene families appears to play an important role in sequential gene expression. In the case of the globin gene for example, the closely linked globin genes are arranged in the order of their developmental expression (Hardison *et al.*, 1979; Jahn *et al.*, 1980; Lacy *et al.*, 1979; Lauer and Maniatis, 1980; Proudfoot *et al.*, 1980). Defining the linkage relationships between actin genes is thus an important element in elucidating the developmental expression of this gene family. The existence of linkage betwen actin genes is most readily demonstrated experimentally by showing that more than one actin-coding sequence resides on the same cloned fragment of DNA. The individual actin-coding sequences can be identified either by restriction enzyme mapping of the cloned fragment using a radioactive probe containing actin sequences or by heteroduplex analysis (see Section II,C).

Using the aforementioned approaches, linked actin genes have been identified in *Dictyostelium*, as well as the sea urchin species *S. purpuratus* and *S. franciscanis*. In *D. discoideum*, 2 actin genes have been shown to be linked by heteroduplex analysis. These 2 genes have like orientations and are separated by 350 bp of noncoding DNA. There are no data indicating the linkage relationship of the remaining 15 actin genes of this species.

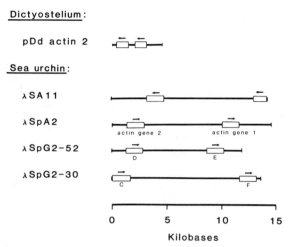

Fig. 1. Known linkages of actin genes. Actin genes are represented by open boxes, and the direction of transcription is indicated with arrows. Note that all these linked actin genes are tandemly arranged.

More extensive linkages have been observed in sea urchins (Fig. 1). In *S. purpuratus,* five different genomic clones isolated in phage have been found to contain two actin genes each (Overbeek *et al.,* 1981; Scheller *et al.,* 1981; Schuler and Keller, 1981). In these five clones, the linked actin genes are separated by 7 to 9 kilobases (kb) of nonactin DNA. Several other clones containing only a single actin gene have been isolated as well. These clones with only one actin gene contain 15–20 kb of genomic DNA, indicating that these particular actin genes are not closely linked to other actin genes.

In a second sea urchin species, *S. franciscanis,* three genomic clones have been isolated that contain two actin genes, and one phage clone was found to contain three actin genes. The actin genes in these four phage clones are linked with an average spacing of only 4–5 kb (Johnson *et al.,* 1983). No clone with only one actin gene has been isolated, indicating that the actin genes of *S. franciscanis* are more tightly clustered than those of *S. purpuratus.*

In *Drosophila,* the *in situ* hybridization data indicated that close linkage of the actin genes does not exist (Fyrberg *et al.,* 1980). In other organisms, definitive data are not available concerning possible linkage. However, in humans eight phage clones containing 11–20 kb of human DNA each have only a single actin gene (Engel *et al.,* 1981).

The presence of multiple actin genes in various organisms suggests that the actin gene has multiplied by some evolutionary mechanism. The close linkage of the actin genes in sea urchins and *Dictyostelium* indicates that some of these actin genes may have duplicated by unequal crossing-over.

C. Intron Position

There are three basic methods for determining the number and location of introns in a gene: visualization of heteroduplexed molecules by electron microscopy, S1 nuclease mapping, and sequencing of DNA and mRNA. Heteroduplex analysis usually involves the hybridization of a cloned gene to the mRNA from which it is derived. Base sequences that occur in the DNA but not in the mRNA appear as looped-out structures or R-loops (Fig. 2). Each of these loops corresponds to an intron in the DNA. The number of introns may be counted and roughly positioned within the gene. However, introns smaller than 100 bp may not be seen (Fyrberg *et al.*, 1981; Nudel *et al.*, 1982). Heteroduplex structures can also be formed between two cloned genes. In this case, introns present in one of the genes and not the other appear as loop structures, while introns common to both appear as bubbles between the two strands of DNA.

The method of S1 nuclease mapping, which is designed to determine the number and size of exons, can be used to infer the number of introns. In this procedure, a cloned gene is uniformly labeled with ^{32}P *in vivo,* and the plasmid containing this gene is then linearized with a restriction enzyme that does not cut in or near the gene sequence. The linearized DNA is denatured and hybridized with mRNA. These hybridized molecules are treated with S1 nuclease to digest single strands of DNA that are not homologous to sequences in the mRNA; intron sequences are thus digested away, while the exons remain intact. The resulting exons are fractionated by gel electrophoresis so that they can be counted and sized. Although this method is useful for quantitating the number of exons and introns as well as the size of the exons, neither the size nor the position of the introns can be elucidated directly (Nudel *et al.*, 1982).

The final method for determining the number and position of introns involves nucleic acid sequencing. Introns in the coding sequence can be identified by comparing the DNA sequence with the amino acid sequence of the encoded protein. Sequencing of only the genomic DNA, however, may result in introns being overlooked in the 5' untranslated region. If large enough, introns in this

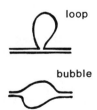

Fig. 2.   Diagram illustrating differences between loop and bubble structures seen in heteroduplex analysis. A loop occurs when intron sequences are present in one molecular and not the other. When both genes share an intron position and intron sequences are not homologous, a bubble structure is formed.

region can be identified by R-loop analysis. Another way to look for introns in the 5' noncoding sequences is to sequence the mature mRNA using a primer extension method (Fornwald *et al.*, 1982). In this procedure, a DNA fragment homologous to the mRNA-coding sequence near the 5' end is hybridized to the mRNA. The DNA fragment serves as a primer and the mRNA then serves as a template for cDNA synthesis using reverse transcriptase and dideoxynucleotides. The resulting sequences are electrophoresed on a sequencing gel, and the sequence for the mRNA determined. A divergence between the mRNA sequence and the genomic sequence indicates the 3' end of an intron. The two sequences again become homologous at the 5' end of the intron. A major limitation to this technique is the necessity for having a DNA primer that only hybridizes to the specific mRNA for which it codes, and not to other closely related mRNAs.

Using the previously discussed general approaches, introns have been identified in at least some of the actin genes of *Drosophila,* sea urchin, yeast, chicken, rat, human, and soybean (Table II). We shall next consider some of these data in detail because they have yielded interesting information concerning the evolutionary origin of introns.

The first actin gene to have its complete nucleotide sequence determined was the yeast actin gene (Gallwitz and Sures, 1980; Ng and Abelson, 1980). An intron of 309 bp was found within the codon for the fourth amino acid. To date, this is the only nuclear gene in yeast known to have an intron.

All six actin genes of *Drosophila,* designated λ*DmA1* through λ*DmA6,* have been partially sequenced. One intron each has been found in four of the genes. From R-looping experiments, genes λ*DmA2* and λ*DmA4* were found to have introns near their 5' ends, whereas λ*DmA6* was shown to have an intron nearer the 3' end. Upon sequencing, the intron in λ*DmA2* was located in the untranslated region eight nucleotides upstream from the ATG initiation codon. The introns of λ*DmA4* and λ*DmA6* on the other hand were found to interrupt glycine codons located at positions 13 and 307, respectively. A small, 60-bp intron at position 307, which had not been seen by R-looping analysis, was discovered in λ*DmA1* upon nucleotide sequencing of this region of the gene. The two remaining actin genes in *Drosophila* appear to have no introns.

Actin genes in the sea urchin *S. purpuratus* exhibit at least two different patterns of intron arrangement. In each of these two patterns, introns are present at amino acid positions 121 and 204. One of the patterns involves two additional introns at amino acid positions 41 and 267 (Cooper and Crain, 1982). The second intron pattern involves an additional intron in the 5'-untranslated region of the gene (Schuler *et al.,* 1982; Zeigler *et al.,* 1983). Another possible intron pattern occurs in those actin genes that are located downstream from another closely linked actin gene. Such actin genes exhibit the second pattern of intron placement, except that no 5'-splice junction for the intron in the untranslated region is identifiable in DNA sequence data (Schuler *et al.,* 1982; Zeigler *et al.,* 1983). In

TABLE II

Intron Positions in Actin Genes Isolated from Various Organisms

Organism	Intron position (amino acid number)												Ref.[b]
	5'UT[a]	3	13	17	41/42	121	150	203/204	267/268	307	327/328	353/354	
Strongylocentrotus purpuratus													
1	+					+		+					1,2,3
2						+		+					1,2
3						+		+					3
S. franciscanis					+	+		+	+				4
Chicken													
α-actin	+				+		+	+	+		+		5
β-actin					+	+			+		+		5
Rat													
α-actin	+				+		+	+	+		+		6
β-actin	+				+	+			+		+		5
Human													
α-actin	+				+		+	+	+		+		7
Drosophila melanogaster													
1										+			8
2	+												8
4			+										8
6										+			8
Saccharomyces cerevisiae		+					+						9,10
Soybean				+								+	11

[a] 5'-untranslated region.

[b] 1, Zeigler et al. (1983); 2, Schuler et al. (1982); 3, Cooper and Crain (1982); 4, Johnson et al. (1983); 5, Fornwald et al. (1982); 6, Zakut et al. (1982); 7, Hamada et al. (1982); 8, Fyrberg et al. (1981); 9, Gallwitz and Sures (1980); 10, Ng and Abelson (1980); and 11, Shah et al. (1982).

the closely related sea urchin species *S. franciscanis*, introns have been found at positions 121 and 204, but the 5'-untranslated region has not yet been sequenced (Foran *et al.*, 1983).

Introns in these same five positions (5'-untranslated, 41, 121, 204, and 267) also occur in bird and mammalian actin genes, although not all in the same gene. The complete nucleotide sequence of an α-actin gene from chicken places six introns in the following positions: 12 nucleotides upstream from the ATG initiation codon, following codon 41, within codon 150, and following codons 204, 267, and 327 (Fornwald *et al.*, 1982). The α-actin of the rat was shown to contain introns in each of these same locations by partial nucleotide sequencing in conjunction with R-loop analysis (Nudel *et al.*, 1982). These findings have since been confirmed with determination of the complete nucleotide sequence of this gene (Zakut *et al.*, 1982). Sequencing of the human cardiac α-actin gene has revealed the conservation of these intron positions in a human α-actin gene (Hamada *et al.*, 1982).

The β-actin and α-actin genes of birds and mammals share introns in the 5'-untranslated region and at codons 204, 267, and 327. The β-actin gene lacks two introns which the α-actin gene contains at amino acid 150 and 204. Unique to the β-actin gene is an intron at position 121 (Fornwald *et al.*, 1982).

In *Dictyostelium,* 7 of the estimated 17 actin genes have been isolated and analyzed by R-looping and nucleotide sequencing (Firtel *et al.*, 1979; McKeown and Firtel, 1981a). R-looping revealed no introns in any of the actin genes. These genes have also been sequenced from the 5'-untranslated regions through the nucleotides corresponding to amino acid 150. Sequences from amino acid 290 extending to the 3' noncoding regions have likewise been determined. Again, no introns were found. The possible presence of introns within the central 400 nucleotides, which have not been sequenced, cannot be rigorously excluded.

The gene shuffling hypothesis first proposed by Gilbert (1978) suggested that genes evolve from modules of coding sequence information, termed exons, separated by noncoding introns. The ancestral actin gene would thus have had several introns separating each of the original exons. As the actin gene evolved it may have lost some introns. Analysis of the actin gene family in view of this hypothesis leads to some interesting speculation about actin gene evolution.

The most general relationship that can be seen in the analysis of intron position is that the deuterostomes (sea urchin, chicken, rat, and human) have maintained more introns than nondeuterostomes (*Drosophila, Dictyostelium, Oxytricha,* and yeast). Most actin genes of nondeuterostomes have no introns, although some have a single intron. Within the deuterostomes, intron positions have been conserved between birds and mammals for actin genes of the same type, just as occurs in the globin gene family (Efstratiadis *et al.*, 1980). α-Actin and β-actin genes share some but not all intron positions. These differences in intron position indicate that the α-actin genes became distinct from the β-actin genes before mammals diverged from birds. The five intron positions present in the sea urchin actin genes are a subset of the seven different intron positions found in the α- and β-actin genes of birds and mammals. This suggests that α- and β-actin became distinct from one another after radiation of the chordates.

The most interesting speculation about the evolution of the actin gene family draws on the observation that a correlation appears to exist between the number of introns in the actin genes of an organism and the organism's level of complexity. Conservation of introns maintains exons as discrete packets which may continue to be shuffled to create new associations of exons. The organisms of higher complexity thus maintain a certain plasticity in their genetic information. This plasticity would enable an organism to evolve more rapidly than one whose genetic information is fixed in larger information packets or in genes without introns.

D. Mapping of Transcriptional Units

Another important structural feature of the actin gene is the location of the sites for initiation and termination of transcription. Of the various methods that have been employed in identifying the 5′ and 3′ boundaries of the actin transcriptional unit, end mapping with S1 nuclease has been most extensively used. In this procedure, mRNA is hybridized to a cloned actin gene in the presence of high concentrations of formamide and at temperatures a few degrees higher than the T_m of DNA–DNA duplexes. Under these conditions DNA–DNA duplexes are unstable, but RNA–DNA hybrids can still form. If the cloned DNA contains 5′- and 3′-flanking sequences extending beyond the 5′ and 3′ boundaries of the mRNA, both terminal positions of the transcript can be defined by single-strand-specific digestion of the overhanging DNA sequences using S1 nuclease.

In both *Dictyostelium* and sea urchin this end-mapping procedure has been carried out using DNA restriction fragments located near the 5′ end of the actin gene. The DNA fragments are first incubated with [^{32}P]ATP and polynucleotide kinase to label the 5′ end of the DNA strand complementary to actin messenger RNA. The resulting DNA probe includes N-terminal actin-coding sequences and 5′-flanking sequences containing putative mRNA leader sequences and the transcriptional start site. When actin mRNA is hybridized to this DNA probe, homologous sequences are protected from single-strand-specific digestion by S1 nuclease. The S1-resistant RNA–DNA hybrid can then be denatured, and the DNA size fractionated by gel electrophoresis to determine the position of the 5′ end of the actin gene. Identification of the initiation nucleotide can be accomplished by running the S1-digested fragments adjacent to a sequencing ladder of the same DNA probe.

In *Dictyostelium*, 5′ ends of four actin genes have been mapped in this way (McKeown and Firtel, 1981b). All four actin genes give rise to several bands at the 5' end, suggesting either multiple initiation sites or nibbling by S1 nuclease in this A–T-rich region. The mRNA leader sequence (delimited at the 3' end by the translation start site and at the 5' end by the transcription start site) for all four actin genes was A–T-rich, heterogeneous relative to the other actin genes, and ranged in size from 20 to 42 nucleotides. Nucleotides tentatively identified as initiation sites included not only the A residue typical of eukaryotic start sites, but C and T residues as well.

In the sea urchin *S. purpuratus,* 5'-end-mapping of two actin genes has revealed the presence of an intron in the 5'-untranslated region of the two genes (Zeigler and Kleinsmith, 1982; Zeigler *et al.,* 1983). Hybridization of actin mRNA to a ^{32}P-labeled DNA probe derived from actin gene *pSA16* (a plasmid subclone of the actin gene in λ*SA16*) followed by S1 nuclease digestion indicates that the 5' end of the mRNA is located 26 nucleotides upstream from the translation start site. However, *in vitro* transcription of *pSA16* in a cell-free HeLa cell extract shows that initiation occurs approximately 280–300 nucleotides upstream from the translation start site. The apparent disparity in these results could be reconciled if an intron exists in the 5'-untranslated region. An examination of this region reveals the existence of two putative, 3'-splice junctions located 25 and 140 bases upstream from the translation start site. Based upon the hybridization/S1 nuclease reactions, the 3'-splice junction located at position −25 appears to be the functional 3'-splice site. A putative 5'-splice junction was located 212 bases upstream from the translation start site. If this 5'-splice junction is functional, then the leader sequence of this actin gene comprises approximately 85 nucleotides with an A–T-rich region located between the translation start site and the 3'-splice junction.

Sequence analysis of the N-terminus and 5'-flanking sequences of a second actin clone, *pSA11-1* (a plasmid subclone of one of the actin genes present in λ*SA11*), has revealed the presence of a 3'-splice site 28 bases upstream from the translation start site. Sequences located between the 3'-splice site and the translation initiation site are homologous to sequences found in the same region of *pSA16*. No obvious 5'-splice site is present for at least 900 nucleotides upstream from the 3'-splice junction. It is thus possible that an intron of 1 kb or more exists in this gene. Alternatively, *pSA11*-1 may represent a pseudogene lacking the transcriptional signals necessary for expression. Since a 3'-splice site is present in *pSA11-1* and this gene is located downstream from another actin gene, it is also possible that this downstream actin gene is transcribed from the promoter/initiation signals located in the 5'-flanking sequences of the upstream actin gene. Northern blot hybridization using actin gene probes has revealed the presence of 4–6-kb RNA transcripts to actin gene probes (G. T. Merlino, personal communication; Cooper and Crain, 1982). Whether these transcripts represent an unprocessed mRNA bearing these two linked actin genes remains to be determined.

The 5' N-terminal coding and flanking sequences for three other genes have also been sequenced (Cooper and Crain, 1982; Schuler *et al.*, 1982). Two of these actin genes, *pSpG17* and actin gene 2, are closely homologous to *pSA16*. However, three putative splice junctions are present in actin gene 2, one of which is not found in *pSA16*. Based on sequence analysis, actin gene 1 possesses the conserved 3'-splice junction 26 nucleotides upstream from the translational start site, but no other significant homologies are found in the 5'-untranslated region when compared with *pSA16*, *pSpG17*, *pSA11-1*, and actin gene 2.

In both *Dictyostelium* and sea urchins, TATA-like sequences have been identified upstream from putative transcription start sites. In the sea urchin actin gene *pSA16*, a likely TATA signal 100 nucleotides upstream from the putative transcription-initiation site has been identified by its sequence similarity to the TATA box. The position of this sequence is atypical when compared with locations of TATA sequences 25–30 nucleotides upstream from transcription-initiation sites found in several eukaryotes (Tsujimoto *et al.*, 1981; Tsai *et al.*, 1981; Vogeli *et al.*, 1981). In *pSpG17*, actin gene 2, and actin gene 1 putative transcriptional start sites were arbitrarily assigned to positions located 25 bases downstream from TATA sequences. No independent verification for these assignments is available, however.

In *Dictyostelium*, TATA-like sequences for four actin genes have been identified 25–30 nucleotides upstream from the transcription-initiation site revealed by S1 nuclease digestion. Two of the genes, actin 8 and actin M6, possess atypical TATA sequences located in the analogous position. Another conserved sequence found in the 5'-untranslated region is a T-rich region of DNA immediately following the transcriptional start site. Though it has also been seen in other *Dictyostelium* genes, the role of this sequence in gene expression is not known.

Primer extension analysis has been used to map the 5'-transcription terminus of the chicken α-actin gene (Fornwald *et al.*, 1982). Genomic DNA fragments 5'-end-labeled in the N-terminal coding region were hybridized to actin mRNA. The DNA fragment was then employed as a primer for synthesis of cDNA by reverse transcriptase. When this reaction is carried out in the presence of dideoxynucleotide terminators, a sequencing ladder can be generated that includes the mRNA leader sequence and transcription initiation site. By this procedure it was determined that the α-actin gene contains a 5'-flanking intron of 111 nucleotides and a leader sequence of 73 nucleotides. The putative initiator nucleotide was identified as a C residue based on primer extension analysis.

R-loop analysis of rat cytoplasmic actin demonstrated the presence of an 820-base, double-strand loop located in the 5'-untranslated region of that gene (Nudel *et al.*, 1982). The exon 5' to this intron could not be observed by electron microscopy, indicating that it must be quite small. This procedure for mapping transcriptional units thus lacks the accuracy afforded by primer extension and S1 nuclease digestion, and so the location of the transcriptional start site cannot be accurately identified (Kinniburgh *et al.*, 1978). However, rat skeletal actin, which also contains an intron in the 5'-untranslated region, has been found by S1 mapping to contain a 55-bp exon upstream from this intron.

End mapping of actin genes at the 3' terminus using S1 nuclease and primer extension analysis has not yet been carried out. However, sequence analysis of the 3'-untranslated regions of *Dictyostelium* and sea urchin actin genes has defined transcription terminator and poly(A) addition signals for a number of these genes (see Section II,E).

E. Actin Gene Homologies

The complete nucleotide sequence of two actin genes and the partial sequence of three other actin genes from *S. purpuratus* have been determined (Cooper and Crain, 1982; Schuler *et al.*, 1982; Zeigler *et al.*, 1983). As expected from their similar restriction enzyme patterns, these five actin genes exhibit considerable sequence homology in both coding and noncoding sequences. Comparison of the actin gene *pSpG17*, sequenced by Cooper and Crain (1982), with actin gene 2 sequenced in Keller's laboratory (Schuler *et al.*, 1982) reveals only 0.8% difference in sequences; of the 9 bp changes, all but 2 are silent mutations. Each of these genes has introns located at amino acid residues 121 and 204. These two introns show homology both intergenically and intragenically. Comparing the two introns in *pSpG17* with each other reveals that the 3' one-third of the intron located at position 121 is 68% homologous to the 3' portion of the intron located at position 204. The remainder of the intron is highly divergent. Intergenically, the intron located at position 121 of *pSpG17* and the analogous intron of actin gene 2 show 94% homology, whereas comparison of the introns located at position 204 reveals only 83% homology. The introns at position 121 and 204 thus appear to have diverged at slightly different rates.

Actin gene 1, which is separated from actin gene 2 by only 7 kb, is 2.2% different in base sequence from actin gene 2. This means that actin gene 2 has diverged more from its closely linked neighbor (actin gene 1) than it has from *pSpG17*, an actin gene of unknown linkage. When comparing the sequences of actin genes 1 and 2, 15 out of 19 bp changes in the coding sequences are found to be silent mutations. The introns, however, show a great deal of divergence. Examination of the intron sequences of actin gene 1 and actin gene 2 reveals the presence of both deletions and single-base changes. If these two genes arose by gene duplication, it follows that they once held all sequences in common. Assuming that the introns have only lost sequences since their duplication, the length of the intron at the time of duplication can be estimated by identifying deleted sequences and adding them back to the intron from which they were deleted. For example, for actin gene 1 the intron located at amino acid 121 has undergone 11 deletions totaling 47 bp. Adding these 47 bp to the present length of this intron (207 bp) yields a total length of 254 bp for the ancestral intron. Likewise, for the intron located at amino acid 204, there have been 5 deletions in actin gene 1 accounting for a total of 10 bp. Since the length of this intron in actin gene 1 is now 224 bp, the ancestral intron is estimated to have been 224 + 10 = 234 bp. If comparable calculations are carried out for actin gene 2, the same results are obtained. The majority of deletions in these introns have occurred between short, repeated sequences (Schuler *et al.*, 1982), an observation compatible with a model for the deletion mechanism that utilizes the occurrence of repeated sequences (Efstratiadis *et al.*, 1980; Streisinger *et al.*, 1966).

When sequences not deleted in either of the genes are compared for point mutations, the intron located at position 121 in actin gene 1 shows 64% homology to the analogous intron of actin gene 2, whereas the intron located at amino acid position 204 in actin gene 1 retains 77% homology to the analogous intron of actin gene 2. Thus, the intron located at position 121 has diverged faster than the intron located at position 204. This is the opposite of what is observed for the intron sequences of *pSpG17* and actin gene 2, in which the intron at position 121 exhibits more homology than the intron located at position 204. These observations suggest that the observed variation in sequence divergence rates in these introns is random. It is also clear that whereas the sequences of the coding portions of these genes have remained relatively constant, the intron sequences have diverged greatly.

The overall homology of the 5'-untranslated sequences of the three actin genes *pSA16* (Zeigler *et al.*, 1983), actin gene 2 (Schuler *et al.*, 1982), and *pSpG17* (Cooper and Crain, 1982) is quite substantial. The two genes most divergent are *pSA16* and *pSpG17*, which show a 12% difference in 350 bp of sequence. Actin gene 2 is 6% different from *pSA16* and *pSpG17*. These three actin genes all show conserved consensus sequences for two 3'-splice junctions in their 5'-untranslated sequences. These splice junction sites occur 25 and 140 bases upstream from the translation initiation codon. Only a single 5'-splice junction sequence has been identified upstream from these 3'-splice sites.

Sequencing of the actin genes of *Dictyostelium* has led to the conclusion that at least 6 of the 15 actin genes code for the same amino acid sequence. The coding sequence of these genes corresponds to the amino acid sequence of *Dictyostelium* actin as determined by Vandekerckhove and Weber (1980). In pairwise comparisons of the coding sequence of these genes, the differences in base sequence ranged from 1.8–5.9% (McKeown and Firtel, 1981a). Three other *Dictyostelium* actin genes code for variant amino acid sequences. M6 and pDd actin 3 code for actins that differ in two and four amino acids, respectively, from the protein sequence as determined by Vandekerckhove and Weber (1980). A third variant, pDd actin 2-sub 2, differs only 7% in DNA sequence from pDd actin 2-sub 1, but 13% in amino acid sequence. The large divergence in the encoded amino acid sequence has led to the speculation that pDd actin 2-sub 2 is a pseudogene

(McKeown and Firtel, 1981a). In contrast to the strong conservation of coding sequences, analysis of the 3'-untranslated sequences reveals considerable variation but enough homology to suggest the existence of three groups of *Dictyostelium* actin genes. No sequence homologies are discernable in the 3'-untranscribed DNA flanking sequences (McKeown and Firtel, 1981a). The 5'-untranslated sequences of the *Dictyostelium* actin genes are rich in AT (Firtel *et al.*, 1979; McKeown and Firtel, 1981b). The 5'-untranslated sequences include two conserved regions which seem to be important for transcription (see Section II,D). These two regions are not conserved in pDd actin 2-sub 2, which further supports the conclusion that it is a pseudogene.

The DNA sequences for α-actin genes of human (Hamada *et al.*, 1982), rat (Zakut *et al.*, 1982), and chicken (Fornwald *et al.*, 1982) have also been recently determined. Homologies existing between the coding sequences of these and other actin genes are summarized in Table III.

F. Adjacent Repeat Sequences

The presence of highly repetitious sequences adjacent to actin genes has been investigated by both Southern blot analysis and DNA-sequencing techniques. To identify repeated sequences in a cloned DNA fragment, the DNA is digested with restriction enzymes, fractionated by gel electrophoresis, and transferred to a support medium. The cloned DNA is then probed for highly repetitious sequences with radioactive genomic DNA. The filter hybridization is subjected to

TABLE III

Homology of Actin-Coding Sequences

Actin genes compared	Base substitutions (%)	Comments
Sea urchin (pSpG17) / Sea urchin (actin gene 2)	0.8	
Sea urchin (actin gene 2) / Sea urchin (actin gene 1)	2.2	0.3% change in amino acid sequence
Sea urchin (pSpG17) / Chicken (pGa-actin 1)	18	
Chicken (pGa-actin 1) / Rat (pAC15.2)	11	All silent mutations
Sea urchin (pSpG17) / *Oxytricha* (pOFACT)	35	
Oxytricha (pOFACT) / *Dictyostelium* (pDd actin 5)	16	

autoradiography for lengths of time that will only produce observable signals from sequences repeated at least 100 times in the genome.

Using these approaches, highly repeated sequences have been identified near or within the actin genes of ciliated protozoans, fruit flies, sea urchins, and humans. In the macronucleus of the ciliated protozoan *Oxytricha fallax*, the DNA exists in linear fragments with an average size of 3.2 kb (Kaine and Spear, 1980). The actin-containing fragment was found to terminate with sequence GGGGTTTT. Sequencing of total macronuclear DNA revealed the presence of this sequence on all macronuclear DNA molecules (Kaine and Spear, 1982). Similar terminal sequences have been identified in other ciliates (Klobutcher *et al.*, 1981). Terminal sequences from *Tetrahymena* have been shown to stabilize linear DNA molecules in yeast (Szostack and Blackburn, 1982).

Within the fourth intron of the human cardiac α-actin gene a sequence of 22 alternating G's and T's was identified by DNA sequencing. By Southern blot analysis this sequence was estimated to be repeated 10^5 times in the human genome (Hamada and Kakunaga, 1982). Because of this sequence's alternating purine and pyrimidine structure, it is capable of assuming a left-handed helix or Z-form DNA structure (Vorlickova *et al.*, 1982; Zimmerman *et al.*, 1982). Although Z-form DNA has been demonstrated to exist in the polytene chromosomes of *Drosophila* (Nordheim *et al.*, 1981), it is not known if the repeated GT sequence of human DNA assumes a Z confirmation *in vivo*. Further Southern blot analysis of the phage clone containing the α-actin gene failed to identify the presence of any member of the Alu repeat family (Hamada and Kakunaga, 1982), which is estimated to be present in the human genome of 10^5 copies (Houck *et al.*, 1979).

A single repeated element has been identified in the 3′ noncoding region of the *Drosophila* actin gene λDmA2 (Fyrberg *et al.*, 1980) by Southern blot hybridization. Its reiteration frequency was determined by cot analysis to be about 100 copies per haploid genome. Sequences homologous to this repeat were found in poly(A) RNA, indicating that at least some members of this repeat family are transcribed.

The most extensive study of repeated sequences associated with actin genes has been carried out on *S. purpuratus*, where several repeated sequences have been identified. In a clone containing actin genes 1 and 2, a repeat of poly(dG-dA) was located 3′ to each of the actin-coding sequences as determined by Southern blot hybridization (Schuler and Keller, 1981). This repeat has also been identified at the 3′ end of the histone genes of sea urchin (Sures *et al.*, 1978; Schaffner *et al.*, 1978). In a separate study, seven different repeated sequences were identified near the actin genes (Scheller *et al.*, 1981). These repeated sequences are found both 5′ and 3′ to actin genes, and are also located between linked actin genes. The role of these repeated sequences in expression in actin genes and in the evolutionary duplication of actin gene sequences is yet to be determined.

III. DEVELOPMENTAL EXPRESSION OF ACTIN GENES

The recent isolation and characterization of actin genes from a variety of eukaryotes have provided an opportunity for studying how the expression of this multigene family is regulated in higher organisms. Studies on sea urchins, *Dictyostelium,* and *Drosophila* have delineated some basic features of actin gene expression, though none of these studies have yet yielded definitive information about the exact mechanism of actin gene regulation. The data currently available are compatible with several possible levels of regulation, including (1) transcription, (2) posttranscriptional processing, and (3) translation of mature mRNA. In the following sections we shall first summarize the major experimental tools that have been used to monitor the control of actin gene expression and then describe the major findings derived from such studies.

A. Experimental Approaches

The synthesis of actin is generally monitored using [^{35}S]methionine as a radioactive tracer, followed by one- or two-dimensional polyacrylamide gel electrophoresis. Two-dimensional electrophoresis is preferable because it allows resolution of actin into its three isoforms, α, β, and γ. One-dimensional electrophoresis cannot resolve these actin forms from each other and in some cases may not even be able to separate actin from comigrating proteins of the same size. One way around this problem is to prepurify actin by DNase I affinity chromatography, a procedure highly selective for actin (MacLeod *et al.,* 1980).

Actin mRNA has been detected in several ways. Isolated mRNA can be assayed by translation in either a wheat germ or rabbit reticulocyte protein-synthesizing system. The labeled translation products can then be analyzed to measure the fraction of synthesized protein that is actin, as well as which forms of actin are synthesized. It should be kept in mind, however, that heterologous translation systems may not be comparable in translational efficiency and specificity to the *in vivo* system. Particular care in the analysis of data is required when mRNA is purified by oligo(dT) affinity chromatography, since mRNAs containing short poly(A) tails (< 40 nucleotides) do not bind efficiently to such columns. For this reason, RNA not binding to oligo(dT) columns as well as bound RNA should be examined for the presence of actin mRNA activity.

Actin mRNA sequences can also be detected by hybridization to actin-specific DNA probes. When dealing with a family of genes it is possible that each member of the family codes for a different mRNA, resulting in heterogeneity in the 5' leader sequences of the mRNAs. Heterogeneity may also be found at the 3'-untranslated end of the mRNA. These heterologous 5' and 3' sequences can be used to design specific DNA probes for determining whether given members of the actin gene family are differentially expressed at various developmental stages. The specificity of such probes assumes that coding sequences are highly conserved, while noncoding sequences are sufficiently nonhomologous to ensure detection of specific actin mRNAs.

One such type of actin DNA probe would contain a short, 5'-coding sequence and contiguous 5'-heterologous leader sequences. Two fragments would be generated from a heterogeneous population of actin mRNAs hybridized to such a DNA probe. The larger fragment, which includes the 5' leader as well as the coding sequence, would result from hybridization of the DNA probe to its respective mRNA. The shorter fragment would derive from hybridization of the DNA probe to other actin mRNAs in which only the coding region is homologous. If the hybridization is carried out in DNA excess so that virtually all actin mRNA present reacts with the DNA probe, it should be possible to quantitate the amount of a specific actin gene transcript present relative to the total population of actin gene transcripts (McKeown and Firtel, 1981b).

B. Slime Mold (*Dictyostelium discoideum*)

As was previously noted, at least 15 distinct actin genes have been identified in *Dictyostelium*. Much of the study of actin gene expression during the developmental cycle of this organism has dealt with an analysis of actin protein synthesis. During vegetative growth of the ameboid cells, four isoforms of actin have been identified by two-dimensional electrophoresis (MacLeod *et al.*, 1980). However, amino acid sequencing of actin isolated from these cells reveals the presence of a single protein constituting greater than 95% of the total actin protein (Vandekerckhove and Weber, 1980). This heterogeneity in isoelectric point without a difference in primary sequence may be due to partial acetylation of the protein, producing two isoforms with the same primary sequence (Zulauf *et al.*, 1981). Two of the isoforms are minor spots and may in fact constitute only 5% of the total actin protein. On the basis of sequence analysis, six actin genes code for identical proteins whose amino acid sequence is in agreement with the known actin protein sequence (McKeown and Firtel, 1981a; Vandekerckhove and Weber, 1980).

Actin in vegetative cells accounts for 8% of total *in vivo* incorporation of [^{35}S]methionine. Margolskee and Lodish (1980b) showed that induction of the vegetative cells into the developmental cycle by amino acid starvation results in a threefold increase in the rate of actin synthesis, accompanied by a threef increase in the synthetic rate of actin mRNA relative to the vegetative level hr of development the relative synthetic rates of actin and actin mR' fallen to 60 and 30%, respectively. The synthetic rates remain relatively constant until 16 hr, when a second decline in synthetic rates occurs. By 24 hr, actin mRNA and protein synthesis have fallen to approximately 1% of the vegetative level.

A regulatory relationship seems to exist between the second decrease in actin synthesis and the ability of developmental mutants to form aggregates. Unlike wild-type cells, developmental mutants which were unable to form aggregates also showed no decline in rates of actin mRNA synthesis after 16 hr of development (Margolskee and Lodish, 1980b).

The correlation in the rates of synthesis of both actin and actin mRNA through development suggests that transcriptional and/or posttranscriptional control might be responsible for actin gene regulation in *Dictyostelium*. It has been concluded that these modulations in relative rates of actin mRNA synthesis reflect transcriptional control of actin gene expression because most of the nuclear poly(A) RNA appears in the cytoplasm, and little size reduction is seen in nuclear poly(A) RNA when compared to cytoplasmic RNA. This implies that posttranscriptional processing is not a factor in the control of actin gene expression in *Dictyostelium*.

MacLeod and associates (1980) found that spores produced at the end of the developmental cycle contain little actin mRNA and carry out little actin protein synthesis. Upon germination of these spores, actin protein synthesis accounts for less than 0.01% of total protein synthesis through the first hour of germination. During the third hour of development actin accounts for 2–3% of total protein synthesis, and by 6 hr actin protein is being synthesized at vegetative levels. It is interesting to note that although no detectable synthesis of actin occurs during the first hour of germination as measured by *in vivo* labeling, *in vitro* translation of poly(A) RNA isolated from germinating spores yields significant amounts of actin synthesis. These findings suggest that translational control of actin gene expression may be operative at this developmental stage. The kinetics of the increase in actin synthesis and of the accumulation of actin mRNA during germination is consistent with actin genes being turned on at the vegetative rate.

While the aforementioned studies provide a generalized profile of actin gene expression in *Dictyostelium,* McKeown and Firtel (1981b) examined the expression of individual members of the actin gene family and found evidence for differential expression of individual actin genes. This study made use of gene-specific probes containing N-terminal actin-coding sequences and heterologous 5′ noncoding sequences. These ^{32}P-end-labeled probes were hybridized to poly(A) RNA isolated at various developmental stages, S1 nuclease digested, size fractionated on polyacrylamide gels, and quantitated as previously described (see Section III,A).

Two genes, actin 5 and actin M6, are expressed at low levels throughout the developmental cycle, showing no fluctuations in their expression relative to the total actin population. The actin mRNA transcribed from pDd actin 2-sub 2 accounts for no more than 1% of the total actin mRNA at any developmental stage. As described earlier (see Section II,E) sequence information suggests this gene may be a pseudogene. Actin genes 6 and 8 are modulated in their expression relative to other actin genes. Through 8 hr of development, actin 8 increases from 18 to 27% of the total actin mRNA. By 13 hr, the percentage of actin mRNA represented by actin 8 had returned to its normal vegetative level.

Actin gene 6 exhibits the most interesting fluctuations. During the first 3 hr of development the fraction of actin mRNA represented by actin gene 6 decreases from 23 to 9%. After 8 hr of development there is slight increase in the relative amount of actin 6 mRNA, followed by another decrease to 2% of total actin messages by 20 hr of development. The investigators concluded that this decline represents a selective turning off of actin gene 6 at the beginning of starvation. If actin gene 6 is turned off at the beginning of starvation and the actin 6 mRNA is assumed to have a half-life of 3 hr (Margolskee and Lodish, 1980a), then the fraction of the total actin mRNA represented by actin 6 mRNA would be 4% after 3 hr of development. However, the data indicate that after the first 3 hr of development, actin 6 mRNA decreases to only 9% of total actin mRNA, suggesting actin 6 continues to be transcribed at vegetative levels. The decrease in percentage of actin 6 mRNA relative to total actin mRNA during the period when total actin mRNA synthesis is declining suggests that actin 6 mRNA decays more quickly than the average actin mRNA.

C. Fruit Fly (*Drosophila melanogaster*)

As previously described in this review, *Drosophila* actin genes comprise a multigene family of six genes located at six scattered chromosomal loci. These genes encode three actin isoforms, I, II, and III; II and III are synthesized in relative abundance from early embryogenesis, whereas actin I does not become abundant until larval myogenesis (Storti *et al.*, 1978; Fyrberg and Donady, 1979; Horovitch *et al.*, 1979). To examine what regulatory mechanisms might be responsible for this tissue- and stage-specific actin synthesis, poly(A) RNA was isolated from embryos, larvae, and pupae, and hybridized to actin-coding DNA probes by the Northern blot procedure. Embryos and larvae contain three size classes of actin mRNA: 1.65, 1.95, and 2.3 kb, whereas pupae only contain actin mRNA of 2.3 kb. When DNA probes containing transcribed 3'-flanking sequences but no actin-coding sequences were hybridized to poly(A) RNA obtained from embryos, larvae, and pupae, only the 1.95-kb species present in the larval stage was detected (Fyrberg *et al.*, 1980; Zulauf *et al.*, 1981).

Similar experiments were performed to analyze actin mRNA synthesis during the first 19 hr of embryonic development in *Drosophila*. Probes containing actin-coding sequences were hybridized to total RNA and poly(A) RNA isolated from embryos at various stages. A 2.2-kb actin mRNA was detected throughout embryogenesis, whereas a 1.77-kb transcript began to show appreciable synthesis at 12 through 19 hr (Sodja *et al.*, 1982). These results suggest that differential expression of actin genes is occurring during early development. Whether this differential expression is regulated at the transcriptional, post-transcriptional, or translational level has not been determined.

D. Sea Urchin (*Strongylocentrotus purpuratus*)

The most thorough investigations of actin gene expression in sea urchin have involved detection of actin-encoding RNAs by the Northern blot procedure (Merlino *et al.*, 1981; Crain *et al.*, 1981). These studies have revealed that two size classes of actin mRNA, 1.8 and 2.2 kb, are synthesized in the developing sea urchin and expressed differentially both temporally and spatially. Both actin mRNAs have been found to accumulate substantially during early development. Minimal amounts of the 2.2-kb mRNA are present in the unfertilized egg, whereas the 1.8-kb mRNA is virtually undetectable. Between 12 and 24 hr there is a dramatic increase in the accumulation of these two mRNAs. Results summarized in Fig. 3 reveal that there is a significant difference in the pattern of accumulation of the 1.8 and 2.2-kb actin mRNAs. The 2.2-kb mRNA is more prevalent in early stages of development (2–5 hr) and peaks at 36 hr, whereas the 1.8-kb mRNA predominates between 18 and 36 hr, peaking at 24 hr. The 2.2-kb actin mRNA predominates once again at 48 hr. In addition to this difference in the temporal pattern of accumulation of the two actin mRNAs, fractionation of

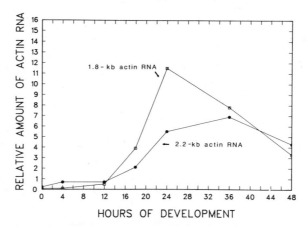

Fig. 3.   Changes in the relative amounts of actin-specific RNAs during sea urchin development. The relative amounts of the 1.8-kb (□) and 2.2-kb (●) actin mRNAs were obtained from densitometric tracing of an autoradiogram of a Northern blot in which a labeled actin cDNA probe was hybridized to electrophoresed total RNA (Merlino *et al.*, 1981).

sea urchin embryos into different tissue types has revealed the existence of a spatial difference in their localization. The 1.8-kb actin mRNA is preferentially localized in the ectoderm, whereas the 2.2-kb actin mRNA is more highly concentrated in the endoderm/mesoderm fraction (Peters *et al.,* 1982).

The accumulation of actin messenger RNAs described in the preceding paragraph appears to be correlated with an increase in the rate of actin synthesis during embryogenesis. Analysis of *in vivo* labeled proteins by gel electrophoresis has revealed a qualitative increase in the amount of actin synthesized (Crain *et al.,* 1981), with quantitative estimates suggesting that a sevenfold increase in the rate of actin synthesis occurs between 12 and 24 hr of development (Peters and Kleinsmith, 1983). Relative amounts of α, β, and γ-actin synthesis vary through development (Infante and Heilman, 1981); the synthesis of β-actin appears to remain relatively constant, whereas the synthesis of α- and γ-actins changes markedly. α-Actin, which is not synthesized in the egg, increases to 50% of total actin synthesis by 25 hr, whereas γ-actin synthesis, which accounts for 90% of total actin in the egg, declines to 38% at 25 hr of development.

The above data raise the question as to what mechanisms are involved in controlling actin synthesis and actin mRNA accumulation during early development in the sea urchin. When considering the general profile of actin mRNA accumulation, it has been shown that the rapid increase of the two RNAs in total RNA is paralleled by rapid increases of the two actin mRNAs in cytoplasmic RNA and polysomal RNA. Thus, overall translational control does not appear to be a major factor. On the other hand, the marked accumulation of actin-specific transcripts suggests that some transcriptional and/or posttranscriptional regulation of actin gene expression occurs during early development, resulting in differential accumulation of the two actin transcripts.

To investigate the possibility that selective gene activation is occurring, the biological system under study must be considered in detail. The developing sea urchin can be taken to approximate a closed system because its mass is constant and it does not feed through the pluteus stage of development (about 72 hr for *S. purpuratus*). The most obvious change that occurs in the developing embryo is its rapid increase in cell number. This increase in cells per embryo is of great importance for understanding the control of gene expression because it produces an increase in the number of gene templates available for transcription.

To determine whether the rapid increase in actin mRNA between 12 and 24 hr of development can be accounted for by the increase in cell number, the number of cells per embryo can be plotted relative to the amount of actin mRNA (Fig. 4). The increase in actin RNA between 12 and 24 hr is greater than the observed increase in number of cells per embryo. This implies that the steady state level of actin RNA *per cell* has increased over this time period.

This analysis, if carried no further, would indicate that the rate of actin gene transcription must have increased in order to account for the rise in steady state

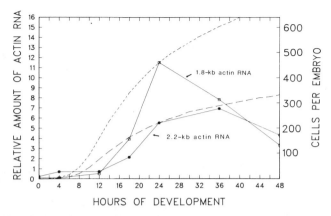

Fig. 4. Comparison of changes in actin RNA levels to changes in the number of cells per embryo in developing sea urchins. The top dashed line represents the number of cells in a sea urchin embryo as a function of development. Derivation of the equation describing this function will be described elsewhere (Peters and Kleinsmith, 1983). The bottom dashed line shows what this same curve looks like when plotted on a scale that attempts to superimpose the curve on the data for the 2.2-kb RNA. The shapes of the curves for the cells per embryo and for the amount of actin RNA are dissimilar, demonstrating no direct relationship.

amounts of actin RNA. There are, however, problems with the analysis as it stands. First, a developing embryo by its nature is far removed from anything approximating a steady state. This is again most vividly demonstrated by the exponential increase in the number of cells per embryo at the onset of development. Thus, it is misleading to speak of steady state levels of any components of the embryo that are dependent on cell number. Second, the incompleteness of this argument can best be demonstrated by considering the units of a rate of transcription. The units are mass of RNA (in picograms) per unit of time per cell. If this is multiplied by cell number, the result is not a mass of RNA, but a mass of RNA per unit of time:

$$\frac{\text{pg RNA}}{\text{cell-hour}} \times \text{cell} = \frac{\text{pg RNA}}{\text{hour}}$$

Thus, the rate of transcription must be multiplied by cell-hours to obtain the correct result of mass of RNA:

$$\frac{\text{pg RNA}}{\text{cell-hour}} \times \text{cell-hour} = \text{pg RNA}$$

The cell-hour is analogous to the man-hour and is thus a unit of work. To evaluate the amount of actin mRNA synthesized over a particular period of development, the rate of transcription of the actin genes is multiplied by the number of cell-hours of work performed by the embryo. The cell-hour function of the embryo is obtained by integrating the cell number function of the embryo with respect to time.

If a graph of the cell-hour function is compared with a graph of actin mRNA plotted as a function of developmental time, it is found that these curves are superimposable through 24 hr of development (Fig. 5). The rapid increase of actin mRNA could thus be accounted for by a constant rate of transcription, assuming negligible degradation of the actin mRNA through 24 hr of development. It is clear that after 24 hr, cell-hours no longer reflects the behavior of actin mRNA. Reasons for the decrease of actin mRNA after 24 hr undoubtedly include actin mRNA degradation, but the rate of actin mRNA synthesis may decrease as well.

The strong correlation observed between the cell-hour function and the increase in the actin mRNA does not verify the assumption of a constant rate of actin gene transcription during early sea urchin development, but this type of analysis demonstrates the necessity of considering the dynamics of the biological system under investigation when interpreting data on mRNA accumulation. The equations developed for the description of the cell number and cell-hour in the sea urchin can be used to analyze the expression of other genes in this organism. In addition, these same concepts can be applied to other developmental systems in which the cell number changes rapidly.

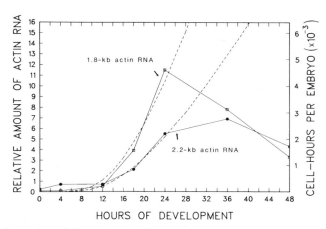

Fig. 5. Comparison of changes in actin RNA levels to the cell-hours of work performed by the embryo. The top dashed line represents the number of cell-hours of work the embryo performs as it develops. Derivation of the equations describing this function are described elsewhere (Peters and Kleinsmith, 1983). The bottom dashed line shows what the cell-hour curve looks like when superimposed on the data for the 2.2-kb RNA. The curves for cell-hours and actin RNA accumulation through 24 hr of development are superimposable, suggesting that the actin genes can be expressed at a constant rate and still account for the rapid accumulation.

ACKNOWLEDGMENTS

Those studies carried out in the authors' laboratory were supported in part by NSF grant PCM78-15300. N.K.P. is supported by NIH training grant 1-P32-GM 07544-05. We thank Valerie Kish for reading over the manuscript.

REFERENCES

Allen, D. (1981). Cell motility. *J. Cell Biol.* **91,** 148s–155s.

Cleveland, D. W., Lopata, M. A., McDonald, R. J., Cowan, N. J., Rutter, W. J., and Kirschner, M. W. (1980). Number and evolutionary conservation of α- and β- tubulin and cytoplasmic β- and γ-actin genes using specific cloned cDNA probes. *Cell* **20,** 95–105.

Cooper, A. D., and Crain, W. R., Jr. (1982). Complete nucleotide sequence of a sea urchin actin gene. *Nucleic Acids Res.* **10,** 4081–4092.

Crain, W. R., Jr., Durica, D. S., and Van Doren, K. (1981). Actin gene expression in developing sea urchin embryos. *Mol. Cell. Biol.* **1,** 711–720.

Durica, D. S., Schloss, J. A., and Crain, W. R., Jr. (1980). Organization of the actin gene sequences in the sea urchin: Molecular cloning of an intron-containing DNA sequence coding for a cytoplasmic actin. *Proc. Natl. Acad. Sci. U.S.A.* **77,** 5682–5687.

Efstratiadis, A., Posakong, J. W., Maniatis, T., Lawn, R. M., O'Connell, C., Spritz, R. A., DeRiel, J. K., Forget, B. G., Weissman, S. M., Slighton, J. L., Blechl, A. E., Smithies, O., Barralle, F. E., Shoulders, C. C., and Proudfoot, N. J. (1980). The structure and evolution of the human β-globin gene family. *Cell* **21,** 653–668.

Engel, J. N., Gunning, P. W., and Kedes, L. (1981). Isolation and characterization of human actin genes. *Proc. Natl. Acad. Sci. U.S.A.* **78,** 4674–4678.

Engel, J. N., Gunning, P. W., and Kedes, L. (1982). Human cytoplasmic actin proteins are encoded by a multigene family. *Mol. Cell. Biol.* **2,** 674–684.

Firtel, R. A., Timm, R., Kimmel, A. R., and McKeown, M. (1979). Unusual nucleotide sequences at the 5' end of the actin genes in *Dictyostelium discoideum. Proc. Natl. Acad. Sci. U.S.A.* **76,** 6206–6210.

Foran, D. R., Johnson, P. G., and Moore, G. P. (1984). In preparation.

Fornwald, J. A., Kuncio, G., Peng, I., and Ordahl, C. P. (1982). The complete nucleotide sequence of the chicken α-actin gene and its evolutionary relationship to the actin gene family. *Nucleic Acids Res.* **10,** 3861–3876.

Franzini-Armstrong, C., and Peachey, L. D. (1981). Striated muscle-contractile and control mechanisms. *J. Cell Biol.* **91,** 166s–188s.

Fyrberg, E. A., and Donady J. J. (1979). Actin heterogeneity in primary embryonic culture cells from *Drosophila melanogaster. Dev. Biol.* **68,** 487–502.

Fyrberg, E. A., Kindle, K. L., and Davidson, N. (1980). The actin genes of *Drosophila:* A dispersed multigene family. *Cell* **19,** 365–378.

Fyrberg, E. A., Bond, B. J., Hershey, N. D., Mixter, K. S., and Davidson, N. (1981). The actin genes of *Drosophila:* Protein coding regions are highly conserved but intron positions are not. *Cell* **24,** 107–116.

Gallwitz, D., and Seidel, R. (1980). Molecular cloning of the actin gene from yeast *Saccharomyces cerevisiae. Nucleic Acids Res.* **8,** 1043–1059.

Gallwitz, D., and Sures, I. (1980). Structure of a split yeast gene: Complete nucleotide sequence of the actin gene in *Saccharomyces cerevisiae. Proc. Natl. Acad. Sci. U.S.A.* **77,** 2546–2550.

Gilbert, W. (1978). Why genes in pieces? *Nature (London)* **271,** 501–508.

Goldman, R., Pollard, T., and Rosenbaum, J., eds. (1976). "Cell Motility." Cold Spring Harbor Lab., Cold Spring Harbor, New York.

Hamada, H., and Kakunaga, T. (1982). Potential Z-DNA forming sequences are highly dispersed in the human genome. *Nature (London)* **298**, 396–398.

Hamada, H., Petrino, M. G., and Kakunaga, T. (1982). Molecular structure and evolutionary origin of human cardiac muscle actin gene. *Proc. Natl. Acad. Sci. U.S.A.* **79**, 5901–5905.

Hardison, R. C., Butler, E. T., III, Lacy, E., Maniatis, T., Rosenthal, N., and Efstratiadis, A. (1979). The structure and transcription of four linked rabbit β-like globin genes. *Cell* **18**, 1285–1297.

Horovitch, R. V., Stoti, R. V., Rich, A., and Pardue, M. L. (1979). Multiple actins in *Drosophila melanogaster. J. Cell Biol.* **82**, 86–92.

Houck, C. H., Rinehart, F. P., and Schmid, C. W. (1979). A ubiquitous family of repeated DNA sequences in the human genome. *J. Mol. Biol.* **132**, 289–306.

Infante, A. A., and Heilman, L. J. (1981). Distribution of messenger ribonucleic acid in polysomes of sea urchin embryos: Translational control of actin synthesis. *Biochemistry* **20**, 1–8.

Jahn, C. L., Hutchinson, C. A., III, Phillips, S. J., Weaver, S., Haigwood, N. L., Voliva, C. F., and Edgell, M. E. (1980). DNA sequence organization of the β-globin complex in BALB/c mouse. *Cell* **21**, 159–168.

Johnson, P. J., Foran, D. R., and Moore, G. P. (1983). Organization and evolution of the actin gene family in sea urchins. *Mol. Cell. Biol.* **3**, 1824–1833.

Kaine, B. P., and Spear, B. B. (1980). Putative actin genes in the macronucleus of *Oxytricha fallax. Proc. Natl. Acad. Sci. U.S.A.* **77**, 5336–5340.

Kaine, B. P., and Spear, B. B. (1982). Nucleotide sequence of a macronuclear gene for actin in *Oxytricha fallax. Nature (London)* **295**, 430–432.

Kindle, K. L., and Firtel, R. A. (1978). Identification and analysis of *Dictyostelium* actin genes, a family of moderately repeated genes. *Cell* **15**, 763–778.

Kinniburgh, A. J., Mertz, J. E., and Ross, J. (1978). The precursor of mouse β-globin messenger RNA contains two intervening RNA sequences. *Cell* **14**, 681–693.

Klobutcher, L. A., Shawton, M. T., Donini, P., and Prescott, D. M. (1981). All gene-size DNA molecules in four species of hypotrichs have the same terminal sequence and an unusual 3' terminus. *Proc. Natl. Acad. Sci. U.S.A.* **78**, 3015–3019.

Lacy, E., Hardison, R. C., Quon, D., and Maniatis, T. (1979). The linkage arrangement of four rabbit β-like globin genes. *Cell* **18**, 1273–1283.

Lauer, J., and Maniatis, T. (1980). The chromosomal arrangement of human α-like globin gene: Sequence homology and α-globin gene deletions. *Cell* **20**, 119–130.

McKeown, M., and Firtel, R. A. (1981a). Evidence for sub-families of actin genes in *Dictyostelium discoideum* as determined by comparison of 3' end sequences. *J. Mol. Biol.* **151**, 593–606.

McKeown, M., and Firtel, R. A. (1981b). Differential expression and 5' end mapping of actin genes in *Dictyostelium. Cell* **24**, 799–807.

McKeown, M., Taylor, W. C., Kindle, K. L., Firtel, R. A., Bender, W., and Davidson, N. (1978). Multiple heterogeneous actin genes in *Dictyostelium. Cell* **15**, 789–800.

MacLeod, C., Firtel, R. A., and Papkoff, J. (1980). Regulation of actin gene expression during spore germination in *Dictyostelium discoideum. Dev. Biol.* **76**, 263–274.

Margolskee, J. P., and Lodish, H. F. (1980a). Half-lives of messenger RNA species during growth and differentiation of *Dictyostelium discoideum. Dev. Biol.* **74**, 37–49.

Margolskee, J. P., and Lodish, H. F. (1980b). The regulation of the synthesis of actin and two other proteins induced early in *Dictyostelium discoideum* development. *Dev. Biol.* **74**, 50–64.

Merlino, G. T., Water, R. D., Chamberlain, J. P., Jackson, D. A., El-Glewely, M. R., and Kleinsmith, L. J. (1980). Cloning of sea urchin actin gene sequences for use in studying the regulation of actin gene transcription. *Proc. Natl. Acad. Sci. U.S.A.* **77**, 765–769.

Merlino, G. T., Water, R. D., Moore, G. P., and Kleinsmith, L. J. (1981). Change in expression of the actin gene family during early sea urchin development. *Dev. Biol.* **85,** 505–508.

Ng, R., and Abelson, J. (1980). Isolation and sequence of the gene for actin in *Saccharomyces cerevisiae. Proc. Natl. Acad. Sci. U.S.A.* **77,** 3912–3916.

Nordheim, A., Pardue, M. L., Lafer, E. M., Moller, A., Stollar, B. D., and Rich, A. (1981). Antibodies to left-handed Z-DNA bind to interband regions of *Drosophila* polytene chromosomes. *Nature (London)* **294,** 417–422.

Nudel, U., Katcoff, D., Zakut, R., Shani, M., Carmon, Y., Finer, M., Czosneck, H., Ginsburg, I., and Yaffe, D. (1982). Isolation and characterization of rat skeletal muscle and cytoplasmic actin genes. *Proc. Natl. Acad. Sci. U.S.A.* **79,** 2763–2767.

Overbeek, P. A., Merlino, G. T., Peters, N. K., Cohn, V. H., Moore, G. P., and Kleinsmith, L. J. (1981). Characterization of five members of the actin gene family in the sea urchin. *Biochem. Biophy. Acta* **656,** 195–205.

Peters, N. K., and Kleinsmith, L. J. (1983). Kinetic model for the study of gene expression in the developing sea urchin. *Develop. Biol.* (in press).

Peters, N. K., Moore, G. P., and Kleinsmith, L. J. (1982). Differential expression of two forms of actin-specific messenger RNA in developing sea urchin embryos. *J. Cell Biol.* **91,** 365a (abstr.).

Pollard, T. P. (1981). Cytoplasmic contractile proteins. *J. Cell Biol.* **91,** 156s–165s.

Pollard, T. P., and Weihing, R. R. (1974). Actin and myosin and cell movement. *CRC Crit. Rev. Biochem.* **2,** 1–65.

Proudfoot, N. J., Shander, M. H., Manley, J. L., Gefter, M. L., and Maniatis, T. (1980). Structure and *in vitro* transcription of human globin genes. *Science* **209,** 1329–1336.

Schaffner, W., Kunz, G., Daetwyler, H., Telford, J., Smith, H. O., and Birnstiel, M. L. (1978). Genes and spacers of cloned sea urchin histone DNA analyzed by sequencing. *Cell* **14,** 655–672.

Scheller, R. H., McAllister, L. B., Crain, W. R., Jr., Durica, D. S., Posakany, J. W., Thomas, T. L., Britten, R. J., and Davidson, E. C. (1981). Organization and expression of multiple actin genes in the sea urchin. *Mol. Cell. Biol.* **1,** 609–628.

Schuler, M. A., and Keller, E. B. (1981). The chromosomal arrangement of two linked actin genes in the sea urchin *S. purpuratus. Nucleic Acids Res.* **9,** 591–604.

Schuler, M. A., McOsker, P., and Keller, E. B. (1982). DNA sequence of two linked actin genes of the sea urchin. *Mol. Cell. Biol.* (in press).

Shah, D. M., Hightower, R. C., and Meagher, R. B. (1982). Complete nucleotide sequence of a soybean actin gene. *Proc. Natl. Acad. Sci. U.S.A.* **79,** 1022–1026.

Sodja, A., Arking, R., and Zafar, R. (1982). Actin gene expression during embryogenesis of *Drosophila melanogaster. Dev. Biol.* **90,** 363–368.

Southern, E. M. (1975). Detection of specific sequences among DNA fragments separated by gel electrophoresis. *J. Mol. Biol.* **98,** 503–517.

Storti, R. V., Horovitch, S. J., Scott, M. P., Rich, A., and Pardue, M. L. (1978). Myogenesis in primary cell cultures from *Drosophila melanogaster:* Protein synthesis and actin heterogeneity during development. *Cell* **13,** 589–598.

Streisinger, G., Okada, Y., Emrich, J., Newton, J., Tsugita, A., Terzaghi, E., and Inouye, M. (1966). Frameshift mutations and the genetic code. *Cold Spring Harbor Symp. Quant. Biol.* **31,** 77–84.

Sures, I., Lowry, J., and Kedes, L. H. (1978). The DNA sequence of sea urchin (*S. purpuratus*) H2A, H2B and H3 histone coding and spacer regions. *Cell* **15,** 1033–1044.

Szostack, J., and Blackburn, E. H. (1982). Cloning yeast telomeres on linear plasmid vectors. *Cell* **29,** 245–255.

Thomas, P. S. (1980). Hybridization of denatured RNA and small DNA fragments transferred to nitrocellulose. *Proc. Natl. Acad. Sci. U.S.A.* **77,** 5201–5205.

Tsai, S. Y., Tsai, M., and O'Malley, B. W. (1981). Specific 5' flanking sequences are required for faithful initiation of *in vitro* transcription of the ovalbumin gene. *Proc. Natl. Acad. Sci. U.S.A.* **78,** 879–883.

Tsujimoto, Y., Hirose, S., Tsuda, M., and Suzuki, Y. (1981). Promoter sequence of fibroin gene assigned by *in vitro* transcription system. *Proc. Natl. Acad. Sci. U.S.A.* **78,** 4838–4842.

Vandekerckhove, J., and Weber, K. (1978a). Mammalian cytoplasmic actins are the products of at least two genes and differ in primary structure in at least 25 identified positions from skeletal muscle actins. *Proc. Natl. Acad. Sci. U.S.A.* **75,** 1106–1110.

Vandekerckhove, J., and Weber, K. (1978b). At least six different actins are expressed in a higher mammal: An analysis based on the amino acid sequence of the amino-terminal tryptic peptide. *J. Mol. Biol.* **126,** 783–802.

Vandekerckhove, J., and Weber, K. (1980). Vegetative *Dictyostelium* cells containing 17 actin genes express a single major actin. *Nature (London)* **284,** 475–477.

Vogeli, G., Ohkubo, H., Sobel, M. E., Yamada, Y., Pastan, I., and DeCrombrugghe, B. (1981). Structure of the promoter for chicken α2 type I collagen gene. *Proc. Natl. Acad. Sci. U.S.A.* **78,** 5334–5338.

Vorlickova, M., Kypr, J., Stokrova, S., and Sponar, J. (1982). A Z-like form of poly(dA-dC)·poly(dG-dT) in solution? *Nucleic Acids Res.* **10,** 1071–1080.

Water, R. D., Pringle, J. R., and Kleinsmith, L. J. (1980). Identification of an actin-like protein and its messenger ribonucleic acid in *Saccharomyces cerevisiae. J. Bacteriol.* **144,** 1143–1151.

Zakut, R., Shani, M., Givol, D., Neuman, S., Yaffe, D., and Nudel, U. (1982). Nucleotide sequence of the rat skeletal muscle actin gene. *Nature (London)* **298,** 857–859.

Zeigler, M. E., and Kleinsmith, L. J. (1982). Cell-free transcription of a cloned sea urchin actin gene. *Fed. Proc., Fed. Am. Soc. Exp. Biol.* **41,** 1293 (abstr.).

Zeigler, M. E., Kish, V. M., and Kleinsmith, L. J. (1983). In preparation.

Zimmerman, C., Tymen, S., Marck, C., and Guschlbauer, W. (1982). Conformational transitions of poly(dA-dC)·poly(dG-dT) induced by high salt or in ethanolic solution. *Nucleic Acids Res.* **10,** 1081–1096.

Zulauf, E., Sanchez, R., Tobin, S. L., Rdest, U., and McCarthy, B. J. (1981). Developmental expression of a *Drosophila* actin gene encoding actin I. *Nature (London)* **296,** 556–558.

12

Functional Architecture at Telomeres of Linear DNA in Eukaryotes

EDWARD M. JOHNSON, PETER BERGOLD,[1] **AND GERALD R. CAMPBELL**[2]

The Rockefeller University
New York, New York

I. INTRODUCTION: CHROMOSOMAL AND EXTRACHROMOSOMAL LINEAR DNA

One of the primary distinctions between the genomes of prokaryotic and eukaryotic cells is found in the simple geometric configuration of their chromosomes. Whereas the genomes of most (if not all) bacteria consist of a single circular chromosome, it is now generally recognized that the multiple chromosomes of higher organisms are predominantly in the form of discretely replicating and segregating linear entities. There are many references to the discrete and linear organization of chromosomes in lower eukaryotes (cf. Kavenoff and Zimm, 1973;

[1]Present address: Memorial Sloan-Kettering Cancer Center, 1275 York Avenue, New York, New York 10021.
[2]Present address: Department of Biochemistry, New York University Medical Center, New York, New York 10016.

RECOMBINANT DNA AND
CELL PROLIFERATION

Petes *et al.*, 1973; Lauer *et al.*, 1977). Much of the evidence from higher organisms has centered on morphological or genetic observations that pose topological problems most readily resolved by postulating linearity in chromosomal DNA. For example, observations of sister chromatid exchanges in mammalian cells are considered strong evidence for the organization of metaphase chromatids as single, linear DNA duplexes (Taylor, 1958; Wolff and Perry, 1975; Wolff, 1977). It is likely that eukaryotes have evolved highly specific mechanisms for the efficient replication and, possibly, functional associations of chromosome ends, and that these are reflected in conservation of sequences or patterns of sequences at termini. Several linear DNA molecules have been isolated from genomes of eukaryotes and shown to replicate as independent linear entities. These molecules include gene-sized fragments produced during macronuclear formation in ciliated protozoa and extrachromosomal rDNA and mitochondrial DNA molecules of several lower eukaryotes. Based on evolutionary considerations there may be reason to suspect that sequences at telomeres* of all of these molecules and of chromosome telomeres will have similar characteristics (Cavalier-Smith, 1974). Sequencing of the ends of palindromic extrachromosomal rDNA molecules from *Tetrahymena* (Blackburn and Gall, 1978), *Dictyostelium* (Emery and Weiner, 1981), and *Physarum* (Bergold *et al.*, 1983) reveals that sequences at termini are strikingly different but have an intriguing pattern of organizational similarities. In this chapter we review known sequences at ends of cellular DNA molecules and attempt to evaluate this information in terms of models for telomere replication and function.

II. DNA TERMINI RESULTING FROM GENOMIC FRAGMENTATION IN CILIATED PROTOZOA

There are now several known instances in which chromosomes of eukaryotes are systematically cleaved into smaller linear pieces which replicate independently. It has long been known that chromosomes in somatic cells of the roundworm *Ascaris* are fragmented, and that this process is accompanied by a considerable loss of chromatin (Wilson, 1925). More recently, genomic fragmentation has been especially well characterized for the ciliated protozoa. These organisms possess two differently functioning nuclei within each cell: a diploid micronucleus which functions as a germinal nucleus in conjugation and a polyploid macronucleus which functions as a locus for gene activation and transcription.

---

*The term "telomere" is usually taken to mean the end of a eukaryotic chromosome. In molecular terms this definition is vague. A more appropriate definition in view of recent observations is as follows: Telomeres are the termini of a linear cellular DNA molecule that support complete replication of the molecule followed by resolution and proper segregation of replicants.

The macronucleus is formed *de novo* after sexual conjugation as a division product of the new zygotic micronucleus. Details of macronuclear development appear to vary widely between the classes of hypotrichous and holotrichous ciliates, but in general the process involves fragmentation and selective amplification of functional DNA sequences accompanied by degradation of much repetitive and nonfunctional DNA (for reviews, see Prescott *et al.*, 1973; Gorovsky, 1980).

In hypotrichous ciliates, including *Oxytricha*, *Stylonychia*, and *Euplotes*, macronuclear development involves polytenization of chromosomes followed by cleavage at interbands and degradation of as much as 90% of the micronuclear DNA. In the vegetative, transcriptionally active macronucleus, DNA exists as an array of molecules ranging in size from about 0.5 to 20 kilobases (kb), with an average of about 2.2 kb (Lawn *et al.*, 1978; Swanton *et al.*, 1980). Kaine and Spear (1982) sequenced a complete *Oxytricha* macronuclear fragment homologous to a yeast actin gene and showed that this linear piece of 1.56 kb contains a single gene.

The sequences at ends of all macronuclear DNA molecules in a given hypotrich are virtually invariant. Sequencing is therefore a relatively simple matter of isolating total macronuclear DNA, end labeling at either 3' or 5' positions, and sequencing by the procedure of Maxam and Gilbert (1980). In certain *Oxytricha* and *Stylonychia* species the terminal sequences are reportedly based on a tandem C_4A_4 repeat as follows (Klobutcher *et al.*, 1981):

$$5'\text{-}\ C_4\ A_4\ C_4\ A_4\ C_4.\ .\ .$$
$$3'\text{-}\ G_4\ T_4\ G_4\ T_4\ G_4\ T_4\ G_4\ T_4\ G_4.\ .\ .$$

(See Table I for a summary of terminal sequences.) This configuration includes a 16-nucleotide, 3'-terminal "overhang" which cannot exist in a hairpin configuration. The terminal sequence in *Euplotes aediculatus* is similar but not identical (Klobutcher *et al.*, 1981):

$$5'\text{-}\ C_4\ A_4\ C_4\ A_4\ C_4\ A_4\ C_4\ (X)_{17}\ T\ T\ G\ A\ A.\ .\ .$$
$$3'\text{-}\ G_2\ T_4\ G_4\ T_4\ G_4\ T_4\ G_4\ T_4\ G_4\ T_4\ G_4\ (X)_{17}\ A\ A\ C\ T\ T.\ .\ .$$

In this case the 3' terminal is 14 nucleotides long. In both of these cases 3' and 5' ends were separately sequenced by the Maxam and Gilbert (1980) procedure following end labeling. Evidence for a 3' protruding end is essentially that the 3' sequence is 14 or 16 nucleotides longer than the 5' sequence. Other configurations are equally conceivable; for example, an internal $G_4T_4G_4T_4$ unpaired loop in the *Oxytricha* terminus would leave flush ends on the molecule. Recent reports strongly opt in favor of an overhang. After 3' end labeling, brief treatment with S1 nuclease (specific for single strands) yields a ladder of gel bands consistent with a G and T single-strand repeat extending to the end of the molecule (Pluta and Spear, 1981). The sequences at each end of a hypotrich macronuclear DNA

Table I

Repeat Sequences at Linear DNA Termini in Eukaryotes

Source	Sequence		Sequence determined from cloned terminus	Single-strand nicks or gaps	Hairpin configuration at terminus
Macronuclear DNA in ciliated protozoa					
Stylonychia, Oxytricha[a]	$5'$- C_4 A_4 C_4 A_4 C_4 ··· $3'$- G_4 T_4 G_4 T_4 G_4 T_4 G_4 ···		—	$(-)^h$	—
Euplotes aediculatus[a]	$5'$- C_4 A_4 C_4 A_4 C_4 A_4 C_4 $(X)_{17}$ T T G A A ···· $3'$- G_4 T_4 G_4 T_4 G_4 T_4 G_4 $(X)_{17}$ A A C T T ····		—	$(-)^h$	—
Glaucoma chattoni[b]	$5'$- C A A $(C_4A_2)_n$ ···· $n \geq 38$		—	$(-)^h$	—
Tetrahymena pyriformis[c]	$5'$- ··· $(C_4A_2)_n$ ···· $n = 20$–70		—	$(-)^h$	—

Episomal rDNA
molecules

T. pyriformis[d]	5'-... (C$_4$A$_2$)$_n$ ... n = 20–70	–	+	(+)[g]
Dictyostelium discoideum[e]	5'-... (C$_n$T)$_m$ ... [29-bp repeat] n = 1–8 m = 18–34	+	?	(+)[g]
D. discoideum rDNA 29-bp tandem repeat[e]	5'-... T C A T C T T A G C C A C C A T G G A G C C A A A A A T T ...	+	?	(+)[g]
Physarum polycephalum 140-bp tandem repeat[f]	5'-... (see sequence below)	+	+	+

Physarum polycephalum 140-bp tandem repeat[f]:

```
                10                20
5'-...C C C G G A T C G A T G C A T A G C G A
           30                40
    T T C A A A C A G G T G C T G G G G G C A
           50                60
    G C G C C T T T T C C A T G T C G T C
           70                80
    T G C C C A G T T C T G C C T C T T T C
           90               100
    T C T T C A C G G G G C G A G C T G C T G
          110               120
    G T A G T G A C G C G C C C C A G C T C T
          130               140
    G A G C C T C A A G A T C G A T T C G T
    G T G G ...
```

[a]Klobutcher et al. (1981).
[b]Katzen et al. (1981).
[c]Yao and Yao (1981).
[d]Blackburn and Gall (1978).
[e]Emery and Weiner (1981).
[f]Bergold et al. (1983). This sequence is repeated 6–10 times with significant degeneracy.
[g]A hairpin structure has been inferred, but no molecular details have been presented.
[h]These DNA termini may possess a protruding single-strand sequence.

molecule are inverted terminal repetitions (Oka *et al.,* 1980; Klobutcher *et al.,* 1981). Denaturation and reannealing can therefore result in formation of single-strand DNA circles.

Macronuclear development in holotrichous ciliates, including *Paramecium* and *Tetrahymena,* does not appear to involve formation and fragmentation of polytene chromosomes. Instead, there is a gradual process of selective gene amplification accompanied by degradation of micronuclear sequences. In these organisms, in contrast to the hypotrichs, nearly 90% of micronuclear DNA is retained in the macronucleus (Yao and Gorovsky, 1974). The average size of macronuclear DNA molecules in the holotrich *Tetrahymena* is much larger than that in macronuclei of hypotrichs (Preer and Preer, 1979), and it has been difficult to determine with certainty whether this DNA is site-specifically fragmented, as is the case in hypotrichs.

Katzen *et al.* (1981) have reported that macronuclear DNA in the holotrichous ciliate *Glaucoma chattoni* is site-specifically fragmented. Differences in genome reorganization between hypotrichs and holotrichs may thus differ primarily in the extent to which such fragmentation occurs rather than in the question of whether fragmentation occurs at all. Sequences at macronuclear ends in *Glaucoma* are similar to those reported for hypotrichous ciliates. Katzen and co-workers have reported that at least 38 repeats of the sequence CCCCAA are present at each 5' terminus. Each 5' terminus has the configuration (Katzen *et al.,* 1981):

$$5'- C \ A \ A \ (C_4A_2)_n \ \ldots \ 3'$$
$$n \geq 38$$

At this point publications have not clarified whether or not a protruding sequence exists at 3' ends in *Glaucoma* as has been reported for hypotrichs (Klobutcher *et al.,* 1981). It is notable that in *Glaucoma* macronuclei, a discrete 9.3-kb class of molecule has been identified as monomeric, nonpalindromic, linear rDNA molecules (Katzen *et al.,* 1981). The ends of these multiple rDNA copies also possess the tandem CCCCAA repeat.

What is the origin of the tandem C and A repeats at the ends of fragmented macronuclear DNA? It has been suggested that CCCCAA is recognized during site-specific chromosome fragmentation in *Glaucoma* (Katzen *et al.,* 1981). The question then arises as to whether or not repeats of such a sequence are present, separating all chromosomally integrated copies of macronuclear genes that exist in micronuclei. There are reported instances where this is not the case. In *O. fallax* Dawson and Herrick (1982) have used labeled, terminal-repeated sequences to probe a micronuclear DNA fragment that carries the ends of two adjacent macronuclear fragments and a short, nonretained stretch in between; they detected no hybridization. (Note that a single C and A unit, such as that reportedly next to integrated *Tetrahymena* rDNA [King and Yao, 1982], would not be detected.) Similarly, Boswell *et al.* (1982) have reported that the intact, repeated C_4A_4 sequence is not present at the ends of macronuclear sequences as they exist in the micronuclear chromosomes of *O. nova*. It is therefore conceivable that these short, repetitive sequences are attached to DNA molecules during or after fragmentation (cf. Boswell *et al.*, 1982). Are there then genomic regions of C and A or G and T repeats that can be excised and attached to other gene fragments? Experiments of Dawson and Herrick (1982) are roughly consistent with this view. They detected micronuclear DNA sequences homologous to the macronuclear DNA termini repeats in *O. fallax*. These micronuclear sequences are present at only one-sixth their level in macronuclei, and they are embedded in high-molecular-weight DNA segments. It has been speculated that some rearrangement mechanism could place these sequences adjacent to appropriate macronuclear genes during chromosome fragmentation (Yao *et al.*, 1981; Dawson and Herrick, 1982). One other possibility to be considered is that the multiple, short, simple-sequence repeats are attached *de novo* and enzymatically to DNA molecules following fragmentation. Further studies should illuminate the problem of the origin of the macronuclear terminal repeats and help assess their functional role. Clearly, these ends play a role in replication of the linear DNA fragments in macronuclei. It is evident that replication of linear DNA must involve special mechanisms for the completion of replication of 5' nucleotides (Watson, 1972). It is conceivable that a percentage of molecules with 3' protruding ends could represent incomplete replication products. Presently, there is no clear indication as to how ends of the macronuclear DNA molecules can be completed. Several proposed mechanisms for terminal completion involve hairpin formation or concatamerization, processes not readily compatible with A and C or G and T repeats with 3' overhangs.

III. CLONED TERMINI OF PALINDROMIC rDNA MOLECULES

The ribosomal genes of several lower eukaryotes are located in pairs of inverted transcription units on palindromic DNA molecules that replicate as linear episomes. These rDNA palindromes have been especially useful in analyzing sequences important for DNA replication. Sequences at termini of three of these rDNA molecules have been determined: those of *Dictyostelium discoideum* (Emery and Weiner, 1981), *Tetrahymena pyriformis* (Blackburn and Gall, 1978), and *Physarum polycephalum* (Bergold *et al.*, 1983). In the case of *Dictyostelium* and *Physarum*, the rDNA ends have been cloned in vectors propagated in *E. coli*.

A. *Dictyostelium* rDNA Termini

Palindromic rDNA in the cellular slime mold *Dictyostelium* is of unusually large size (Cockburn *et al.*, 1978). Molecules 87 kb long contain not only genes for 17, 5.8, and 25 S rRNAs but also, distal to these, genes coding for 5 S rRNA (Maizels, 1976; Frankel *et al.*, 1977). The two larger rRNA transcription units are separated by a central, nontranscribed spacer of about 50 kb. A terminal, nontranscribed spacer of about 5 kb exists at each end (Frankel *et al.*, 1977; Cockburn *et al.*, 1978). Emery and Weiner (1981) have cloned and sequenced a terminal restriction fragment of *Dictyostelium* rDNA. The rDNA terminal *Hin*dIII fragment was treated with S1 nuclease to remove any possible blocking hairpin sequences and cloned in pBR322 *Bam*HI and *Eco*RI sites after attachment of *Bam*HI linkers and further digestion with *Bam*HI and *Eco*RI. The resulting cloned *Eco*RI terminal fragment was then sequenced by the Maxam and Gilbert (1980) procedure. Sequences of five cloned termini were presented.

Each cloned rDNA terminus contains an irregular satellite sequence consisting primarily of C and T copolymer. The most distal regions of the *Eco*RI fragments are heterogeneous with respect to length. This heterogeneity is accounted for by variable tracts of a sequence with the general formula $[C_nT]_m$ (Emery and Weiner, 1981), where n varies from 1 to 8 and m varies from 18 to 34. Immediately proximal to the C_nT satellite are six nearly perfect repeats of a 29-base pair (bp) sequence, presented in Table I.

The single-sequence repeat at *Dictyostelium* rDNA ends bears certain similarities to sequences at ends of macronuclear DNA in ciliated protozoa. For example, both consist of strands with several hundred nucleotide stretches composed of two nucleotides. In many details, however, similarities between DNA termini in these two instances are tenuous. The terminal C_nT repeat in *Dictyostelium* rDNA is highly irregular, whereas the terminal repeat in fragmented macronuclear DNA is highly regular. There is little to compare in the actual 5′ sequences other than that they are C-rich (see Table I). It is notable that the S1 nuclease treatment used in cloning ends of *Dictyostelium* rDNA would have removed any protruding single-strand sequence, such as that reported for macronuclear DNA in ciliated protozoa. Furthermore, if hairpin formation at the rDNA terminal involves a non-base-paired loop, then the terminal nucleotides may have been removed before cloning. Restriction mapping demonstrates that the C_n T repeat does extend to very near the actual terminus if not the very end. The irregularity of the C_nT repeat and its heterogeneity in different rDNA terminal clones suggest the possibility of some form of recombination or gene conversion involving extrachromosomal DNA termini. In this regard it is interesting that the C_nT repeat seems to be limited to rDNA molecules. Emery and Weiner (1981) did not detect significant hybridization of a labeled, cloned rDNA terminal fragment to genomic DNA in blots containing nuclear DNA restriction digests.

B. *Tetrahymena* rDNA Termini

1. Sequences at the Ends of the Extrachromosomal rDNA

Because the extrachromosomal rDNA molecules of *Tetrahymena* have been extensively characterized, they provide a particularly useful system for analyzing the functional significance of terminal sequences. The ribosomal genes in *Tetrahymena* micronuclei exist in a single, chromosomally integrated copy of a 4.8-kb transcription unit together with associated transcribed and nontranscribed spacers (Yao and Gall, 1977). During macronuclear development this integrated copy is amplified to form nearly 10,000 copies per cell of a 20-kb palindromic version in which two transcription units are arranged in inverse orientation on either side of a 2-kb central spacer (Engberg *et al.*, 1976; Karrer and Gall, 1976; Yao and Gall, 1977). Pan and Blackburn (1981) have detected single mac-ronuclear copies of extrachromosomal rDNA half-palindromes and proposed that these are intermediates in formation of the full palindromes. Yao and Gall (1977) have proposed a replication mechanism whereby palindromic rDNA copies are amplified directly from the single micronuclear half-palindrome. At this point the molecular details of either the excision of rDNA or the mechanism of ampli-fication are not known, and it is conceivable that extrachromosomal rDNA copies arise through more than one mechanism.

The palindromic linear rDNA copies replicate independently in the mac-ronucleus (Karrer and Gall, 1976; Yao and Gall, 1977). Replication originates in the central spacer of the molecule at a site slightly displaced from the center of symmetry and proceeds outward to the ends (Cech and Brehm, 1981). It is notable that macronuclear half-palindromic rDNA copies are also capable of independent replication (Pan and Blackburn, 1981). In this case the replication origin is apparently near one end of the molecule (Cech and Brehm, 1981).

The *Tetrahymena* palindromic rDNA termini are blocked from 5' end labeling and are assumed to be in a hairpin configuration, although little is known about the molecular details of the terminal cross-link (Blackburn and Gall, 1978; Szostak and Blackburn, 1982). Blackburn and Gall (1978) have reported the presence of single-strand discontinuities at the rDNA ends. These nicks or gaps allow selective labeling of the rDNA ends by brief nick translation, demonstrat-ing the presence of free 3'-OH groups internally on single strands near termini. Such labeling at the termini is highly specific for the sequence CCCCAA. It has been shown that this hexanucleotide sequence [the same as that repeated at ends of macronuclear DNA in *G. chattoni* (Katzen *et al.*, 1981)] is repeated 20–70 times at each rDNA terminus (Blackburn and Gall, 1978). Discontinuities within this sequence are present every 2 to 4 repeats and are most likely short gaps, perhaps of just a single nucleotide, occurring after the second A in the C_4A_2 unit. Are hairpin structures at ends necessary for autonomous replication of linear rDNA? Pan and Blackburn (1981) have demonstrated that one end of the half-palindromic, extrachromosomal rDNA molecule possesses CCCCAA repeats whereas the other end, that corresponding to the center of the palindromic mole-cule, possesses a 0.3-kb genomic DNA sequence not found in the palindrome. This 0.3-kb segment, the sequence of which is not yet known, is not cross-linked in a hairpin configuration. Since the half-palindromic molecule reportedly repli-cates autonomously (Pan and Blackburn, 1981), it is possible that the hairpin configuration is not necessary for replication of this end. It is also possible that hairpin structures are formed transiently at this end and could not be detected. It is notable that this nonhairpin rDNA end is very near the origin of rDNA replication (Cech and Brehm, 1981), and it is conceivable that it serves a function in replication different from that of ends with hexanucleotide repeats and hairpins.

2. Use of Tetrahymena rDNA Termini in the Cloning of Chromosome Telomeres in Yeast

The termini of the palindromic rDNA molecule have been used to clone yeast chromosome telomeric sequences on linear plasmid vectors. Szostak and Blackburn (1982) have prepared a DNA vector, based on an *E. coli* circular plasmid, that carries the yeast ars 1 function to allow autonomous replication (Stinchcomb *et al.,* 1979) and the *LEU2* gene to allow selection after transformation of a LEU2 yeast strain (Ratzkin and Carbon, 1977). This circular vector was converted to a linear DNA vector by restriction cleavage and ligation to 1.5-kb *Bam*HI terminal fragments of the *Tetrahymena* rDNA palindrome (Szostak and Blackburn, 1982). This linear vector (pSZ216) was used to transform yeast, and several transformants were found to contain the full-length, 12-kb plasmid.

Several unusual aspects of the *Tetrahymena* rDNA ends were found to be maintained in linear plasmids recovered from transformant yeast. These include the multiple CCCCAA repeats, the specific single-strand gaps within these repeats, and the cross-linked, presumably hairpin terminus (Szostak and Blackburn, 1982).

Cloning of yeast telomeres was performed by selecting for yeast chromosomal DNA restriction fragments that were able to substitute functionally for one of the *Tetrahymena* ends of the linear plasmid. Following an asymmetric restriction cleavage of the linear plasmid, the fragment containing the LEU2 sequence and a *Tetrahymena* rDNA end was ligated to yeast chromosomal DNA fragments. The ligation mixture was used to transform yeast. Several transformants were recovered with a plasmid smaller than the original pSZ216, suggesting that one end was derived from yeast DNA. The putative yeast end from one of the transformant plasmids was isolated and used to probe yeast DNA genomic blots. Hybridization to 30–40 yeast restriction bands was observed—the approximate number expected if the probe is homologous to telomeres of all 17 yeast chromosomes. A ring chromosome III from a haploid strain does not have chromosome III telomeric sequences and did not hybridize with the cloned probe from the linear plasmid. These results are suggestive, if not proof, that yeast telomeres have been cloned. In any case, the cloned yeast segments are functional telomeres by the criterion that they support replication of a linear DNA molecule. Recent observations indicate that overall length of linear plasmid vectors is an important factor for maintaining mitotic and meiotic stability in yeast (Murray and Szostak, 1983; Dani and Zakian, 1983).

Complete sequences for cloned yeast telomeres have not yet been reported, but several features of their structure have been investigated. The yeast ends of transformant linear plasmids possess single-strand discontinuities that allow labeling by nick translation (Szostak and Blackburn, 1982). In addition, the yeast ends are cross-linked in an apparent hairpin configuration, as are the *Tetrahymena* ends. In further experiments Szostak and Blackburn (1982) addressed the question of whether a terminal hairpin configuration alone was sufficient to support replication of the linear plasmid. A synthetic hairpin terminus was substituted for one of the *Tetrahymena* ends. The resulting plasmid was found to replicate, but in all cases replicating molecules were trapped in the form of circular replication intermediates. It may be concluded that some properties of telomeric sequences are important for proper regeneration of complete hairpin termini.

3. Additional Comparisons of Telomeres of rDNA and Yeast Chromosomes

Preliminary sequencing results indicate that yeast telomeres possess a satellite sequence consisting of C and A in an irregular repeat (J. Szostak and E. Blackburn, personal communication). This irregularity is similar to that of the C and T repeat found at *Dictyostelium* rDNA ends (Emery and Weiner, 1981). Within this repeat are apparently stretches of alternating A and C, since a poly(dGdT·dCdA) probe anneals to the yeast telomeres cloned on a linear plasmid vector (Walmsley *et al.*, 1983). In addition, the poly(GT) probe annealed with *Tetrahymena* rDNA ends after they had been propagated on the linear vector in yeast, but did not anneal with the C_4A_2 repeat of these ends as isolated from rDNA. It thus seems that poly(GT) tracts are added to the *Tetrahymena* telomeres when the linear vectors are cloned in yeast. The poly(GT) tracts could be added either enzymatically or by recombination of *Tetrahymena* ends with yeast telomeres.

The yeast chromosome telomeres may possess a sequence complexity greater than that of their counterparts in the fragmented genomes of ciliated protozoa. Chan and Tye (1983) report that DNA sequences adjacent to telomeres, and extending from about 12 to 30 kb, contain repeats that are conserved and possess multiple origins of replication. The repeat elements are 0–3.75, 1–1.5, and 5.2 kb in length. These repeats are conserved within *Saccharomyces cerevisiae* strains but show striking divergence in different yeast species. The presence of multiple, tandem, complex, divergent repeats is also a characteristic of the rDNA telomeres of *Physarum polycephalum* (Johnson, 1980; Bergold *et al.*, 1983).

C. Termini of the rDNA Minichromosome of *Physarum polycephalum*

The ribosomal genes of the slime mold *P. polycephalum* are arranged as inverse pairs on a giant (61-kb) palindromic rDNA molecule that is present at

160–300 copies per diploid nucleus (Molgaard *et al.*, 1976; Vogt and Braun, 1976; Campbell *et al.*, 1979; see Fig. 1). This rDNA comprises virtually all of the DNA in interphase nucleoli (Bradbury *et al.*, 1973), and its ease of isolation has made it a particularly good system for studies of ribosomal gene chromatin (Johnson *et al.*, 1978, 1979; Prior *et al.*, 1980, 1983). The *Physarum* rDNA is maintained as an episomal, linear DNA molecule at all stages of the life cycle including those of haploid spores and amoebae (Hall and Braum, 1977; Affolter and Braun, 1978). No chromosomally integrated rDNA copy has thus far been detected, and there is no evidence for rDNA amplification at any stage. Replication of the rDNA originates near the center of the molecule (Vogt and Braun, 1977). *Physarum* plasmodia are diploid and possess 46 chromosomes in strain a × i (Mohberg *et al.*, 1973). There are no data to suggest that the *Physarum* genome is fragmented or that the rDNA is excised from a larger genomic segment. Thus, with respect to many properties of replication and distribution the *Physarum* rDNA acts as a functional minichromosome (Seebeck *et al.*, 1979).

Restriction mapping of the rDNA molecule has revealed a significant length heterogeneity at termini (Campbell *et al.*, 1979); terminal fragments generated by a single restriction enzyme vary by $\pm$ 0.6 kb. Electron microscopically, it was seen that after partial denaturation, the rDNA ends anneal to form multiple foldback loops (Campbell *et al.*, 1979; Hardman *et al.*, 1979). The length and position of these foldback segments were mapped and found to comprise a series of inverted repeats of about 100 bp extending for an average of 600 bp at or near the rDNA ends (Johnson, 1980). Length heterogeneity is due to differences in both the number of these terminal repeats and the sequences of the individual repeats. Brief nick translation of the rDNA ends selectively labels a sequence beginning CCCTA, indicating the presence of single-strand discontinuities with free 3'-OH groups (Johnson, 1980). Following CCCTA, the labeled sequence becomes heterogeneous, although labeling is preferential for the inverted repeat sequences. The inability of these discontinuities to be removed by treatment with T4 DNA ligase suggests that they are gaps rather than nicks, and filling in these breaks suggests that they are short, possibly one nucleotide long (Johnson, 1980).

1. Cloning and Sequencing of Inverted Repeats at rDNA Termini

The *Physarum* rDNA termini were cloned by a modification of the method of Emery and Weiner (1981) as used for *Dictyostelium* rDNA ends. Brief S1 nuclease treatment of isolated rDNA was employed to unblock possible hairpin termini and cleave at single-strand gaps. Following this treatment, synthetic *Eco*RI linkers were ligated to the terminal *Eco*RI restriction fragment, and the resulting linked segments cloned in phage Charon 13 (Bergold *et al.*, 1983). This procedure was designed to circumvent problems of cloning DNA ends with

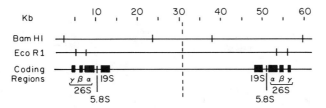

Fig. 1.   The palindromic rDNA molecule of *Physarum polycephalum*.

single-strand gaps and hairpin structures. Cloning of the *Physarum* rDNA ends is necessary in order to carry out sequencing since length and sequence heterogeneity of uncloned ends rules out sequencing by directly end labeling isolated cellular DNA, as could be done with macronuclear fragments from ciliated protozoa.

Several full-length or nearly full-length rDNA termini have been cloned using the aforementioned procedure. The terminal rDNA restriction fragment averages approximately 5 kb in length (Campbell *et al.*, 1979; Gubler *et al.*, 1979) and is the only rDNA fragment of a size convenient for cloning in Charon 13, which accommodates DNA *Eco*RI fragments of 2.9–16.7 kb (Blattner *et al.*, 1977). Restriction mapping of cloned rDNA termini reveals a series of tandem *Hae*III repeats of approximately 140 bp with occasional *Hae*III sites at 20 and 50 bp within these repeats (Bergold *et al.*, 1983). This *Hae*III ladder agrees with *Hae*III digests of uncloned rDNA ends, which also yield a terminal ladder pattern (Johnson, 1980; Bergold *et al.*, 1983). The cloned rDNA terminal insert from one clone, designated PrD 229b, was denatured and reannealed for electron microscopy. The foldback regions of this cloned insert occupy the same position and are of the same approximate extent of those in uncloned rDNA, confirming that we have cloned the terminal sequence of the rDNA molecule and that this sequence consists of a series of foldback repeats.

The sequence of more than 800 nucleotides at the *Physarum* rDNA termini, as deduced from clone PrD 229b, has been presented (Bergold *et al.*, 1983). This sequence consists of six tandem repeats of a unit bounded by *Hae*III sites and averaging 140 bp in length but varying from 136 to 144 bp. The sequence of the most distal repeat unit is shown in Table I, and its most energetically favorable hairpin configuration (from Bergold *et al.*, 1983) is presented in Fig. 2. This terminal repeat unit is somewhat atypical of the more proximal *Hae*III repeat units in that it has no internal *Hae*III sites and no CCCTA potential single-strand gap sites.

2. Novel Aspects of the Physarum rDNA Telomere

There are several features of the complex inverted repeat series at rDNA ends that render these ends unusual in comparison to ends of other known linear

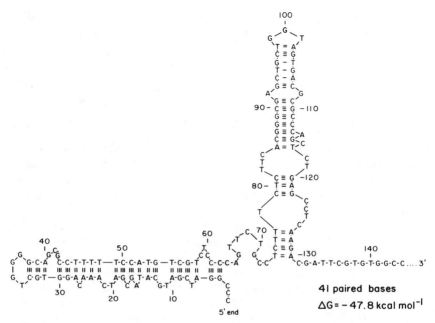

Fig. 2.   DNA sequence and possible secondary structure of an rDNA telomeric repeat. The sequence shown is the most distal example of a unit, averaging 140 nucleotides, repeated with degeneracy at least six times at the termini of *Physarum* rDNA. The hairpin structure shown is the most energetically favorable of secondary structure possibilities. (Courtesy of Bergold *et al.*, 1983. Copyright © 1983 by M.I.T. Press.)

DNAs in eukaryotes. First of all, the *Physarum* rDNA ends are the first sequenced DNA ends from eukaryotes that do not consist of a single sequence made up of essentially two nucleotides. Some caution must be exercised here, however. The S1 nuclease treatment employed prior to cloning could have systematically removed a short terminal sequence from rDNA ends. Therefore, it is conceivable that the end of the terminal sequence presented is not actually the 5′ "terminal nucleotide" of the rDNA. The concept of any such terminal nucleotide is, however, rendered virtually inapplicable by the array of hairpin structures with single-strand gaps found at the rDNA ends.

a. Hairpin Structure and Position of Single-Strand Gaps. The *Physarum* rDNA ends consist of a series of inverted repeat sequences capable of forming complex hairpin secondary structures (Bergold *et al.*, 1983). We asked whether or not these hairpin structures are actually formed *in vivo* in *Physarum* nuclei. We isolated rDNA from nucleoli by procedures designed to minimize any possible partial denaturation at termini, and we spread the rDNA molecules directly for visualization. We found that in growing plasmodia a significant fraction of the rDNA termini could be seen as an array of multiple hairpin loops although nearly 80% of the ends had no readily visible secondary structure. These results suggest that formation of secondary structures at rDNA termini does occur *in vivo* but that this structure does not permanently occupy each end. This finding is consistent with the possible involvement of transiently formed hairpins in completion of terminal replication or in association of termini with other chromosomal components.

In addition to observing native rDNA termini in hairpin configurations, we observed a small but significant fraction of rDNA ends in juxtaposition to other, noncontiguous DNA strands in what appear to be recombination intermediates (Bergold *et al.*, 1983). Many of the complex structures observed result from interaction of sequences in the region of rDNA terminal secondary structure with internal sequences on unidentified DNA duplexes. It is attractive to speculate that this recombination is initiated by insertion of a single-strand hairpin loop into a homologous, or closely homologous, DNA duplex. Recombination initiated in similar fashion in *E. coli* is mediated by the protein product of the *recA* gene (Cunningham *et al.*, 1979; Wu *et al.*, 1982). A protein complex is bound very tightly, possibly covalently, to the inverted repeat sequences at *Physarum* rDNA termini (Cheung *et al.*, 1981). The functional role of this complex is not known, but it could be involved in transitions in secondary structure at the termini or in formation of single-strand discontinuities.

The sequence, CCCTA, labeled by nick translation at terminal single-strand gaps, is strategically located in or near apex loops formed by hairpins (Bergold *et al.*, 1983). This pentanucleotide sequence and its homolog TAGGG are among several elements that are relatively conserved in the 140-bp repeats at rDNA termini. It can be seen in Table I that the most distal repeat has neither of these sequences. However, nucleotides 44–46 correspond to the position of the CCCTA site in most other *Hae*III repeats, and this region contains a permuted version of this, CCT, in the terminal repeat. Of the six terminal *Hae*III inverted repeats in clone PrD 229b, four possess the sequence CCCTA in approximately the same position, whereas one repeat in addition to the terminal one does not. This repeat also has a permuted version of a gap site, CGCCTA. Two repeats possess the homology TAGGG positioned eight or nine nucleotides 5' to the gap sequence. The positions of the potential gap sequences found at the terminus in clone PrD 229b are shown in Table II. In the column describing the type of overhang in Table II, we refer to the most energetically favorable single-strand protrusion resulting from removal of the first C in the sequence CCCTA occurring in a possible hairpin formed as a foldback of that particular repeat. It is clear, however, that there are a myriad of foldbacks that can be formed among the terminal inverted repeats. Table II shows that gapping in these secondary structures could lead to formation of either 5' or 3' single-strand overhangs. Such overhanging single-strand sequences would have the effect of polarizing the

Table II

Potential Gap Positions in *Physarum* rDNA Terminal Repeats

Repeat	Nucleotides	Gap site homolog	Position	Potential gap site	Position	Type of overhang
1	1–144	(TGGG)	34–37	(CCT)	44–46	—
2	170–265	—		CCCTA	207–211	5′
3	266–399	TAGGG	295–299	CCCTA	308–312	5′
4	400–539	(TCAGGG)	439–444	(CCTA)	453–456	—
5	540–676	(TGGG)	583–586	CCCTA	595–599	3′
6	677–808	TAGGG	716–720	CCCTA	730–734	3′

direction of strand transfer if they were involved in initiation of recombination as described above and by Bergold *et al.* (1983).

b. Structural and Functional Aspects of Conserved Sequences in Terminal Inverted Repeats. The 140-nucleotide *Hae*III inverted repeats found at *Physarum* rDNA termini all have a pyrimidine-rich hexanucleotide sequence containing at least four T residues (Bergold *et al.*, 1983). This sequence is seen at nucleotides 46–51 in the terminal *Hae*III repeat presented in Table I. In most repeats this T-rich sequence is followed a few nucleotides downstream by the nucleotides TGTCPu. This arrangement is highly reminiscent of the sequence at the termination point of the *Physarum* rRNA transcription unit (Kukita *et al.*, 1981). This termination sequence is located at the 3′ end of the 26 S rRNA gene. Downstream from the transcription termination site there exists a potential hairpin stem and loop configuration of 33 nucleotides (Kukita *et al.*, 1981). This inverted repeat sequence bears a significant homology to nucleotides 59–94 in the rDNA terminal repeat sequence presented in Table I, which are slightly downstream from the terminator-like sequence at nucleotides 46–57 (presented in Table III). Table III shows that the *Physarum* rDNA terminator sequence of TTTTTTTTGTCGG has a counterpart sequence at rRNA termination sites in *Xenopus* and yeast. In particular, the hexanucleotide TGTCNG is present in all of these eukaryotes. In *Physarum* and *Xenopus* this sequence is TGTCGG and immediately follows a run of T residues. In yeast the hexanucleotide is TGTCTG and precedes a run of T residues (Table III). At the rDNA ends, four of the seven *Hae*III repeat sequences possess the tetranucleotide TGTC following a T-rich sequence. Five of the seven repeats possess the approximate TGT, and six of seven the appropriate TG.

The terminator-like sequences repeated at the *Physarum* rDNA ends can each be part of a stem and loop configuration base paired with an A-rich sequence seen at nucleotides 19–28 in the *Hae*III repeat sequence presented here (Bergold

Table III

Sequences Resembling rRNA Transcription Terminators Repeated near *Physarum* rDNA Ends

Source	Sequence[a]
Physarum polycephalum	T T T T T T G T C G G ↓[b]
Xenopus laevis	C T T T T G T↓ C G G[c]
Saccharomyces cerevisiae	T G T C T G A T T T G T T T T T T A T ↓[d]
Escherichia coli rrn B operon	T T T C G T T T T A T C T G[e]
Physarum rDNA terminal repeats	
1	T T T T T C C A T G T C G T C[f]
2	C C T T T T T G A A T C G
3	T C T T A C T T A G C T A G C
4	T G T T C C C C T G T C G G G
5	T T T T C C A A T G T C A T C
6	T A T T T A A C A A A T G T A C G G
	T C T T A T A T T G T C A T G

[a] Arrows (↓) denote transcription termination sites where known.

[b] Kukita et al. (1981).

[c] Sollner-Webb and Reeder (1979); Bakken et al. (1982).

[d] Veldman et al. (1980).

[e] Brosius et al. (1981).

[f] Bergold et al. (1983).

et al., 1983). This inverted repeat configuration resembles the palindromic terminator sequences found in rRNA operons of *E. coli* (Young, 1979; Sekiya *et al.*, 1980; Brosius *et al.*, 1981) and in the λ 6 S terminator (Brosius *et al.*, 1981). However, in these prokaryotes the conserved hexanucleotide following a run of T residues is of the general sequence PyPuTCTG, and in no known case does it match the sequences of the *Physarum* rRNA terminator.

Nowhere in the telomeric rDNA repeats are sequences found that resemble sequences in the 26 S rRNA gene or any other rRNA-coding region. [The genes for 5 S and tRNAs have not been detected on the *Physarum* rDNA molecule (Hall and Braun, 1977).] It is likely that the rDNA telomeric sequences are not transcribed since labeled rRNA transcripts from isolated nuclei do not hybridize with the rDNA terminal restriction fragments (Sun *et al.*, 1979). Why then do the telomeric inverted repeats possess apparent termination signals? It is conceivable that the repeats at the ends of the rDNA molecule have evolved to possess and conserve termination signals to certify the selective advantage of not allowing transcription downstream from these repeats. Transcription through the rDNA ends and downstream could only occur if the rDNA were at some point integrated with other DNA sequences. Such integration with chromosomal DNA would imply that the adjacent sequences are not transcribed by RNA polymerase I (pol I), and it would therefore be advantageous to insure that pol I transcription of the rRNA genes would not extend to adjacent sequences.

It is also conceivable that requirements for similar chromatin structural domains could determine certain homologies between transcription terminator regions and DNA telomeres. For example, it may be advantageous to have nucleosomes (or related chromosomal subunits) phased in a particular order at both termination sites and DNA telomeres. In both cases, the phasing could be related to interaction of exposed DNA sequences with noncontiguous DNA molecules or with the nuclear matrix. Sequences at terminator or terminator-like regions could be involved in nucleosome phasing. This is suggested by the observations that poly(T) sequences are present upstream from the TGTC sequence in *Physarum* (Kukita *et al.*, 1981) and *Xenopus* (Sollner-Webb and Reeder, 1979), but downstream from it in yeast (Veldman *et al.*, 1980). The poly(T) sequences could have long-range effects on nucleosome phasing—by, for example, being energetically less favorably wound into a core configuration—and thus help to expose DNA sequences, either upstream or downstream, involved in more specific interactions with chromosomal components.

The notion that a particular DNA three-dimensional configuration is required at telomeres gains support from experiments of Walmsley *et al.* (1983). These workers report hybridization of the alternating copolymer poly(dGdT·dCdA) to restriction fragments corresponding to yeast telomeres. In addition, tracts of this copolymer are added to the ends of the extrachromosomal rDNA molecules of *Tetrahymena* when cloned in yeast. Poly(dGdT·dCdA) can exist *in vitro* as Z-form, left-hand helical DNA—as opposed to prevalent B-form, right-hand helical DNA (Zimmerman, 1982). Thus, the possibility exists that certain copolymer stretches at telomeres function through their effects on DNA helical structure. Again, these effects could be long-range with respect to nucleosome phasing or interactions with the nuclear matrix and other chromosomal components.

c. Proteins at rDNA Termini. The rDNA molecule of *Physarum*, as isolated directly from nucleoli with care taken to minimize protease activity, possesses a protein complex bound very tightly near the termini (Cheung et al., 1981). This complex has been detected by treating rDNA with the ^{125}I-labeled Bolton-Hunter reagent, which selectively reacts with N-terminal or lysine ε-amino groups.

When either rDNA or isolated rDNA termini labeled this way are digested with DNase I, two discrete protein bands of 5,000 and 13,000 molecular weight are detected (the latter is most likely two bands) as well as larger bands of 40,000–45,000 molecular weight that may be aggregation products of protein and DNA. The structural relationship of these protein bands is not known. They may be subunits of a larger protein–DNA complex or they may be structurally and functionally distinct. Proteins of the same molecular weights are detected on rDNA whether or not nucleoli are treated with phenol immediately after isolation. Thus, proteolysis is probably not a factor in these experiments although this cannot be ruled out entirely. The protein at rDNA ends is not removed by treatment of the rDNA with SDS, phenol, guanidinium chloride, CsCl, formamide, or urea, but is removed by treatment with 0.1 M NaOH or proteinase K (Cheung et al., 1981). Specific activities of iodination alone do not allow calculation of the number of protein complexes per rDNA molecule. However, in additional experiments protein at termini could be visualized by electron microscopy after reacting rDNA with dinitrofluorobenzene followed by treatment with anti-[DNP-protein] antibody (Cheung et al., 1981). In some electron micrographs we could see more than one complex per terminus, suggesting multiple binding sites for proteins. The nature of the chemical bond between protein and rDNA has yet to be determined. It has been observed that proteins linked to termini of herpes simplex virus DNA segments are apparently not covalently bound even though they are SDS resistant (Wu et al., 1979). In the present case, resistance to various chemical treatments suggests a covalent linkage, and lability to NaOH would be consistent with a phosphoester linkage, although other bonds are possible.

In adenovirus a single bound protein of 55,000 molecular weight is linked directly to the 5′ terminus of each strand and may serve as a primer for initiation of replication (Challberg et al., 1980). This protein is present on a high percentage of adenovirus molecules and is attached through a phosphodiester linkage between the β-OH of a serine residue and the 5′-OH of deoxycytidine (Desiderio and Kelly, 1981). Several other eukaryotic viruses possess terminal proteins that serve a primer function (see Wimmer, 1982, for review). It is unlikely that rDNA-bound protein in *Physarum* serves an identical primer function because rDNA replication initiates in the center of the molecule and proceeds toward the ends. The terminal protein of *Physarum* may, however, be involved in rDNA replication—possibly in initiation of recombination to effect strand transfer (Bergold *et al.*, 1983).

It is conceivable that covalent attachment of proteins occurs during formation of single-strand gaps. Several reports suggest that eukaryotic DNA type I topoisomerases function via formation of covalent DNA enzyme intermediates (Depew *et al.*, 1978; Been and Champoux, 1980; Tse *et al.*, 1980; Champoux, 1981). For example, cleavage of a DNA phosphodiester bond by *E. coli* topoisomerase I is accompanied by covalent linkage of the enzyme to the 5' phosphoryl group of the terminal nucleotide at the cleavage site (Tse *et al.*, 1980). It is attractive to speculate that the tightly bound protein detected at *Physarum* rDNA ends functions in similar fashion to catalyze formation of single-strand breaks. It may now be possible to test this hypothesis. The availability of cloned repeat units containing single-strand gap recognition sequences should allow a search for enzymes that specifically form nicks or gaps at the sequence CCCTA.

Blackburn and Chiou (1981) have reported that the tandemly repeated CCC-CAA sequence at *Tetrahymena* rDNA ends is packaged in chromosomal complexes different from standard nucleosomes. When macronuclear chromatin was digested with staphylococcal nuclease and the resulting DNA fragments separated by gel electrophoresis and probed with labeled CCCCAA repeats, a band pattern was obtained that differed from the characteristic nucleosome repeat ladder. The average protected subunit length was >300 bp instead of the standard 144 bp. It is possible that the rDNA terminal repeat is in this case complexed with proteins other than histones. It is also possible that the CCCCAA DNA sequence imposes constraints on packaging by histones that result in an unusual chromatin subunit structure. For example, the CCCCAA periodicity could induce a particular phasing of chromatin subunits with respect to the underlying DNA so that staphylococcal nuclease digestion yields a nonstandard pattern. The phasing of chromatin subunits could be instrumental in exposing specific DNA regions to the action of enzymes that function at telomeres. Phasing could also be instrumental in allowing particular inverted repeat sequences to form hairpin secondary structures. Packaging by nucleosomes imposes a negative supercoil on a given sequence that has a helix-destabilizing effect (Prunell *et al.*, 1979; Wang, 1982); thus, phasing on a regularly repeated sequence could significantly alter the energy requirements for changing the secondary structure configuration of that sequence. In the *Physarum* and *Dictyostelium* rDNA termini, the repeated segments of multiple T or C and T residues could have the effect of phasing nucleosomes because it has been demonstrated that nucleosomes form poorly on poly(dA)·poly(dT) tracts (Kunkel and Martinson, 1981). In *Physarum* rDNA termini such phasing oriented by pyrimidine-rich segments could significantly affect the transitions in secondary structure proposed for the complex, 140-bp, inverted repeat units (Johnson, 1980; Bergold *et al.*, 1983).

IV. MODELS FOR TELOMERE FUNCTION

In order to qualify as a telomere under the definition proposed in the Section I a DNA terminus must satisfy certain functional prerequisites. One requirement, as discussed previously, is that the ends of a linear molecule support complete replication of both strands of the DNA duplex. In addition, the ends should allow for regeneration by daughter molecules of the original, autonomous, linear configuration of the parent molecule. The ends of eukaryotic chromosomes must allow regeneration of the autonomous configuration necessary for independent segregation at meiosis and mitosis. Any functional consideration of chromosome telomeres should also account for the genetic stability of these ends relative to the enhanced ability of broken ends of chromosomes to recombine randomly (McClintock, 1941, 1942). There are almost certainly many additional telomeric functions of which we presently know very little. For example, it has been proposed that during meiosis homologous telomeres pair by moving along the nuclear membrane before formation of the synaptonemal complex (Gillies, 1975). Binding of telomeres to the nuclear membrane has been reported (Diaz and Lewis, 1975).

Several models for telomere function in replication have been based on features of DNA viruses that replicate as linear entities. Watson (1972) outlined a mechanism for replication of phage T7 DNA in *E. coli* that relies on catenation of incomplete duplexes by base pairing at complementary 3' tails. A structural requirement for this model—satisfied by linear T7 DNA—is that the molecule possesses redundant DNA termini. This particular structural feature is not a universal characteristic of self-replicating, linear DNA molecules in eukaryotes. For example, the ends of macronuclear DNA molecules in hypotrichous ciliates are inverted terminal repetitions rather than being redundant (Klobutcher *et al.*, 1981; Oka *et al.*, 1980).

Several viruses that replicate in eukaryotes also have ends that are good subjects for models of DNA terminal replication. These include parvoviruses (Astell *et al.*, 1979), vaccinia (Garon *et al.*, 1978; Wittek and Moss, 1980), and herpes viruses (cf. Roizman, 1979). In vaccinia the viral DNA termini possess an inverted repeat sequence (Garon *et al.*, 1978), and this terminal segment contains multiple tandem repeats of 70 bp (Wittek and Moss, 1980; Moss *et al.*, 1981; Barondy and Moss, 1982). No nicks or gaps have been detected in these repeats in isolated viral DNA. The ends of vaccinia DNA are in an apparent hairpin configuration in virions (Geshelin and Burns, 1974) but not in cells (Pogo, 1977). In herpes simplex virus (HSV) both U_L and U_S segments of the viral DNA are flanked by inverted repeat sequences (see Roizman, 1979). Single-strand gaps, possibly selectively located, have been detected in HSV-1 DNA (Frenkel and Roizman, 1972; Wadsworth *et al.*, 1976). In addition to terminal inverted repeats and single-strand breaks, both HSV segments are reported to possess a terminal protein complex bound very tightly but not covalently (Wu *et al.*, 1979). The resemblance, noted in these several characteristics, between HSV and *Physarum* rDNA termini may reflect similarities in replicating ends of these linear DNAs. The protein complex at termini of adenovirus 2 DNA is evidently involved in priming DNA synthesis, which in this case originates at the ends (Rekosh *et al.*, 1977; Challberg *et al.*, 1980). This means of beginning replication thus far has no known counterpart in cellular DNA.

Cavalier-Smith (1974) first proposed a model in which terminal inverted repeat sequences at chromosome telomeres were given a role in completion of replication. The model hypothesizes hairpin formation at incomplete ends, followed by DNA synthesis primed at the hairpins, and finally by ligation and single-strand cleavage. Bateman (1975) has simplified this model by proposing that both ends of a chromosome DNA duplex are hairpins and that the chromosome is essentially a single, continuous DNA strand. The notion of a replicative role for hairpin sequences at termini received some support from observations of Astell and co-workers (1979), who sequenced 3' ends of DNA from four rodent parvoviruses. The parvoviruses possess linear, single-strand DNA genomes. Astell *et al.* (1979) found that the 3' ends of each of these molecules comprised a 115- or 116-nucleotide inverted repeat that forms a Y-shaped hairpin secondary structure. They proposed a terminal completion mechanism that involves DNA synthesis, primed by hairpin formation and accompanied by strand displacement. As in the Cavalier-Smith (1974) and Bateman (1975) models, specific (or at least localized) nick formation and ligation are necessary aspects for completion.

A model for completion of ends of a linear mitochondrial DNA molecule in *Tetrahymena* has been presented by Goldback *et al.* (1979). In this case recombination occurs at multiple, tandemly repeated terminal sequences and involves annealing of an unfinished 3' end with an internally displaced DNA single strand produced at a nick or gap. This mechanism does not require inverted repeats at termini, nor does it necessarily regenerate a hairpin terminus. It may therefore be most useful in understanding replication of ends that are flush or possess protruding single strands, such as those of macronuclear DNA in hypotrichs. It is interesting with regard to this model that displaced single DNA strands of 2–19 nucleotides have been found associated with single-strand breaks in the circular DNA genome of cauliflower mosaic virus (Franck *et al.*, 1980). A similar model for completion that does not require inverted repeats or hairpins has been proposed by Heumann (1976). This model also proposes pairing of terminal sequences with internal repeats. A model has been put forward to explain observations regarding the growth of chromosome ends in multiplying trypanosomes (Bernards *et al.*, 1983). Certain genes for the variant surface glycoproteins of these organisms are located near a discontinuity in the DNA, presumably a chromosome end. Sequences of varying length 3' to the glycoprotein genes and near the DNA ends are termed "barren regions" since they are devoid of known restriction sites (Van der Ploeg *et al.*, 1982). Thus far, these regions have not been cloned or sequenced. In multiplying trypanosomes the DNA fragments containing these apparent telomeres increase in length at a rate of about 10 bp per division. The proposed explanation for this increase assumes replication of an intact hairpin followed by nicking, DNA unpairing, and filling in of gaps. It is further proposed that eventually these lengthened ends will be shortened through recombination. The details of any such recombination process are not known, but exchange of sequences at telomeres may be involved in selective activation of different variant surface glycoprotein genes (Bernards *et al.*, 1983).

Terminal completion models based on synthesis primed at hairpin structures are capable of explaining autonomous replication of linear DNA without resorting to catenation or telomeric fusion to form "superchromosomes" (Cavalier-Smith, 1974). However, several characteristics of abnormal mammalian chromosomes are best explained by considering telomere fusion as part of the replicative process. For example, Benn (1976) found that more than 2% of telomeres in senescing human primary cell lines are fused end-to-end in the form of dicentric chromosomes. Dutrillaux *et al.* (1978) have reported that nearly 1% of cultured cells from a patient with Thiberge-Weissenbach syndrome are fused at telomeres. Dancis and Holmquist (1979) have proposed a model for terminal completion that relies on recombination at hairpin sequences at termini to temporarily eliminate chromosome ends. This terminal fusion is proposed to occur before replication; after replication, hairpin ends are regenerated by nicking, unpairing, and ligation. This model assumes that any pair of telomeres can be involved in a particular fusion and consequently, that all telomeres are of similar sequence.

The terminal completion model of Dancis and Holmquist (1979) may be unnecessarily rigid in requiring recombination at telomeres to occur prior to replication. We have recently proposed a model for telomeric recombination and terminal completion that is based on both sequencing and electron microscopic observation of *Physarum* rDNA ends (Bergold *et al.*, 1983). In this model, shown in Fig. 3, recombination is initiated with strand invasion by a single-strand hairpin loop into a closely homologous DNA duplex. Terminal completion is effected by strand transfer and subsequent repair and ligation. The model assumes the existence in cells of enzymes making specific, single-strand nicks or gaps. It also assumes a mechanism for allowing reversible transitions between fully duplex and hairpin secondary structures. A somewhat different model im-

I. Hairpin Formation at Termini

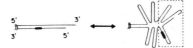

II. Synopsis

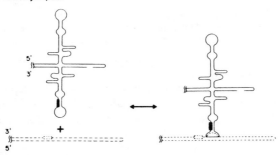

III. Specific Gapping and Strand Transfer

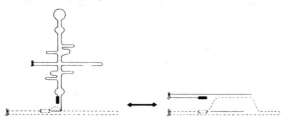

IV. Strand Scission and Separation

V. Internal DNA Synthesis and Repair

Fig. 3. A model for the mechanism of completion of replication at a linear rDNA end. (I) An incompletely replicated DNA strand (solid lines) may undergo secondary structure formation at the terminus. (II) The single-stranded regions at the end of the hairpin loops invade a homologous sequence on another DNA molecule (dotted lines). (III) Specific gapping at CCCTA sequences (indicated by a dark box) near the hairpin loop initiates a strand transfer event, giving rise to recombination intermediates. (IV) Recombination intermediates are resolved by strand scission, allowing the DNA molecules to separate. See Fig. 4 for visualization of structures consistent with this step. Letters refer to those in Fig. 4. (V) Internal DNA synthesis and repair of any mismatches complete the new rDNA terminus and regenerate the homologous sequence. (Courtesy of Bergold *et al.*, 1983. Copyright © 1983 by M.I.T. Press.)

plicating DNA cruciform structures in telomere replication has been proposed by McFadden and Morgan (1982). In postulating fusion at telomeres the model of Bergold *et al.* (1983) is similar to that of Dancis and Holmquist (1979). In the Bergold model, however, recombination can be initiated at any of a number of tandemly repeated hairpins that need not be at the very end of the chromosome in order to effect completion. This recombination may occur after the bulk of replication. Such recombination could result in the formation of full, dicentric chromosomes as observed by others (Benn, 1976; Dutrillaux *et al.*, 1978) if mechanisms for strand scission and separation of the fused chromosomes are faulty. With regard to the *Physarum* rDNA ends, a recombinational mechanism for completion helps explain both the length and sequence heterogeneity observed in the region of terminal inverted repeats. Upon electron microscopic visualization of native rDNA molecules, a significant number of complex terminal structures are seen that are consistent with our model of recombination at telomeres. One such structure is shown in Fig. 4 with an interpretative drawing referring to the model in Fig. 3.

Recombination at the ends of the rDNA can also provide a mechanism for evolutionary stabilization of sequences among the array of extrachromosomal rDNA copies (cf. Dover, 1982) via processes of unequal crossing-over (Arnheim *et al.*, 1980) or gene conversion (Slightom *et al.*, 1980).

A segment of *Drosophila melanogaster* DNA containing multiple tandem repeats has been localized to all telomeres of polytene chromosomes by *in situ* hybridization (Rubin, 1978). It is not known whether this sequence constitutes the molecular end of the chromosome. Several clones of *D. melanogaster* DNA have been found to hybridize to telomere regions (Young *et al.*, 1983). Most of these clones do not possess the precise Rubin repeat although there are shorter regions of homology. Hybridization of one of these clones (λ T-A) closely follows the morphology of extreme ends of chromosomes 2L and 3L, which under particular spreading conditions consist of a tightly-constricted band and loose ruffle, respectively. These *Drosophila* telomeric sequences apparently do not contain short, gapped C + A repeats (Rubin, 1978; Young *et al.*, 1983). The λ T-A telomeric clone shares complex DNA sequences with regions of pericentric heterochromatin, suggesting some similar function for sequences at telomeres and sequences near centromeres. There is presently little or no evidence for association of sequences at telomeres and centromeres. Perhaps similarity in sequence may be related to the presence of multiple replication origins in these chromosome regions. Clustering of replication origins has been localized within a series of tandem repeats near yeast telomeres (Chan and Tye, 1983).

A comparison of the known sequences at DNA termini (see Table I) suffices to point out the risks of attempting to generalize about mechanisms of telomeric replication over a broad evolutionary range of organisms. It is clear that multiply repeated sequences are present at ends of all cellular DNAs thus far sequenced,

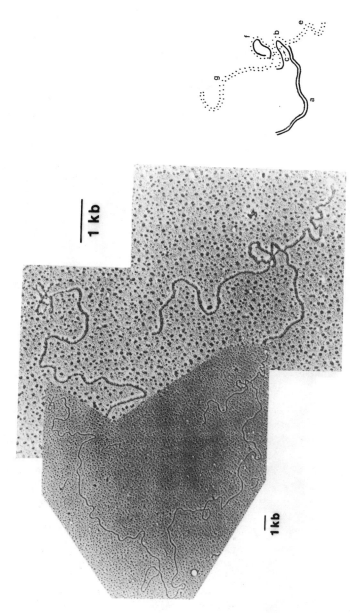

Fig. 4. Hairpin structures and possible recombination forms at *Physarum* rDNA telomeres. This is an electron micrograph of one 61-kb rDNA palindrome. The molecule was spread after isolation under conditions calculated to minimize partial denaturation. At one end multiple foldback structures are visible, as magnified at top right. At the other end a complex structure is seen that may represent recombination between the rDNA telomere and a fragmented segment from chromosomal DNA. Letters on the drawing at right refer to DNA segments similarly lettered in Fig. 3(IV) of the model for recombination presented by Bergold *et al.* (1983). ═, rDNA; ∶∶∶, noncontiguous molecule of recombining DNA. (Courtesy of Bergold *et al.*, 1983. Copyright © 1983 by M.I.T. Press.)

but the configurations of these repeats differ considerably in different cells. For example, hairpin repeats have been sequenced for *Physarum* rDNA (Bergold *et al.*, 1983) and are also likely in *Tetrahymena* (Blackburn and Gall, 1978) and *Dictyostelium* (Emery and Weiner, 1981). Yet the C and A repeat at ends of macronuclear DNA in hypotrichous ciliates (Klobutcher *et al.*, 1981), despite similarities to the *Tetrahymena* rDNA repeat (Blackburn and Gall, 1978), is unlikely to be present as a hairpin. The presence of single-strand discontinuities at telomeres seems to be a unifying characteristic of several disparate terminal repeat sequences. The finding of such nicks or gaps in *Tetrahymena* (Blackburn and Gall, 1978), *Physarum* (Johnson, 1980; Bergold *et al.*, 1983), and *Dictyostelium* (Emery and Weiner, 1981) and the observation that these breaks at *Tetrahymena* rDNA ends are maintained in yeast (Szostak and Blackburn, 1982) strongly suggest the existence of a similar gap-generating enzyme apparatus in lower eukaryotes, if not in higher organisms. The complexity of the terminal hairpin repeat in *Physarum* rDNA (Bergold *et al.*, 1983) may presage a burgeoning of telomeric functions in higher eukaryotes, extending beyond occupation with DNA replication. We shall be able to understand these functions better after telomeres of several mammalian cells have been cloned and sequenced. We shall at that point also better understand processes of telomere replication, including the replication-dependent generation of certain chromosome abnormalities.

REFERENCES

Affolter, H.-U., and Braun, R. (1978). Ribosomal DNA in spores of *Physarum polycephalum*. *Biochim. Biophys. Acta* **519**, 118–124.

Arnheim, N., Seperack, P., Banerji, J., Lang, R. B., Miesfeld, R., and Marcu, K. B. (1980). Mouse rDNA non-transcribed spacer sequences are found flanking immunoglobulin C_H genes and elsewhere throughout the genome. *Cell* **22**, 179–185.

Astell, C. R., Smith, M., Chou, M. B., and Ward, D. C. (1979). Structure of 3'-hairpin termini of 4 rodent parvovirus genomes: Nucleotide sequence homology at origins of replication. *Cell* **17**, 691–703.

Bakken, A., Morgan, G., Sollner-Webb, B., Roan, J., Busby, S., and Reeder, R. H. (1982). Mapping of transcription initiation and termination signals on *Xenopus leavis* ribosomal DNA. *Proc. Natl. Acad. Sci. U.S.A.* **79**, 56–60.

Barondy, B. M., and Moss, B. (1982). Sequence homologies of diverse length tandem repetition near ends of vaccinia virus genome suggest unequal crossing over. *Nucleic Acids Res.* **10**, 5673–5680.

Bateman, A. J. (1975). Simplification of palindromic telomere theory. *Nature (London)* **253**, 379.

Been, M. D., and Champoux, J. J. (1980). Breakage of single-stranded DNA by rat liver nicking-closing enzyme with the formation of a DNA-enzyme complex. *Nucleic Acids Res.* **8**, 6129–6142.

Benn, P. A. (1976). Specific chromosome aberrations in senescent fibroblast cell lines derived from human embryos. *Am. J. Hum. Genet.* **28**, 465–473.

Bergold, P., Campbell, G. R., Littau, V. C., and Johnson, E. M. (1983). Sequence and hairpin structure of an inverted repeat series at termini of the *Physarum* extrachromosomal rDNA molecule. *Cell* **32,** 1287–1299.

Bernards, A., Michels, P. A. M., Lincke, C. R., and Borst, P. (1983). Growth of chromosome ends in multiplying trypanosomes. *Nature* **303,** 592–597.

Blackburn, E. H., and Chiou, S.-S. (1981). Non-nucleosomal packaging of a tandemly-repeated DNA sequence at termini of extrachromosomal DNA coding for rRNA in *Tetrahymena. Proc. Natl. Acad. Sci. U.S.A.* **78,** 2263–2267.

Blackburn, E. H., and Gall, J. G. (1978). A tandemly-repeated sequence at termini of the extrachromosomal ribosomal genes in *Tetrahymena. J. Mol. Biol.* **120,** 33–53.

Blattner, F., Williams, B. G., Blechl, A. E., Denniston-Thompson, K., Faber, H. F., Furlong, L.-A., Grunwald, D. J., Kiefer, D. O., Moore, D. D., Schumm, J. W., Sheldon, E. L., and Smithies, O. (1977). Charon phages: Safer derivatives of bacteriophage lambda for DNA cloning. *Science* **196,** 161–169.

Boswell, R. E., Klobutcher, L. A. and Prescott, D. M. (1982). Inverted terminal repeats are added to genes during macronuclear development in *Oxytricha nova. Proc. Natl. Acad. Sci. U.S.A.* **79,** 3255–3259.

Bradbury, E. M., Matthews, H. R., McNaughton, J., and Molgaard, H. (1973). Subnuclear components of *Physarum polycephalum. Biochim. Biophys. Acta* **335,** 19–29.

Brosius, J., Dull, T. J., Sleeter, D. D., and Noller, H. F. (1981). Gene organization and primary structure of a ribosomal RNA operon from *Escherichia coli. J. Mol. Biol.* **148,** 107–127.

Campbell, G. R., Littau, V. C., Melera, P., Allfrey, V. G., and Johnson, E. M. (1979). Unique sequence arrangement of ribosomal genes in the palindromic rDNA molecules of *Physarum polycephalum. Nucleic Acids Res.* **6,** 1433–1447.

Cavalier-Smith, T. (1974). Palindromic base sequences and replication of eucaryote chromosome ends. *Nature (London)* **250,** 467–470.

Cech, T. R., and Brehm, S. L. (1981). Replication of the extrachromosomal ribosomal RNA genes of *Tetrahymena thermophila. Nucleic Acids Res.* **9,** 3531–3543.

Challberg, M. D., Desiderio, S. V., and Kelly, T. J., Jr. (1980). Adenovirus DNA replication in vitro: Characterization of a protein covalently linked to nascent DNA strands. *Proc. Natl. Acad. Sci. U.S.A.* **77,** 5105–5109.

Champoux, J. J. (1981). DNA is linked to the rat liver DNA nicking-closing enzyme by a phosphodiester band to tryosine. *J. Biol. Chem.* **256,** 4805–4809.

Chan, C. S. M., and Tye, B.-K. (1983). Organization of DNA sequences and replication origins at yeast telomeres. *Cell* **33,** 563–573.

Cheung, M. K., Drivas, D. T., Littau, V. C., and Johnson, E. M. (1981). Protein tightly bound near the termini of the *Physarum* extrachromosomal rDNA palindrome. *J. Cell Biol.* **91,** 309–314.

Cockburn, A. F., Taylor, W. C., and Firtel, R. A. (1978). *Dictyostelium* rDNA consists of nonchromosomal palindromic dimers containing 5S and 36S coding regions. *Chromosoma* **70,** 19–29.

Cunningham, R. P., Shibata, T., Das Gupta, C., and Radding, C. M. (1979). Homologous pairing in genetic recombination: Single strands induce recA protein to unwind duplex DNA. *Nature (London)* **281,** 191–195.

Dancis, B. M., and Holmquist, G. P. (1979). Telomere replication and fusion in eucaryotes. *J. Theor. Biol.* **78,** 211–224.

Dani, G. M., and Zakian, V. A. (1983). Mitotic and meiotic stability of linear plasmids in yeast. *Proc. Natl. Acad. Sci. U.S.A.* **80,** 3406–3410.

Dawson, D., and Herrick, G. (1982). Micronuclear DNA sequences of *Oxytricha fallax* homologous to the macronuclear inverted terminal repeat. *Nucleic Acids Res.* **10,** 2911–2924.

Depew, R. E., Lin, L. R., and Wang, J. C. (1978). Interaction between DNA and *Escherichia coli* protein ω. Formation of a complex between single-stranded DNA and ω protein. *J. Biol. Chem.* **253,** 511–518.

Desiderio, S. V., and Kelly, T. J. (1981). Structure of the linkage between adenovirus DNA and the 55,000 molecular weight terminal protein. *J. Mol. Biol.* **145**, 319–337.

Diaz, D., and Lewis, K. R. (1975). Interphase chromosome arrangement in *Anopheles atroparvus*. *Chromosoma* **52**, 27–35.

Dover, G. (1982). Molecular drive: A cohesive mode of species evolution. *Nature (London)* **299**, 111–117.

Dutrillaux, B., Aurias, A., Couturier, J., Croquette, M. F., and Viegas-Pequiguot, E. (1978). Human somatic chromosome chains and rings—preliminary note on end to end fusions. *Cytogenet. Cell Genet.* **20**, 70–77.

Emery, H. S., and Weiner, A. M. (1981). An irregular satellite sequence is found at the termini of the linear extrachromosomal rDNA in *Dictyostelium discoideium*. *Cell* **26**, 411–419.

Engberg, J., Andersson, P., Leick, V., and Collins, J. (1976). Free ribosomal DNA molecules from *Tetrahymena pyriformis* GL are giant palindromes. *J. Mol. Biol.* **104**, 455–470.

Franck, A., Guilley, H., Jonard, K., and Hirth, L. (1980). Nucleotide sequence of cauliflower mosaic virus DNA. *Cell* **21**, 285–294.

Frankel, G., Cockburn, A. F., Kindle, K. L., and Firtel, R. A. (1977). Organization of the ribosomal RNA genes of *Dictyostelium discoideium:* Mapping of the transcribed region. *J. Mol. Biol.* **109**, 539–558.

Frenkel, N., and Roizman, B. (1972). Separation of the herpesvirus deoxyribonucleic acid on sedimentation in alkaline gradients. *J. Virol.* **10**, 565–572.

Garon, C. F., Barbosa, E., and Moss, B. (1978). Visualization of an inverted terminal repetition in Vaccinia virus DNA. *Proc. Natl. Acad. Sci. U.S.A.* **75**, 4863–4867.

Geshelin, P., and Burns, K. I. (1974). Characterization of the naturally occurring crosslinks in Vaccinia virus DNA. *J. Mol. Biol.* **88**, 785–796.

Gillies, C. B. (1975). Synaptonemal complex and chromosome structure. *Annu. Rev. Genet.* **9**, 91–105.

Goldback, R. W., Bollen-de Boer, J. E., Van Bruggen, E. F. J., and Borst, P. (1979). Replication of the linear mitochondrial DNA of *Tetrahymena pyriformis*. *Biochim. Biophys. Acta* **562**, 400–417.

Gorovsky, M. A. (1980). Genome organization and reorganization in *Tetrahymena*. *Annu. Rev. Genet.* **14**, 203–239.

Gubler, U., Wyler, T., and Braun, R. (1979). The gene for 26S rRNA in *Physarum* contains two insertions. *FEBS Lett.* **100**, 347–350.

Hall, L., and Braun, R. (1977). The organization of genes for transfer RNA and ribosomal RNA in amoebae and plasmodia of *Physarum polycephalum*. *Eur. J. Biochem.* **76**, 165–174.

Hardman, N., Jack, P. L., Brown, A. J. P., and McLachlan, A. (1979). Characterization of ribosomal satellite in total nuclear DNA from *Physarum polycephalum*. *Biochim. Biophys. Acta* **562**, 365–376.

Heumann, J. M. (1976). A model for replication at the ends of linear chromosomes. *Nucleic Acids Res.* **3**, 3167–3171.

Johnson, E. M. (1980). A family of inverted repeat sequences and specific single-strand gaps at the termini of the *Physarum* rDNA palindrome. *Cell* **22**, 875–886.

Johnson, E. M., Allfrey, V. G., Bradbury, E. M., and Matthews, H. R. (1978). Altered nucleosome structure containing DNA sequences complementary to 19S and 26S ribosomal RNA in *Physarum polycephalum*. *Proc. Natl. Acad. Sci. U.S.A.* **75**, 1116–1120.

Johnson, E. M., Campbell, G. R., and Allfrey, V. G. (1979). Different nucleosome structures on transcribing and non-transcribing ribosomal gene sequences. *Science* **206**, 1192–1194.

Kaine, B. P., and Spear, B. B. (1982). Nucleotide sequence of a macronuclear gene for actin in *Oxytricha fallax*. *Nature (London)* **295**, 430–432.

Karrer, K. M., and Gall, J. G. (1976). The macronuclear rDNA of *Tetrahymena pyriformis* is a palindrome. *J. Mol. Biol.* **104**, 421–453.

Katzen, A. L., Cann, G. M., and Blackburn, E. H. (1981). Sequence-specific fragmentation of macronuclear DNA in a holotrichous ciliate. *Cell* **24**, 313–320.

Kavenoff, R., and Zimm, B. H. (1973). Chromosome sized DNA molecules from *Drosophila. Chromosoma* **41**, 1–28.

King, B. O., and Yao, M.-C. (1982). Tandemly repeated hexanucleotide at *Tetrahymena* rDNA free end is generated from a single copy during development. *Cell* **31**, 177–182.

Klobutcher, L. A., Swanton, M. T., Donini, P., and Prescott, D. M. (1981). All gene-sized DNA molecules in four species of hypotrichs have the same terminal sequence and an unusual 3′ terminus. *Proc. Natl. Acad. Sci. U.S.A.* **78**, 3015–3019.

Kukita, T., Sakaki, Y., Nomiyama, H., Otsuka, T., Kuhara, S., and Takagi, Y. (1981). Structure around the 3′ terminus of the 26S ribosomal RNA gene of *Physarum polycephalum. Gene* **16**, 309–315.

Kunkel, G. R., and Martinson, H. G. (1981). Nucleosomes will not form on double-stranded RNA or over poly(dA)·poly(dT) tracts in recombinant DNA. *Nucleic Acids Res.* **9**, 6869–6888.

Lauer, G. D., Roberts, T. M., and Klotz, L. C. (1977). Determination of the nuclear DNA content of *S. cerevisiae* and implications for the organization of DNA in yeast chromosomes. *J. Mol. Biol.* **113**, 507–526.

Lawn, R. M., Heumann, J. M., Herrick, G., and Prescott, D. M. (1978). Gene-size DNA molecules in *Oxytricha. Cold Spring Harbor Symp. Quant. Biol.* **42**, 483–492.

McClintock, B. (1941). The stability of broken ends of chromosomes in *Zea mays. Genetics* **26**, 234–282.

McClintock, B. (1942). The fusion of broken ends of chromosomes following nuclear fusion. *Proc. Natl. Acad. Sci. U.S.A.* **28**, 458–463.

McFadden, G., and Morgan, A. R. (1982). DNA cruciform structures: Implications for telomere replication in eucaryotes and instability of long palindromic DNA sequences in procaryotes. *J. Theor. Biol.* **97**, 343–349.

Maizels, N. (1976). *Dictyostelium* 17S, 25S and 5S rDNAs lie within a 38,000 base-pair repeated unit. *Cell* **9**, 431–438.

Maxam, A. W., and Gilbert, W. (1980). Sequencing end-labeled DNA with base-specific chemical cleavages. *In* "Methods in Enzymology" (L. Grossman and K. Moldave, eds.), Vol. 65, pp. 499–560. Academic Press, New York.

Mohberg, J., Babcock, K. L., Haugli, F. B., and Rusch, H. P. (1973). Nuclear DNA content and chromosome number in the myxomycete *Physarum polycephalum. Dev. Biol.* **34**, 228–245.

Molgaard, H. V., Matthews, H. R., and Bradbury, E. M. (1976). Organization of genes for ribosomal RNA in *Physarum polycephalum. Eur. J. Biochem.* **68**, 541–549.

Moss, B., Winters, E., and Cooper, N. (1981). Instability and reiteration of DNA sequences within the Vaccinia virus genome. *Proc. Natl. Acad. Sci. U.S.A.* **78**, 6354–6358.

Murray, A., and Szostak, J. W. (1983). Construction of artificial chromosomes in yeast. *Nature* **305**, 189–193.

Oka, Y., Shiota, S., Nakai, S., Nishida, Y., and Okubo, S. (1980). Inverted terminal repeated sequence in the macronuclear DNA of *Stylonychia pustulata. Gene* **10**, 301–306.

Pan, W.-C., and Blackburn, E. H. (1981). Single extrachromosomal ribosomal RNA gene copies synthesized during amplification of the rDNA in *Tetrahymena. Cell* **23**, 459–466 .

Petes, T. D., Byers, B., and Fangman, W. L. (1973). Size and structure of yeast chromosomal DNA. *Proc. Natl. Acad. Sci. U.S.A.* **70**, 3072–3076.

Pluta, A. F., and Spear, B. B. (1981). The terminal organization of macronuclear DNA in *Oxytricha fallax. J. Cell Biol.* **91**, 138a.

Pogo, B. G. T. (1977). Elimination of naturally-occurring crosslinks in Vaccinia virus DNA after viral penetration into cells. *Proc. Natl. Acad. Sci. U.S.A.* **74**, 1139–1142.

Preer, J. R., and Preer, L. B. (1979). The size of macronuclear DNA and its relationship to models for maintaining genic balance. *J. Protozool.* **26**, 14–18.

Prescott, D. M., Murti, K. G., and Bostock, C. J. (1973). Genetic apparatus of *Stylonychia* sp. *Nature (London)* **242**, 597–600.

Prior, C. P., Cantor, C. R., Johnson, E. M., and Allfrey V. G. (1980). Incorporation of pyrene-labeled histone H3 into *Physarum* chromatin: A system for studying changes in nucleosomes assembled in vivo. *Cell* **20**, 597–608.

Prior, C. P., Cantor, C. R., Johnson, E. M., Littau, V. C. and Allfrey, V. G. (1983). Reversible changes in nucleosome structure and histone H3 accessibility in transcriptionally active and inactive states of rDNA chromatin. *Cell* **34**, 1033–1042.

Prunell, A., Kornberg, R. D., Lutter, L., Klug, A., Levitt, M., and Crick, F. H. C. (1979). Periodicity of deoxyribonuclease I digestion of chromatin. *Science* **204**, 855–858.

Ratzkin, B., and Carbon, J. (1977). Functional expression of cloned yeast DNA in *Escherichia coli*. *Proc. Natl. Acad. Sci. U.S.A.* **74**, 487–491.

Rekosh, D. M. K., Russell, W. C., Bellet, A. J. D., and Robinson, A. J. (1977). Identification of a protein linked to the ends of adenovirus DNA. *Cell* **11**, 283–295.

Roizman, B. (1979). The structure and isomerization of herpes simplex virus genome. *Cell* **16**, 481–494.

Rubin, G. M. (1978). Isolation of a telomeric DNA sequence from *Drosophila melanogaster*. *Cold Spring Harbor Symp. Quant. Biol.* **42**, 1121–1135.

Seebeck, T., Stalder, J., and Braun, R. (1979). Isolation of a minichromosome containing the ribosomal genes from *Physarum polycephalum*. *Biochemistry* **18**, 484–490.

Sekiya, T., Mori, M., Takahashi, N., and Nishimura, S. (1980). Sequence of the distal tRNA, Asp gene and the transcription termination signal in the *Escherichia coli* ribosomal RNA operon rrnF (or G). *Nucleic Acids Res.* **8**, 3809–3827.

Slightom, J. L., Blechl, A. E., and Smithies, O. (1980). Human fetal $^G\gamma$- and $^A\gamma$-globin genes: Complete nucleotide sequences suggest that DNA can be exchanged between these duplicated genes. *Cell* **21**, 627–638.

Sollner-Webb, B., and Reeder, R. H. (1979). The nucleotide sequence of the initiation and termination sites for ribosomal RNA transcription sites for ribosomal RNA transcription in *X. leavis*. *Cell* **18**, 485–499.

Stinchcomb, D. T., Struhl, K., and Davis, R. W. (1979). Isolation and characterization of a yeast chromosomal replicator. *Nature* **282**, 39–43.

Sun, I.Y.-C., Johnson, E. M., and Allfrey, V. G. (1979). Initiation of transcription of ribosomal deoxyribonucleic acid sequences in isolated nuclei of *Physarum polycephalum*: Studies using nucleoside 5'-[γ-S]triphosphates and labeled precursors. *Biochemistry* **18**, 4572–4580.

Swanton, M. T., Heumann, J. M., and Prescott, D. M. (1980). Gene-sized DNA molecules of the macronuclei in three species of hypotrichs: Size distributions and absence of nicks. DNA of ciliated protozoa. VIII. *Chromosoma* **77**, 217–227.

Szostak, J. W., and Blackburn, E. H. (1982). Cloning yeast telomeres on linear plasmid vectors. *Cell* **29**, 245–255.

Taylor, J. H. (1958). Sister chromatid exchanges in tritium-labeled chromosomes. *Genetics* **43**, 515–529.

Tse, Y.-C., Kirkegaard, K., and Wang, J. C. (1980). Covalent bonds between protein and DNA. *J. Biol. Chem.* **255**, 5560–5565.

Van der Ploeg, L. H. T., Bernards, A., Rijsewijk, F. A. M., and Borst, P. (1982). Characterization of the DNA duplication–transposition that controls the expression of two genes for variant surface glycoproteins of *Trypanosoma brucei*. *Nucleic Acids Res.* **10**, 593–609.

Veldman, G. M., Kloopwijk, J., Jonge, P., Leer, R. J., and Planta, R. J. (1980). The transcription termination site of the ribosomal RNA operon in yeast. *Nucleic Acids Res.* **8**, 5179–5192.

Vogt, V. M., and Braun, R. (1976). Structure of ribosomal DNA in *Physarum polycephalum*. *J. Mol. Biol.* **106**, 567–587.

Vogt, V. M., and Braun, R. (1977). The replication of ribosomal DNA in *Physarum polycephalum*. *Eur. J. Biochem.* **80,** 557–466.

Wadsworth, S., Hayward, G. S., and Roizman, B. (1976). Anatomy of herpes simplex virus DNA. V. Terminally repetitive sequences. *J. Virol.* **17,** 503–512.

Walmsley, R. M., Szostak, J. W. and Petes, T. D. (1983). Is there left-handed DNA at the ends of yeast chromosomes? *Nature (London)* **302,** 84–86.

Wang, J. C. (1982). The path of DNA in the nucleosome. *Cell* **29,** 724–726.

Watson, J. D. (1972). The origin of concatameric T7 DNA. *Nature (London), New Biol.* **239,** 197–201.

Wilson, E. B. (1925), "The Cell in Development and Heredity," 3rd ed. Macmillan, New York.

Wimmer, E. (1982). Genome-linked protein of virus. *Cell* **28,** 199–201.

Wittek, R., and Moss, B. (1980). Tandem repeats within the inverted terminal repetition of vaccinia virus DNA. *Cell* **21,** 277–284.

Wolff, S. (1977). Sister chromatid exchange. *Annu. Rev. Genet.* **11,** 183–201.

Wolff, S., and Perry, P. (1975). Insights on chromosome structure from sister chromatid exchange ratios and the lack of both isolabelling and heterolabelling as determined by the FPG technique. *Exp. Cell Res.* **93,** 23–30.

Wu, A. M., Kahn, R., Das Gupta, C., and Radding, C. M. (1982). Formation of nascent heteroduplex structures by RecA protein and DNA. *Cell* **30,** 37–44.

Wu, M., Hyman, R. W., and Davidson, N. (1979). Electron microscopic mapping of proteins bound to herpes simplex virus DNA. *Nucleic Acids Res.* **6,** 3427–3441.

Yao, M.-C., and Gall, J. G. (1977). A single integrated gene for ribosomal RNA in a eucaryote, *Tetrahymena pyriformis*. *Cell* **12,** 121–132.

Yao, M.-C., and Gorovsky, M. A. (1974). Comparison of the sequences of macro- and micronuclear DNA of *Tetrahymena pyriformis*. *Chromosoma* **48,** 1–18.

Yao, M.-C., and Yao, C. H. (1981). Repeated hexanucleotide C-C-C-C-A-A is present near the ends of macronuclear DNA of *Tetrahymena*. *Proc. Natl. Acad. Sci. U.S.A.* **78,** 7436–7439.

Yao, M.-C., Blackburn, E. H., and Gall, J. G. (1981). Tandemly-repeated C-C-C-C-A-A hexanucleotide of *Tetrahymena* rDNA is present elsewhere in the genome and may be related to the alteration of the somatic genome. *J. Cell Biol.* **90,** 515–520.

Young, B. S., Pession, A., Traverse, K. L., French, C., and Pardue, M. L. (1983). Telomere regions in *Drosophila* share complex DNA sequences with pericentric heterochromatin. *Cell* **34,** 85–94.

Young, R. A. (1979). Transcription termination in the *Escherichia coli* ribosomal RNA operon rrnC. *J. Biol. Chem.* **254,** 12725–12731.

Zimmerman, S. B. (1982). The three-dimensional structure of DNA. *Annu. Rev. Biochem.* **51,** 395–427.

III
Overview

13

Recombinant DNA Approaches to Studying Control of Cell Proliferation: An Overview

RENATO BASERGA

Department of Pathology
and Fels Research Institute
Temple University School of Medicine
Philadelphia, Pennsylvania

I. INTRODUCTION

Recently, I have seen stated in the literature several times that we know very little about cell proliferation in animal cells. In fact, one statement said unequivocally that we know nothing about what controls cell proliferation. These statements stem in part from ignorance, and I mean ignorance not as an insult but in the correct sense of the word, that is, lack of knowledge. With the enormous amount of information accumulating in the past 30 years in the biomedical sciences it would be too much to ask from everybody a complete knowledge of all aspects of cell and molecular biology. It is therefore not surprising that many biologists are not aware that *something* is known about cell proliferation. Such

337

statements stem also in part from a comparison between the detailed knowledge we have of λ phage and the imperfect knowledge we have of cell proliferation. Clearly, when compared to what is known about certain simple biological systems, the field of animal cell reproduction is still based on fragmentary knowledge and fraught with ambiguities. However, having followed the field of cell proliferation for about 30 years I can take a more optimistic view. When I began to study the regulation of cell proliferation back in the early 1950s all that was known about cell proliferation were the morphological aspects, that is, mitosis and interphase. In biology books of that time mitosis received a relatively extensive treatment in which the morphological aspects of the mitotic process were described in detail, from prophase to anaphase and metaphase to telophase. The interphase received a cursory treatment since there was very little to say about interphase except that when cells were not in mitosis, they were in interphase. Considering those beginnings, one can see that some progress has been made in elucidating the biochemistry and the molecular biology of cell division.

The study of cell proliferation has indeed progressed *pari passu* with the expansion of classical biology into the areas of biochemistry, cell biology, and molecular biology. Specifically, a considerable amount of information has accumulated in four general areas: (1) growth factors, including nutrients and inhibitory factors; (2) mathematical models and kinetics of cell proliferation; (3) biochemical aspects of cell proliferation; and (4) the genes and gene products that are necessary and specific for cell proliferation.

Of these four topics I limit myself in this review to the last two, briefly indicating my opinions on mathematical models and kinetics of cell proliferation. Growth factors and inhibitory factors have already been discussed in detail (Baserga, 1981).

II. MATHEMATICAL MODELS AND KINETICS OF CELL PROLIFERATION

The cell cycle was discovered by Howard and Pelc in 1951. The concept that cell DNA replication is limited to a discrete period of the interphase preceding mitosis is a concept that today may seem trivial but was immensely valuable for many years. The cell reproductive cycle was subdivided into four phases: G_1, during which the cell prepares for DNA replication; S phase, during which the genetic material is replicated; G_2, during which the cell prepares for mitosis; and mitosis itself. It is important to remember that the terms G_1 and G_2 simply meant, when they were created in 1951, gap 1 and gap 2. Gap 1 was the gap between mitosis (recognizable by morphology) and S phase (recognizable by autoradiography). Gap 2 was the gap between S phase and mitosis. There was no implication in this terminology that G_1 and G_2 actually were essential. They simply were convenient notations to describe the physiological state of the cell. Clearly, cell reproduction can be reduced to two essential events, namely, the doubling of all cellular components and the organization of the mitotic apparatus so that these components can be distributed at mitosis in approximately equal parts to the daughter cells. As will be discussed later, doubling of DNA can be separated from the doubling of the other cell components, i.e., one can actually distinguish growth in size of the cells from cell DNA replication. But even if we were to reduce cell division to three rather than two fundamental processes, that is, growth in size, cell DNA replication, and mitosis, one should remember that it does not matter when the first two processes are carried out provided they are completed before mitosis. So it is not surprising that some cells do not have a G_1 and other cells do not have a G_2. However, some cells have G_1 periods and some have G_2 periods, i.e., gaps between mitosis and DNA synthesis and between DNA synthesis and mitosis. Furthermore and more importantly, there are cells in the animal body that have G_1 and G_2 periods. So these two periods, although not essential, are usually present in dividing cells both *in vitro* and *in vivo*. Thus, it would be a disadvantage to ignore them simply because they are absent in some cell lines.

Students in the field of cell proliferation often unnecessarily argue whether G_1 or G_0 exists. No biologist would maintain that G_0 and G_1 actually do exist just as no biochemist would insist that K_m and V_{max} do exist. G_0 and G_1, K_m and V_{max}, chromosomal maps, and the numbering of nucleotides are just convenient notations that are extremely useful to investigators in their respective fields.

These semantic discussions have generated many papers but very few data. And the same comment applies in some cases to kinetic models although these have greatly facilitated our study of cell proliferations in the past. Models that have been proposed to describe the growth of cell populations have been and still are useful, but one has to realize their limitations. The transition probability model proposed by Smith and Martin in 1974 is an example. Whether it is correct or not (whether there are two, one, or no transition probabilities), the model at best describes the statistical behavior of populations of cells but does not explain the cell cycle in biochemical or molecular terms. One cannot infer the molecular basis of cell proliferation on the basis of pure kinetic models.

The real objectives for students of cell proliferation are the study of the growth factors (including inhibitory factors) and the genes and gene products that regulate cell division. As this volume clearly shows, the time has come in which cell proliferation can be profitably studied at the biochemical or molecular level. It is reasonable to predict that mathematical models and kinetics of cell proliferation will become less and less useful in the future as more detailed questions about the molecules that regulate the division of cells are answered.

III. BIOCHEMISTRY OF CELL PROLIFERATION

In the biochemistry and molecular biology of cell proliferation one is naturally looking for all those genes that are involved in the regulation of cell division, but more specifically, one always hopes to identify the gene that in fact controls whether a cell should enter the cycle. Here, we must first define what we mean by control of cell proliferation. Frequently, we witness the announcement that a new cellular component that "controls" cell proliferation has been discovered. The list is endless: membrane glycoproteins and glycolipids, cell size, nuclear size, cAMP, cyclic GMP, rRNA synthesis and accumulation, Ca^{2+}, Mg^{2+}, deoxynucleotide pool, levels of thymidine kinase, DNA polymerase, ribonucleotide reductase, phosphorylation of histones, dephosphorylation of histones, phosphorylation of non-histone proteins or ribosomal proteins, acetylation or deacetylation of histones, plasminogen activator, tyrosine kinase, ornithine decarboxylase, and calmodulin. The problem as I mentioned is how to define the term "control of cell proliferation." Clearly, a cell deficient in magnesium may have trouble in carrying out a number of enzymatic reactions including some that are necessary for the reproduction of the cell. But that is not to say that magnesium levels control the extent of cell proliferation in tissues. The absence of an amino acid clearly results in arrest of cell proliferation, but it would be rather naïve to say that that amino acid is indeed the component that controls the extent of cell proliferation in animal cells. The confusion here is between cellular components that are necessary for cell division and those that "initiate" the series of processes leading to mitosis. Many genes, gene products, amino acids, ions, and other cellular components are required for an arrested cell to reenter the cell cycle and to enter S phase. But by control of cell proliferation one should actually refer to the gene, or at the most a handful of genes, whose expression initiates the transition from a resting to a growing stage. With this premise let us review the biochemical events in cell proliferation that are important for our understanding of the mechanisms of cell division.

IV. THE ROLE OF RNA POLYMERASE II

The earliest indications of the presence in proliferating cells of transcripts absent in G_0 cells came from experiments with inhibitors of DNA synthesis, for instance, actinomycin D, and their effect on the progression of cells from G_0 or G_1 to S phase (for a review, see Baserga, 1976). These data, though suggestive, were not convincing, because experiments with inhibitors of RNA synthesis are subject to alternative interpretations. However, evidence has been obtained clearly indicating that unique copy gene transcription is necessary for the $G_0 \rightarrow G_1 \rightarrow$ S transition. The most convincing evidence has come from our laboratory and can be summarized (Table I) into three different experiments which were carried out on *ts* mutants of the cell cycle. The *ts* mutants that we used are *ts*AF8 and *ts*13 cells, both of which were originally isolated from BHK cells by Basilico and co-workers (Meiss and Basilico, 1972; Talavera and Basilico, 1977). These

TABLE I

Evidence That Unique Copy Gene Transcription Is Necessary for the $G_0 \rightarrow G_1 \rightarrow S$ Transition

---

1. Cytoplasts from S-phase cells can complement G_1 mutants of the cell cycle, but cytoplasts from G_0 cells cannot
2. tsAF8 cells, a ts mutant of RNA polymerase II, arrest in G_1 at the nonpermissive temperature
3. α-Amanitin, a drug that specifically inhibits RNA polymerase II, arrests cells in G_1

---

cells are bonafide G_1 mutants because: (1) When collected by mitotic detachment and plated at the nonpermissive temperature, the cells do not enter S phase; (2) when made quiescent by serum restriction and subsequently stimulated at the nonpermissive temperature, the cells do not enter S phase; (3) the cells enter S phase at the permissive temperature, whether plated after mitotic detachment or stimulated after nutritional deprivation; and (4) the cells arrest in G_1, and only in G_1, even when they are shifted up to the nonpermissive temperature in other phases of the cell cycle, i.e., if shifted up during S, M, or G_2, they complete their mitotic cycle until they again reach a G_1 point. By using these mutants we have established the following:

1. Cytoplasts from S-phase cells can complement G_1 mutants of the cell cycle, but cytoplasts from G_0 cells cannot (Jonak and Baserga, 1979, 1980). This means that the nucleus is necessary for the cytoplasmic appearance of at least three functions that are required for the $G_0 \rightarrow G_1 \rightarrow S$ transition. It is true that the information provided by S-phase cytoplasts may not be mRNA, or may be mRNA already present in G_0 cytoplasts in an inactive form. Still, a functional nucleus would be necessary for the putative activation. Similar conclusions have been reached also by Smith and Stiles (1981), who stimulated G_0 3T3 cells by fusing them to cytoplasts of cells induced by exposure to the mitogenic platelet-derived growth factor.

2. tsAF8 cells have been shown to be a mutant of RNA polymerase II by biochemical (Rossini and Baserga, 1978; Rossini et $al.$, 1980) and genetic (Shales et $al.$, 1980; Ingles and Shales, 1982) evidence, and more recently, by direct microinjection of RNA polymerase II into tsAF8 cells (Baserga et $al.$, 1982). This was the final and rigorous demonstration that these cells are a mutant of RNA polymerase II. When tsAF8 cells in G_0 were microinjected with RNA polymerase II and then shifted up to the nonpermissive temperature, 45% of the cells were capable of entering the S phase of the cell cycle. In these experiments, only 10% of the control cells on the same cover slip as the microinjected cells and subjected to exactly the same treatment except the microinjection entered S phase. These results, therefore, complete the demonstration that tsAF8 cells have a defective RNA polymerase II. Since tsAF8 cells arrest in G_1 (Burstin et $al.$,

1974) these experiments also indicate that a functional RNA polymerase II is required for the entry of cells into S whether from G_0 or mitosis.

3. When cells are microinjected with α-amanitin, a specific inhibitor of RNA polymerase II, the cells arrest in G_1, although cells already in S continue to synthesize DNA (Baserga *et al.*, 1982). Since α-amanitin has no other cellular target except the α-amanitin-binding subunit of RNA polymerase II (Ingles, 1978), this is direct evidence that RNA polymerase II transcripts are required for the entry of cells into S.

V. THE SEARCH FOR GENES

Although it has been conjectured that some RNA polymerase II transcripts were necessary for the transition of cells from a resting stage to S phase, it is only with these experiments from our laboratory that a rigorous demonstration has been made that indeed, unique copy gene transcription is required for such a transition. It seems therefore reasonable to look for mRNAs that are present in G_1- or S-phase cells but are absent, or markedly decreased, in G_0 cells. With the recombinant DNA technology presently available this search becomes feasible, and indeed, a number of laboratories are already engaged in the identification of such mRNAs. Clearly, in future directions this is one of the most important tasks that faces biologists involved in the study of cell proliferation. The problem can be reduced to the identification of mRNAs that are present only in a certain physiological state of the cells, i.e., mRNAs that are present in state A and not in state B. However, a number of molecular biologists state that such an approach is not feasible because of the complexity of the mammalian genome, and because many mRNAs are present only in a few copies. Yet, its feasibility has already been demonstrated in at least four instances by: Kramer and Anderson (1980) in yeast growing in either low or high phosphate; Mangiarotti *et al.* (1981) in *Dictyostelium* either growing or in a differentiated state; Crampton *et al.* (1980) in lymphocytes and fibroblasts; and Lee *et al.* (1981) in K12 cells incubated at either permissive or nonpermissive temperatures. Indeed, the feasibility of this approach is clearly born out by the findings of Sierra *et al.* (1982) who, while looking for histone mRNAs, stumbled accidentally on a G_1-specific mRNA.

Of course, this approach will not allow the identification of all genes that control cell proliferation. It would simply identify some genes whose mRNAs are present in G_1 phase or in S phase and absent in G_0. Perhaps this group of genes will include the genes such as those that kick quiescent cells into the cell cycle. The identification of mRNAs specific for G_1 is in itself of value and constitutes a substantial step toward the final goal. Indeed, even simple quantitative differences in the expression of some genes in resting and in proliferating cells would be of considerable interest. On the basis of the aforementioned

TABLE II

Conditional Mutants of the Cell Cycle

Cell line	Apparent lesion		References
Chinese hamster WG1A (K12)	Late G_1		Roscoe *et al.* (1973)
Syrian hamster BHK21	(*ts*AF8)	G_1	Burstin *et al.* (1974)
	(*ts*13)	G_1	Talavera and Basilico (1977)
	(*ts*HJ4)	G_1	Talavera and Basilico (1977)
	(dna-*ts*BN2) DNA synthesis		Eilen *et al.* (1980)
Chinese hamster CCL39	(BF113)	G_1	Scheffler and Buttin (1973)
Mouse B54		G_1	Liskay (1974)
Hamster HM-1 (*ts*546)	Mitosis		Wang (1974)
Murine leukemia L5178Y	Mitosis and cytokinesis		Shiomi and Sato (1976)
Chinese hamster CHO (CS4-D3)	G_1 (cold-sensitive)		Crane and Thomas (1976)
(*ts*C8)	DNA replication		McCracken (1982)
Mouse L (*ts*A169)	DNA replication		Sheinin (1976)
Mouse BALB/3T3 (*ts*-2)	DNA synthesis		Slater and Ozer (1976)
Mouse FM3A (*ts*85)	Late S/G_2		Yasuda *et al.* (1981)
(*ts*131b)	DNA replication		Hyodo and Suzuki (1982)
Chinese hamster WG1A (H. 3.5)		G_1	Landy-Otsuka and Scheffler (1980)
Syrian hamster BHK21 (*ts*BN75)		G_2 and S	Nishimoto *et al.* (1980)
Mouse FM3A (*ts*T244)	DNA synthesis		Tsai *et al.* (1979)
Syrian hamster BHK21 (*ts*422E)	Cell division		Mora *et al.* (1980)

literature, even with a limited number of cDNA clones, the chances are high that some differences will be found. These data, if nothing else, will also allow us to get a reasonable idea of the genetic complexity of cell cycle regulation, i.e., how many genes are presumably involved in the modulation of cell proliferation.

A second set of genes sought by molecular biologists interested in cell proliferation are those whose products are defective in cell cycle-specific *ts* mutants. A substantial number of *ts* mutants of the cell cycle have been described, and a partial list is given in Table II. These are mutants for every phase of the cell cycle: G_1, S, G_2, and M. A combination of recombinant DNA technology and high efficiency transfection should make it possible to identify and clone the genes in the wild-type cells that are responsible for the cell cycle defects in the mutants. Once these genes are cloned, one should be able to settle the question of whether the expression of cell cycle-related genes is cell cycle-dependent.

VI. THE p53 PROTEIN

As mentioned previously, researchers are very interested in identifying the gene or the gene product that actually controls the proliferation of mammalian cells, i.e., the gene or gene product that regulates the transition of cells from a

TABLE III

Effect of Antibodies Microinjected into the Nuclei of Mammalian Cells Stimulated to Proliferate[a]

Stimulus	Antibody	Cells in DNA synthesis (%)
None	None	10
SV40 DNA	None	70
SV40 DNA	Nonimmune IgG	63
SV40 DNA	Monoclonal anti-T	20
10% Serum	None	83
10% Serum	Monoclonal anti-T	85

[a] Adapted from Floros et al. (1981). ts13 Cells, made quiescent, were stimulated with 10% serum or by microinjection of pSV2G, a recombinant plasmid containing the entire early region of SV40. The antibodies or the preimmune IgG were microinjected directly into the nuclei of cells at the time of stimulation.

resting to a proliferating stage. I will describe in this section another approach that has been developed in my laboratory and consists essentially of the microinjection of a monoclonal antibody against a specific protein. If that protein is involved in the regulation of cell proliferation, the microinjection of the monoclonal antibody will inhibit under certain conditions the proliferation of cells in culture. Before describing our results with a monoclonal antibody against the p53 protein, it is necessary to give the background upon which these experiments are based. These experiments, described in detail by Floros et al. (1981), are summarized in Table III. When quiescent cells are microinjected with SV40 DNA or a cloned fragment of the SV40 early region, stimulation of cell DNA replication occurs. However, if the SV40 early region is comicroinjected with a monoclonal antibody against the SV40 T antigen, there is a marked inhibition of SV40 induced cell DNA synthesis. This could be due to a toxic effect of the microinjected monoclonal antibody. However, when this same monoclonal antibody against the SV40 T antigen is microinjected in serum-stimulated cell DNA replication. Furthermore, when a preimmune IgG is comicroinjected with the SV40 early region in quiescent cells, SV40 induced cell DNA synthesis is not inhibited. These experiments by Floros et al. (1981) conclusively demonstrated that antibodies can be microinjected into the nuclei of quiescent cells without toxicity and with specificity of action.

With this background we were able to investigate the effect of a monoclonal antibody against a p53 protein. The p53 protein is a transformation-related protein, which has been seriously considered as a candidate for the regulation of cell proliferation (Linzer and Levine, 1979; DeLeo et al., 1979; Carroll et al., 1980; Milner and McCormick, 1980; Dippold et al., 1981; Milner and Milner, 1981).

This protein is present in many transformed cells and, in smaller amounts, even in some actively proliferating normal cells. A monoclonal antibody against the p53 protein was described by Dippold *et al.* (1981). This monoclonal antibody immunoprecipitates the p53 protein and gives positive immunofluorescence with a number of transformed mouse cells, regardless of whether they have been transformed with viruses, X-radiation, or chemicals. It is negative for BALB/c 3T3 cells. This protein, also described in primary cultures of 12-day mouse embryos, disappears if the cells are passaged.

Swiss 3T3 cells were made quiescent and were then stimulated by serum, and their ability to enter S phase was determined by standard autoradiographic methods. Some of the cells served as controls, whereas others were microinjected with either the monoclonal antibody against the p53 protein or a monoclonal antibody to an unrelated surface antigen Lyt-2.2. This monoclonal antibody, belonging to the same monoclonal class as the monoclonal against the p53 protein, served as a control. The results are summarized in Table IV (taken from Mercer *et al.*, 1982). Regardless of when it was microinjected into the nuclei of Swiss 3T3 cells, the monoclonal antibody Lyt-2.2 had no effect whatsoever on serum-stimulated cell DNA synthesis. However, when the monoclonal antibody against the p53 protein was microinjected into Swiss 3T3 cells, it inhibited subsequent serum-stimulated DNA synthesis but only when it was microinjected between 2 hr before and 2 hr after serum stimulation. If the cells were serum stimulated for 4 hr or longer and then microinjected with the monoclonal antibody against the p53 protein, there was no inhibition of serum-induced cell DNA synthesis. There was also no inhibition of cell DNA synthesis when the monoclo-

TABLE IV

Effect of a Microinjected Monoclonal Antibody against the p53 Protein on Serum-Stimulated Cell DNA Synthesis[a]

Time of microinjection (hr)	Anti-p53	Anti-Lyt-2.2	Controls
−2	31	64	63
−0.5	23	51	51
+2	46	60	66
+4	64	61	62
+6	45	50	47
+17	65	65	70

[a] Adapted from Mercer *et al.* (1982). Swiss 3T3 cells were made quiescent (2.5% labeling index) and subsequently stimulated with 10% serum. At the times designated in the first column they were microinjected with a monoclonal antibody against the p53 protein or with a monoclonal antibody against an unrelated surface antigen, Lyt-2.2. Cells were labeled for 17 hr, except the last row, where the labeling period lasted only 1 hr.

nal antibody against p53 or the Lyt-2.2 was microinjected into DNA-synthesizing cells. These experiments were repeated severalfold and even with a different monoclonal antibody (Gurney *et al.*, 1980) and always with the same results, although the degree of inhibition varied from one experiment to another. The results were highly specific because the timing of the microinjection of the monoclonal antibody against the p53 protein was so important. Its effect, therefore, cannot be attributed to toxicity. Furthermore, when an antibody against RNA polymerase I was microinjected into the nuclei of quiescent cells, serum-stimulated DNA synthesis was not inhibited at all, although nucleolar RNA synthesis was inhibited for several hours. We have therefore conclusively demonstrated that the microinjection of a monoclonal antibody against a p53 protein inhibits serum-stimulated cell DNA synthesis when the antibody is microinjected between -2 and $+2$ hr of stimulation.

These results yield two conclusions: one firm, and the second arguable. The firm conclusion is that the p53 protein is involved in the regulation of cell proliferation. Inhibition of the p53 protein by monoclonal antibody inhibits, under certain conditions, serum-stimulated DNA synthesis. The second conclusion is arguable because there are alternative explanations. However, one explanation is that the p53 protein is no longer needed after the first few hours after stimulation. If this is true, then the p53 protein could be regarded as at least one of the proteins that regulates the transition of cells from G_0 into the cell cycle.

Clearly, this technique of microinjection of specific monoclonal antibody into the nuclei of cells will also be useful in identifying other proteins that may be involved in the regulation of cell proliferation in mammalian cells.

VII. CELL SIZE AND THE ROLE OF RIBOSOMAL RNA

Several years ago Mitchison (1971) had pointed out that there are two cell cycles in yeast cells, a growth cycle and a cell DNA synthesis or nuclear cycle. It is evident that a prerequisite for cell division is an increasing size of the cell. During balanced growth under physiological conditions "the two daughter cells produced at each division are identical to the parent at the same time in the preceding cycle. This requires that all cell components are doubled during the course of each cell cycle" (Fraser and Nurse, 1978). If the cells did not double in size between one mitosis and the next, they would get smaller and smaller at each division and eventually the cells would vanish. There is no question that under physiological conditions the cells must grow in size, and therefore, they must also increase the amount of rRNA which is a very good indicator of the size of the cell. The question that has often been asked is whether rRNA does control the entry of cells into DNA synthesis.

A very popular theory in many laboratories is that the size of the cell controls the entry of cells into DNA synthesis. We had begun a number of experiments to show more conclusively that the entry of cells into S strictly depended on the amount of rRNA present in the cells. The results were exactly the opposite. There is now a considerable amount of evidence that the size of the cell is not a prerequisite for the entry of cells into S phase, whether the size of the cell is measured as cell mass, amount or synthesis of rRNA, or amount of protein. The supporting evidence is the following:

1. If the production of ribosomes is inhibited cells cannot divide, but they can enter the S phase (Mora *et al.*, 1980). This was done with a temperature-sensitive mutant, derived from BHK cells, called 422E, that does not make ribosomes at the nonpermissive temperature. Shifting of the cells to the nonpermissive temperature does not inhibit the entry of these cells into S phase, although it does inhibit cell division.

2. The doubling time of cells in culture is strictly dependent on the doubling time of the amount of proteins, yet cells can enter S phase even with subnormal amounts of proteins (Ronning *et al.*, 1981).

3. Adenovirus 2 causes cell DNA replication in quiescent cells in culture (Rossini *et al.*, 1979) without a concomitant increase in cell size (Pochron *et al.*, 1980). This means that the adenovirus genome encodes information necessary and sufficient for the induction of cell DNA synthesis but not for the doubling of other cellular components.

4. Similar experiments were done with fragments of SV40 described by Soprano in Chapter 1 of this volume.

5. Zetterberg *et al.* (1982) have shown that an alkaline shock of 2 min can induce cell DNA synthesis in quiescent 3T3 cells. However, the cells do not grow in size or accumulate RNA or proteins. If insulin is added to the medium after the alkaline shock, the cells also grow in size.

These results conclusively demonstrate that cell DNA synthesis is not dependent on the size of the cell. The future here lies in identifying the genes that control the entry of cells into DNA synthesis and keep them distinct from the growth in size that occurs independently from cell DNA replication. Studies with cells infected with adenovirus or transfected with fragments of SV40 could be very useful.

VIII. CONCLUSIONS

Thirty years ago the only thing known about cell proliferation was that cells were either in interphase or in mitosis. Much progress has been made, but even more importantly, the control of cell proliferation can now be elucidated at the molecular level. Much research needs to be done, but the future seems to be promising. The identification of growth factors and inhibitory factors, the identification of the genes and gene products that control the transition of cells from the resting to the proliferative stage, and the posibility of studying the factors that control mitosis rather than cell DNA synthesis are all intriguing possibilities that will be the object of investigation in several laboratories in the near future.

REFERENCES

Baserga, R. (1976). "Multiplication and Division in Mammalian Cells." Dekker, New York.

Baserga, R., ed. (1981). "Tissue Growth Factors." Springer-Verlag, Berlin and New York.

Baserga, R., Waechter, D. E., Soprano, K. J., and Galanti, N. (1982). Molecular biology of cell division. *Ann. N.Y. Acad. Sci.* **397,** 110–120.

Burstin, S. J., Meiss, H. K., and Basilico, C. (1974). A temperature-sensitive cell cycle mutant of the BHK cell line. *J. Cell. Physiol.* **84,** 397–408.

Carroll, R. B., Muello, K., and Melero, J. A. (1980). Coordinate expression of the 48K host nuclear phosphoprotein and SV40 Ag upon primary infection of mouse cells. *Virology* **102,** 447–456.

Crampton, J., Humphries, S., Woods, D., and Williamson, R. (1980). The isolation of cloned cDNA sequences which are differentially expressed in human lymphocytes and fibroblasts. *Nucleic Acids Res.* **8,** 6007–6017.

Crane, M. St. J., and Thomas, D. B. (1976). Cell cycle, cell shape mutant with features of the G_0 state. *Nature (London)* **261,** 205–208.

DeLeo, A. B., Jay, G., Appella, E., Dubois, G. C., Law, L. W., and Old, L. J. (1979). Detection of a transformation-related antigen in chemically induced sarcomas and other transformed cells of the mouse. *Proc. Natl. Acad. Sci. U.S.A.* **76,** 2420–2424.

Dippold, W. G., Jay, G., DeLeo, A. B, Khoury, G., and Old, L. (1981). p53 Transformation-related protein: Detection by monoclonal antibody in mouse-human cells. *Proc. Natl. Acad. Sci. U.S.A.* **78,** 1696–1699.

Eilen, E., Hand, R., and Basilico, C. (1980). Decreased initiation of DNA synthesis in a temperature-sensitive mutant of hamster cells. *J. Cell. Physiol.* **105,** 259–266.

Floros, J., Jonak, G., Galanti, N., and Baserga, R. (1981). Induction of cell DNA replication in G_1 specific ts mutants by microinjection of SV40 DNA. *Exp. Cell Res.* **132,** 215–223.

Fraser, R. S. S., and Nurse, P. (1978). Novel cell cycle control of RNA synthesis in yeasts. *Nature (London)* **271,** 726–730.

Gurney, E. G., Harrison, R. O., and Fenno, J. (1980). Monoclonal antibodies against Simian Virus 40 T antigens: Evidence for distinct subclasses of large T antigen and for similarities among nonviral T antigens. *J. Virol.* **34,** 752–763.

Howard, A., and Pelc, S. R. (1951). Nuclear incorporation of P^{32} as demonstrated by autoradiographs. *Exp. Cell Res.* **2,** 178–187.

Hyodo, M., and Suzuki, K. (1982). A temperature-sensitive mutant isolated from mouse FM3A cells defective in DNA replication at a non-permissive temperature. *Exp. Cell Res.* **137,** 31–38.

Ingles, C. J. (1978). Temperature-sensitive RNA polymerase II mutations in chinese hamster ovary cells. *Proc. Natl. Acad. Sci. U.S.A.* **75,** 405–409.

Ingles, C. J., and Shales, M. (1982). DNA mediated transfer of an RNA polymerase II gene. Reversion of the temperature-sensitive hamster cell cycle mutant tsAF8 by mammalian DNA. *Mol. Cell. Biol.* **2,** 666–673.

Jonak, G. J., and Baserga, R. (1979). Cytoplasmic regulation of G_1 specific temperature-sensitive functions. *Cell* **18,** 117–123.

Jonak, G., and Baserga, R. (1980). The cytoplasmic appearance of three functions expressed during the G_0-G_1-S transition is nucleus dependent. *J. Cell. Physiol.* **105,** 347–354.

Kramer, A. K., and Anderson, N. (1980). Isolation of yeast genes with mRNA levels controlled by phosphate concentration. *Proc. Natl. Acad. Sci. U.S.A.* **77,** 6541–6545.

Landy-Otsuka, F., and Scheffler, I. (1980). Enzyme induction in a temperature-sensitive cell cycle mutant of Chinese hamster fibroblasts. *J. Cell. Physiol.* **105,** 209–220.

Lee, S. E., Delegeane, A., and Scharff, D. (1981). Highly conserved glucose-regulated protein in hamster and chicken cells: Preliminary characterization of its cDNA clone. *Proc. Natl. Acad. Sci. U.S.A.* **78,** 4922–4925.

Linzer, D. I. H., and Levine, A. J. (1979). Characterization of a 54K Dalton cellular SV40 tumor antigen present in SV40-transformed cells and an uninfected embryonal carcinoma cells. *Cell* **17,** 43–52.

Liskay, R. M. (1974). A mammalian somatic "cell cycle" mutant defective in G_1. *J. Cell. Physiol.* **84,** 49–56.

McCracken, A. (1982). A temperature-sensitive DNA synthesis mutant isolated from the Chinese hamster ovary cell line. *Somatic Cell Genet.* **8,** 179–195.

Mangiarotti, G., Chung, S., Zucker, C., and Lodish, H. F. (1981). Selection and analysis of cloned developmentally regulated *Dictyostelium discoideum* genes by hybridization competition. *Nucleic Acids Res.* **9,** 947–963.

Meiss, H. K., and Basilico, C. (1972). Temperature sensitive mutants of BHK 21 cells. *Nature (London), New Biol.* **239,** 66–68.

Mercer, W. E., Nelson, D., LeLeo, A. B., Old, L. J., and Baserga, R. (1982). A microinjected monoclonal antibody to the p53 protein inhibits serum induced DNA synthesis in 3T3 cells. *Proc. Natl. Acad. Sci. U.S.A.* **79,** 6309–6312.

Milner, J., and McCormick, F. (1980). Lymphocyte stimulation: Concanavalin A induces the expression of a 53K protein. *Cell Biol. Int. Rep.* **4,** 663–667.

Milner, J., and Milner, S. (1981). SV40-53K antigen: A possible role for 53K in normal cells. *Virology* **112,** 785–788.

Mitchison, J. M. (1971). "The Biology of the Cell Cycle." Cambridge Univ. Press, London and New York.

Mora, M., Darzynkiewicz, Z., and Baserga, R. (1980). DNA synthesis and cell division in a mammalian cell mutant temperature sensitive for the processing of ribosomal RNA. *Exp. Cell Res.* **125,** 241–249.

Nishimoto, T., Takahashi, T., and Basilico, C. (1980). A temperature-sensitive mutation affecting S-phase progression can lead to accumulation of cells with a G_2 content. *Somatic Cell Genet.* **6,** 465–476.

Pochron, S., Rossini, M., Darzynkiewcz, Z., Traganos, F., and Baserga, R. (1980). Failure of accumulation of cellular RNA in hamster cells stimulated to synthesis DNA by infection with adenovirus 2. *J. Biol. Chem.* **255,** 4411–4413.

Ronning, O. W., Lindmo, T., Pettersen, E. O., and Seglen, P. O. (1981). The role of protein accumulation in the cell cycle control of human NHIK 3025 cells. *J. Cell. Physiol.* **109,** 411–418.

Roscoe, D. H., Robinson, H., and Carbonell, A. W. (1973). DNA synthesis and mitosis in a temperature sensitive Chinese cell line. *J. Cell. Physiol.* **82,** 333–338.

Rossini, M., and Baserga, R. (1978). RNA synthesis in a cell cycle specific temperature-sensitive mutant from a hamster cell line. *Biochemistry* **17,** 858–863.

Rossini, M., Weinmann, R., and Baserga, R. (1979). DNA synthesis in temperature-sensitive mutants of the cell cycle infected by polyoma virus and adenovirus. *Proc. Natl. Acad. Sci. U.S.A.* **76,** 4441–4445.

Rossini, M., Baserga, S., Huang, C. H., Ingles, C. J., and Baserga, R. (1980). Changes in RNA polymerase II in a cell cycle specific temperature-sensitive mutant of hamster cells. *J. Cell. Physiol.* **103,** 97–103.

Scheffler, I. E., and Buttin, G. (1973). Conditionally lethal mutations in Chinese hamster cells. Isolation of a temperature-sensitive line and its investigation by cell cycle studies. *J. Cell. Physiol.* **81,** 199–216.

Shales, M., Bergsagel, G., and Ingles, C. J. (1980). Defective RNA polymerase II in the G_1 specific temperature-sensitive hamster cell mutant tsAF8. *J. Cell. Physiol.* **105,** 527–532.

Sheinin, R. (1976). Preliminary characterization of the temperature-sensitive defect in DNA replication in a mutant mouse L cell. *Cell* **7,** 49–57.

Shiomi, T., and Sato, K. (1976). A temperature-sensitive mutant defective in mitosis and cytokinesis. *Exp. Cell Res.* **100,** 297–302.

Sierra, F., Lichtler, A., Marashi, F., Rickles, R., Van Dyke, T., Clark, S., Wells, J., Stein, G., and Stein, J. (1982). Organization of human histone genes. *Proc. Natl. Acad Sci. U.S.A.* **79,** 1795–1799.

Slater, M. L., and Ozer, H. L. (1976). Temperature-sensitive mutant of BALB/3T3 cells: Description of a mutant affected in cellular and polyoma virus DNA synthesis. *Cell* **7,** 289–295.

Smith, J. C., and Stiles, C. D. (1981). Cytoplasmic transfer of the mitogenic response to platelet-derived growth factor. *Proc. Natl. Acad. Sci. U.S.A.* **78,** 4363–4367.

Smith, J. I., and Martin, L. (1974). Do cells cycle? *Proc. Natl. Acad. Sci. U.S.A.* **70,** 1263–1267.

Talavera, A., and Basilico, C. (1977). Temperature-sensitive mutants of BHK cells affected in cell cycle progression. *J. Cell. Physiol.* **92,** 425–436.

Tsai, Y., Hanaoka, F., Nakano, M. M., and Yamada, M. (1979). A mammalian DNA⁻ mutant decreasing nuclear DNA polymerase α activity at nonpermissive temperature. *Biochem. Biophys. Res. Commun.* **91,** 1190–1195.

Wang, R. J. (1974). Temperature-sensitive mammalian cell line blocked in mitosis. *Nature (London)* **248,** 76–78.

Yasuda, H., Matsumoto, Y., Mita, S., Marunouchi, T., and Yamada, M. (1981). A mouse temperature-sensitive mutant defective in H1 histone phosphorylation is defective in deoxyribonucleic acid synthesis and chromosome condensation. *Biochemistry* **20,** 4414–4419.

Zetterberg, A., Engström, W., and Larsson, D. (1982). Growth activation of resting cells: Induction of balanced and imbalanced growth. *Ann. N.Y. Acad. Sci.* **397,** 130–147.

Index

CELL BIOLOGY: A Series of Monographs

EDITORS

D. E. BUETOW

*Department of Physiology
and Biophysics
University of Illinois
Urbana, Illinois*

I. L. CAMERON

*Department of Anatomy
University of Texas
Health Science Center at San Antonio
San Antonio, Texas*

G. M. PADILLA

*Department of Physiology
Duke University Medical Center
Durham, North Carolina*

A. M. ZIMMERMAN

*Department of Zoology
University of Toronto
Toronto, Ontario, Canada*

G. M. Padilla, G. L. Whitson, and I. L. Cameron (editors). THE CELL CYCLE: Gene-Enzyme Interactions, 1969

A. M. Zimmerman (editor). HIGH PRESSURE EFFECTS ON CELLULAR PROCESSES, 1970

I. L. Cameron and J. D. Thrasher (editors). CELLULAR AND MOLECULAR RENEWAL IN THE MAMMALIAN BODY, 1971

I. L. Cameron, G. M. Padilla, and A. M. Zimmerman (editors). DEVELOPMENTAL ASPECTS OF THE CELL CYCLE, 1971

P. F. Smith. The BIOLOGY OF MYCOPLASMAS, 1971

Gary L. Whitson (editor). CONCEPTS IN RADIATION CELL BIOLOGY, 1972

Donald L. Hill. THE BIOCHEMISTRY AND PHYSIOLOGY OF *TETRA-HYMENA*, 1972

Kwang W. Jeon (editor). THE BIOLOGY OF AMOEBA, 1973

Dean F. Martin and George M. Padilla (editors). MARINE PHARMACOGNOSY: Action of Marine Biotoxins at the Cellular Level, 1973

Joseph A. Erwin (editor). LIPIDS AND BIOMEMBRANES OF EUKARYOTIC MICROORGANISMS, 1973

A. M. Zimmerman, G. M. Padilla, and I. L. Cameron (editors). DRUGS AND THE CELL CYCLE, 1973

Stuart Coward (editor). DEVELOPMENTAL REGULATION: Aspects of Cell Differentiation, 1973

I. L. Cameron and J. R. Jeter, Jr. (editors). ACIDIC PROTEINS OF THE NUCLEUS, 1974

Govindjee (editor). BIOENERGETICS OF PHOTOSYNTHESIS, 1975

James R. Jeter, Jr., Ivan L. Cameron, George M. Padilla, and Arthur M. Zimmerman (editors). CELL CYCLE REGULATION, 1978

Gary L. Whitson (editor). NUCLEAR–CYTOPLASMIC INTERACTIONS IN THE CELL CYCLE, 1980

Danton H. O'Day and Paul A. Horgen (editors). SEXUAL INTERACTIONS IN EUKARYOTIC MICROBES, 1981

Ivan L. Cameron and Thomas B. Pool (editors). THE TRANSFORMED CELL, 1981

Arthur M. Zimmerman and Arthur Forer (editors). MITOSIS/CYTOKINESIS, 1981

Ian R. Brown (editor). MOLECULAR APPROACHES TO NEUROBIOLOGY, 1982

Henry C. Aldrich and John W. Daniel (editors). CELL BIOLOGY OF *PHYSARUM* AND *DIDYMIUM*, Volume I: Organisms, Nucleus, and Cell Cycle, 1982; Volume II: Differentiation, Metabolism, and Methodology, 1982

John A. Heddle (editor). MUTAGENICITY: New Horizons in Genetic Toxicology, 1982

Potu N. Rao, Robert T. Johnson, and Karl Sperling (editors). PREMATURE CHROMOSOME CONDENSATION: Application in Basic, Clinical, and Mutation Research, 1982

George M. Padilla and Kenneth S. McCarty, Sr. (editors). GENETIC EXPRESSION IN THE CELL CYCLE, 1982

David S. McDevitt (editor). CELL BIOLOGY OF THE EYE, 1982

P. Michael Conn (editor). CELLULAR REGULATION OF SECRETION AND RELEASE, 1982

Govindjee (editor). PHOTOSYNTHESIS, Volume I: Energy Conversion by Plants and Bacteria, 1982; Volume II: Development, Carbon Metabolism, and Plant Productivity, 1982

John Morrow. EUKARYOTIC CELL GENETICS, 1983

John F. Hartmann (editor). MECHANISM AND CONTROL OF ANIMAL FERTILIZATION, 1983

Gary S. Stein and Janet L. Stein (editors). RECOMBINANT DNA AND CELL PROLIFERATION, 1984

In preparation

Prasad S. Sunkara (editor). NOVEL APPROACHES TO CANCER CHEMO-THERAPY, 1984

B. G. Atkinson and D. B. Walden (editors). CHANGES IN GENE EXPRESSION IN RESPONSE TO ENVIRONMENTAL STRESS, 1984